SCIENCE
BEYOND BORDERS

International Collaborations
in Basic Sciences

SCIENCE BEYOND BORDERS

International Collaborations in Basic Sciences

Dinesh K Srivastava

V S Ramamurthy

Shailesh Nayak

D Suba Chandran

National Institute of Advanced Studies, India

World Scientific

NEW JERSEY • LONDON • SINGAPORE • GENEVA • BEIJING • SHANGHAI • TAIPEI • CHENNAI

Published by

World Scientific Publishing Co. Pte. Ltd.
5 Toh Tuck Link, Singapore 596224
USA office: 27 Warren Street, Suite 401-402, Hackensack, NJ 07601
UK office: 57 Shelton Street, Covent Garden, London WC2H 9HE

Library of Congress Control Number: 2024044624

British Library Cataloguing-in-Publication Data
A catalogue record for this book is available from the British Library.

SCIENCE BEYOND BORDERS
International Collaborations in Basic Sciences

ISBN 978-981-98-0194-7 (hardcover)
ISBN 978-981-98-0195-4 (ebook for institutions)
ISBN 978-981-98-0196-1 (ebook for individuals)

For any available supplementary material, please visit
https://www.worldscientific.com/worldscibooks/10.1142/14074#t=suppl

Desk Editors: Kannan Krishnan/Joseph Ang

Typeset by Stallion Press
Email: enquiries@stallionpress.com

Dedication

"सह नाववतु।

सह नौ भुनक्तु।

सह वीर्यं करवावहै।

तेजस्वि नावधीतमस्तु

मा विद्विषावहै।

शान्तिः शान्तिः शान्तिः ।।"

"May we be protected. May we be nourished. May we work together with great energy (perform great deeds together). *May our intellect be sharpened. May there be no animosity between us. May this bring peace to us, peace to our nations, and peace to the entire universe."*

— Taittiriya Upanishad (2.2.2)

Foreword

Homo sapiens became the dominant human species, outlasting and absorbing all other human species, because they were better at knowledge sharing and collaboration. As our species spread over the continents and civilisations evolved, human society was fragmented into different ethnicities, languages, cultures, religions, etc. The history of human societies is replete with conflicts between the diverse groupings that people identified themselves with. These are bolstered by the conflict between nations, viewing the world through their perceptions of individual national interests.

Nature and natural phenomena, however, do not recognise borders or boundaries. Airstreams, rivers, and oceans flow according to the laws of nature, unaffected by the artificial boundaries created by humans to delineate nation-states. These also include the spread of diseases across continents, exemplified by the COVID-19 pandemic. Cooperation of the entire humanity is necessary to tackle and solve the problems threatening the very existence of life on Earth due to climate change, pandemics, natural disasters, etc. We have seen how people across nations, notwithstanding their differences, cooperated to a large extent during the COVID-19 pandemic to bring it under control. Vaccine development and other modalities of tackling the disease were possible due to scientific cooperation and sharing the fruits of research.

At first glance through the chapter headings of *Science Beyond Borders* by Profs. D. K. Srivastava, V. S. Ramamurthy, Shailesh Nayak, and D. Suba Chandran, a reader may wonder what the diverse topics like bird migration, coffee and tea, education, steel, tsunamis, astronomy, or scientific institutions like CERN or ITER can have in common. But as I started going through it, I was fascinated by the skill the authors display in recounting the story of developments in each chapter, and through this process, they show that the sharing of knowledge and cooperation lead to positive outcomes for all stakeholders. They show how science has played a key role in the spread of knowledge throughout the ages and has given us a way to overcome the problems that confront us at present.

Throughout this book, the authors trace the history of the two parallel strands of the behaviour of human society, *viz.*, conflict and cooperation through the ages, with many examples throughout the book. Although many of the case histories are drawn from the Indian subcontinent, the examples of the interplay of conflict and cooperation from across the world are presented in a very fluid manner without any technical jargon that the readers would find difficult to follow.

The book starts with how the Homo sapiens benefitted from the exchange of ideas and the spread of knowledge from prehistoric to modern times across nations through trade, travel, wars, invasion, and occupation of lands of the vanquished. Even the heinous practice of slavery gave rise to the transmission of knowledge from the native lands of the slaves to those of the slave traders. Of course, in the latter cases, the colonisers were the main beneficiaries, but even the natives in some cases were unintended beneficiaries of the conquests. Coffee and tea, the two beverages most of us cannot do without, are shown to have played an important role in bringing people together to create new knowledge. At the same time, we should not forget that commerce in these beverages has fuelled the colonisation of large parts of the world.

The history of the development of paper is traced, and the way it helped spread knowledge from one part of the world to another through merchants, travellers and religious monks is discussed. Also recounted are the accounts of enormous losses that civilisations had to bear due to the burning of books by conquering armies. The authors make the important point that even though electronic media transferring information at lightning speed have usurped the role of paper, it still is necessary to preserve knowledge on paper for posterity.

The authors demonstrate through examples how the spirit of scientific cooperation, even between nations at war, can overcome the odds for the greater good of humanity and possibly reveal a way in which cooperation can triumph over conflict in human society. They strengthen their argument with the examples of many successful international projects in different branches of science. These is turn have brought about unprecedented breakthroughs in improving our understanding of the universe and led to better health and prosperity for millions. Here, the authors prove their expertise in participating in and managing several international collaboration projects in diverse scientific fields such as Physics, Astronomy, Earth Sciences, Space, and Education. The specific institutions of CERN, ITER, ICTP, FAIR, and those in space sciences and earth sciences are truly outstanding examples of international scientific collaborations involving thousands of scientists and engineers from many nations working towards specific common goals.

The chapter on the green revolution is a prime example of such international cooperation and the transmission of scientific knowledge in transforming India from a food-importing nation, living "from ship to mouth," to a net food exporter despite the huge population increase in the intervening period. The chapters on tsunamis, poles, and meteorology drive home the point that these are worldwide phenomena affecting all the people on the planet, and the challenge of the climate crisis must be tackled through intense international cooperation. The main threat to cooperation comes from the narrow self-interests of communities and nations that blind us from the fact that the Earth is a lonely planet that is home to all life. We must all remember what Mahatma Gandhi said: "The world has enough for everyone's needs but not enough for everyone's greed."

Climate change is by any reasoning the biggest challenge that humanity is facing right now. It is threatening the entire life cycle of our planet Earth and needs to be tackled urgently (I avoid the term "war footing" in keeping with the spirit of the book). Only scientific cooperation on an unprecedented scale respecting the human rights of all people and fairness to all holds the key to the solution to the global crisis of climate change. The authors make the strong case that the battle for global poverty elimination, health for all, and mitigating the adverse effects of climate change can be won with global participation and collaboration in sciences. This is a message of hope and redemption for humans. The authors have done a wonderful job of driving home this point in this book. I hope that their message gets the attention it deserves.

Prof. Amit Roy
Former Director, Inter-University Accelerator Centre, New Delhi,
Ex-Raja Ramanna Fellow,
Variable Energy Cyclotron Centre, Kolkata,
Advisor, Manipal Centre for Natural Sciences,
Manipal Academy of Higher Education

Preface

"The only thing which will redeem mankind is cooperation."

— Bertrand Russel, Scientist, Mathematician, and Philosopher

Humanity has always thrived on knowledge sharing and collaborations, often across national borders and cultures. By pooling resources, expertise, and perspectives, it could not only solve problems at a much larger scale but also accelerate discoveries. This culture of collaboration and cooperation encompassed not only fields like education, science, technology, and medicine but also economic growth and social development. The benefits were, obviously, for all humanity. More recently, developments in global weather forecasting, cartography, and astronomy, for example, have benefited from collaborations of one kind or another across the boundaries of nations.

The recent COVID-19 pandemic is a glorious example of what can be achieved through knowledge sharing and collaboration. The year 2019 started normally enough. However, within a span of a few months, a rapidly increasing number of people from Wuhan in China started showing symptoms of severe respiratory syndrome that included fever, dizziness, and cough, and some started succumbing to it. Very soon, it spread across the world and turned into a pandemic. The COVID-19 pandemic, which affected every nation in the world, changed it very rapidly and all normal activities came to a complete halt, as country after country went into a state of complete lockdown. We saw that no man or woman was an island

and physical borders or walls provided only partial protection. With no vaccine in sight throughout 2020, the only thing that the nations could do was help each other, reduce the risk of infection, and provide support to those affected by it.

And we saw ample evidence of cooperation at all levels — be it by making personal protection equipment available, supplying sanitisers, supplying oxygen plants and oxygen concentrators, supplying COVID testing kits, distributing food to the needy, and supplying medicines like paracetamol and Remdesivir. Front-line workers all over the world put their duties above their lives, and many made the ultimate sacrifice.

Our faith in honest science was our only hope, even though we knew that vaccine developments often take 10–15 years (the fastest vaccine development till then had been about four years for mumps).

And science did not disappoint us. The genome sequencing of the COVID-19 virus was quickly completed and made publicly available by Chinese scientists on January 12, 2020. By early December 2020, several vaccines had shown excellent results in large trials and many more were showing promise. By December 2020, a vaccine made by Pfizer and BioNTech and another one made by Moderna using mRNA became the first and the second fully tested immunisations to be approved for emergency use. The third vaccine to be approved was produced by AstraZeneca (named COVISHIELD in India) in collaboration with the University of Oxford. By then, several vaccines (including COVAXIN and iNCOVACC developed in India) had been produced and approved for use. People across the world were rapidly (though, sadly, not rapidly enough) vaccinated and the pandemic was controlled and reduced to endemic status.

While these developments in such quick succession benefitted from years of research on the SARS virus, they were made possible by billions of dollars made available by governments and private funding agencies, which allowed the pharmaceutical companies to take financial risks by running some tests at the same time. The vaccine developers could do the preclinical and phase I, II, and III trials and manufacturing in parallel, instead of doing them sequentially. The large funding allowed companies to start producing the vaccines even before they were found to be effective and may not have worked and were to be discarded. This kind of funding was not available for diseases we faced earlier because on this occasion the entire world, including the rich nations, faced devastation.

This unique success story gives us hope and a golden template for a future where we recognise that we — and the planet which sustains us — are interdependent, vulnerable, and much in need of cooperation to survive and sustain ourselves. It also reminds us of the famous quip by Jean-Jacques Rousseau (1712–1778 CE): "Providence created natural disasters to force men to cooperate."

Perhaps we also needed a real crisis to enable us to truly appreciate what Bertrand Russel meant by "*The only thing which will redeem mankind is cooperation.*"

The plan to write this book as a celebration of the indomitable human spirit of international cooperation and collaboration in science and technology was put into motion against the background of these events.

We were also buoyed by the exultation of scientists across the world through the insights they gained and discoveries they made in recent decades as they collaborated across borders — pooling their efforts, funds, and intellectual resources, and riding on the energy and support thus released. Thus, the Large Hadron Collider was commissioned and it discovered the Higgs boson, which had remained elusive for fifty years since it was first postulated; the LIGO collaboration detected gravity waves as predicted by Albert Einstein a century earlier; the construction of the world's largest fusion experiment, ITER, was started; the USSR, USA, European Union, and others collaborated on the International Space Station; the first planets outside the solar system were discovered; and the construction of the Facility of Anti-proton and Ion Research was initiated.

Unfortunately, just when the world was basking in the sunshine of international collaboration to face unforeseen challenges, the Ukraine War broke out. It is continuing as we write this. Even as the world was struggling to cope with it, immense destruction was unleashed in a simmering conflict between Palestine and Israel. And all this is happening against the background of tensions that are blowing hot and cold between the US, Russia, China, North Korea, and others, which have the potential to destabilise the entire world if allowed to get out of control.

The Russia–Ukraine War, in particular, has had terrible consequences for international collaboration in science and technology. Not only does Russia (and Russian scientists) stand isolated from several international scientific collaborations but this has also unleashed a cascade of delays and cost escalations, putting at risk the very viability of some projects.

It is worthwhile recalling what the President of the Royal Society, Joseph Bank, wrote to the French scientist J. J. Houtou de Labillardiere in 1796 when their countries were at war:

> "That the Science may be at Peace while their Politics are at war is an axiom, we have learned from your [French] protection to Capt. Cook & surely nothing is so likely to abate the unjustifiable Rancour that Politicians frequently entertain against each other as to see Harmony and goodwill prevail among their Brethren who cultivate science."

We also recall that Christian Huygens, the famous Dutch scientist, helped establish the French Academy of Sciences along the lines of the Royal Society of London when the Dutch were fighting a war against the French.

This book was planned as a celebration of the scientific achievements of mankind over centuries past and centuries to come. Now, we expect it to additionally become a beacon of hope in troubled times, as we trust mankind to recover from every devastation, nurse its wounds, collect its thoughts, and get together to usher in peace and prosperity, again. This has happened again and again in the past, and we hope that it will happen yet again — for these wars will surely come to an end one day, and the scars will surely heal with time, as wise counsel prevails. It is our fervent hope that international collaborations in basic sciences can act as a soothing balm to hasten this process.

We know that international collaborations in technology can accelerate the process of technology development. This hope stems from the well-known fact that a group of international collaborators can bring the wisdom of the entire world and all its civilisations to the field of exploration. And we know that even the most isolated nations and tribes often have a pearl of wisdom to offer.

To succeed, international collaboration in science and technology must treat all collaborators as equals. CERN, in Geneva, has shown the way for a successful framework for international collaboration, which has endured for close to seventy years now and which is continuously expanding. Its spirit of open access, collaboration, tolerance, and freedom of thought has been used as a golden template for many international collaborations like ITER, FAIR, SKA, and SESAME. The invention (and sharing) of the World Wide Web is considered emblematic of the spirit that drives advances in basic science, which enables open innovation and education and connects the worldwide community through shared values.

We add that CERN is at the forefront of the next generation of High-performance computing (HPC) technology, essential for scientific research.

Supercomputers — capable of performing a billion times billion calculations per second — are on the verge of becoming a reality. This power, coupled with a growing use of machine learning, will be vital in ensuring the success of big science projects like the Square Kilometre Array and CERN's High-Luminosity Large Hadron Collider. The Square Kilometre Array Observatory entered into a cooperation agreement with CERN, GÉANT (the pan-European network and services provider for research and education), and PRACE (the Partnership for Advanced Computing in Europe) in 2020 to overcome challenges related to the use of high-performance computing to support large, data-intensive science projects.

Similarly, the Human Genome Project was a monumental international scientific endeavour that ran from 1990 to 2003. Its main goal was to map and sequence the entire human genome (i.e., identify the order of all the chemical base pairs, A, C, G, and T, that make up our DNA), and identify and map all the genes within the human genome, understanding their function and how they contribute to our biology. It involved international collaboration between scientists from the US, UK, France, Germany, Japan, China, and other countries, and utilised groundbreaking technologies for DNA sequencing and analysis to tackle the massive amount of data involved. The success of the project has revolutionised genetic research and personalised medicine and led to advancements in fields like gene therapy, drug discovery, and diagnostic testing.

In fact, there is no branch of scientific investigation that did not and cannot benefit from open international collaboration. The financial, technical, and intellectual demands for carrying out some of the major scientific investigations are spiralling out of the reach of even the richest countries. In addition, putting a major part of the funds of any country into any one specific area tends to starve other areas of research funds. This restraining irritant can also be tackled most effectively by participating in an international collaboration.

It is understood by all by now that public goods like air, water, biodiversity, education, knowledge, access to the internet, peace, and welfare can be preserved only with the involvement of all the stakeholders. We have already seen that fighting certain types of diseases requires a strong collaboration between the scientific community, governments, and companies producing vaccines. In fields such as meteorology, scientific cooperation is accepted as essential, because without it every country would lose as predictions, warnings, and fruitful mitigatory actions are possible only with a global exchange of data.

The greatest calamity facing mankind today is global climate change, which threatens our very existence. It affects us all. It has been realised that environmental challenges are global in scope and require solutions that integrate knowledge across disciplines, cultures, and organisations. It has been argued that solutions to these challenges will come from diverse teams rather than individuals or single academic disciplines; therefore, graduate students must be trained to work in these diverse teams, with international collaboration in education. It will require the participation of all nations — large and small, industrialised and developing — to develop and adopt climate change mitigation measures.

The Intergovernmental Panel on Climate Change, the Nobel Peace Prize-winning body, has been bringing together thousands of scientists from diverse backgrounds in one of the widest representations and rigorous collaborations to assess the science related to climate change. Its reports have played a crucial role in recommending the global climate policy. We fervently hope that all nations across the globe heed these warnings and implement the recommendations before it is too late.

We recall that culture refers to socially learned behaviours that persist in a population over time. Studies of animal behaviour have revealed that animal culture like human culture can be cumulative and characterised by sequential innovations built on previous ones (and accumulated over historical times: the "ratchet effect"). However, it has been argued that human cumulative culture involves behaviour so complex that it lies beyond the capacity of any individual to discover during their lifetime. After all, we alone in the animal kingdom have created art and literature, socio-political systems that have leveraged large-scale cooperation, and the scientific and technological knowledge to colonise the whole planet, explore space,[1] and make wonderful discoveries in science, which are universal, by definition.

Citizen Scientists

We also recall a wonderful framework of international collaboration: "citizen scientist." It is a fantastic way for anyone, regardless of background or scientific experience, to contribute to real research. Thus, for

[1]A. Mesoudi and A. Thornton, "What is cumulative cultural evolution?", *Proceedings of the Royal Society B: Biological Sciences*, (June 13, 2018).

example, Galaxy Zoo, an online platform, allows volunteers to classify hundreds of thousands of galaxies, helping astronomers understand galaxy formation and evolution. Similarly, eBird, a global birdwatching platform, collects valuable data on bird populations, migrations, and distributions, crucial for conservation and studying climate change impacts. By now, many such efforts are in place, which help accelerate research with comprehensive data collection covering vast areas and analysis. These also foster public understanding of science and its importance. With the online availability of a large body of data, curious citizens have made valuable contributions. Recall the young engineer Shanmuga Subramanian from Chennai who compared frame-by-frame pictures of the moon's surface, available from NASA, before and after the crash of Vikram — the lander of Chandrayaan-2 — and located its debris, while sitting at home. These citizen scientists have identified many new species of birds, insects, and fish. Yes, science and knowledge have progressed, and this progress has been accelerating through international collaboration. We celebrate this continuing adventure of mankind.

Interdisciplinary Collaboration

The growing complexity of global research necessitates the formation of interdisciplinary, international, and inter-organisational research teams. Thus, the marvellous accelerators and detectors which helped decipher the building blocks of matter were built by a collaboration of engineers and physicists. The complex and extremely sensitive telescopes have required the collaboration of astronomers, engineers, physicists, and computer scientists. Research in biology has been scaling new heights with the help of accelerators and extremely powerful computers. The list is endless.

We reiterate that this book further celebrates the curiosity and cooperation of mankind in solving the problems we face and the discoveries we made, which have given us immense intellectual delight, made our lives comfortable, and turned the world into a global village by vastly improved connectivity.

We are thankful to Dr. K. K. Phua and the World Scientific Publishing Company, Singapore, for readily agreeing to publish this book and waiting for us to complete it well beyond the the originally stipulated date of submission. One of us (DKS) is grateful to the Indian National Science Academy, New Delhi, for its support in the form of the INSA Senior

Scientist position and to the National Institute of Advanced Studies, Bengaluru, for hosting him as an Honorary Visiting Professor.

We are grateful to our spouses, Mrs. Rekha Srivastava, Mrs. Raji Ramamurthy, Mrs. Preeti Nayak, and Mrs. Selvi Chandran, for their support and tolerance of our obsession with this book, which led to our neglect of some of our other responsibilities. We often used our children, students, colleagues, and friends as sounding boards for our overenthusiastic and uber-idealistic hopes, and we are thankful to them for bearing with us and acquainting us with hard reality.

Dinesh K. Srivastava, V. S. Ramamurthy, Shailesh Nayak, and D. Suba Chandran.

National Institute of Advanced Studies, Bengaluru 560012.

About the Authors

Dinesh Kumar Srivastava (born 1952) is presently Senior Scientist, Indian National Science Academy and Honorary Visiting Professor at the National Institute of Advanced Studies, Bengaluru. He is now working on energy, environment, climate change, international collaborations, and science outreach while continuing his research on quark–gluon plasma.

He graduated from Allahabad University in 1970 and joined the Training School of the Bhabha Atomic Research Centre, Mumbai. He started working on the Variable Energy Project at the Bhabha Atomic Research Centre in 1971 and retired as Director and Distinguished Scientist at the Variable Energy Cyclotron Centre, Kolkata, in 2016. Later, he continued at the Centre as Raja Ramanna Fellow at the Department of Atomic Energy (DAE) until 2019. He then worked at the National Institute of Advanced Studies, Bengaluru, as a Homi Bhabha Chair Professor from 2019–2023.

Prof. Srivastava is a Fellow of the National Academy of Sciences, India, and the Indian National Science Academy. He has held an Honorary Professorship at Amity University since 2019 and was Emeritus Professor at the Homi Bhabha National Institute.

He was awarded the Indian Nuclear Society's award for "Outstanding Contribution to Teaching of Nuclear Sciences." Prof. Srivastava received the Outstanding Referee Award from the American Physical Society (APS) in 2009 and was presented the Lifetime Achievement Award by

Chitkara University in 2012. He was also awarded the Homi Bhabha Lecturer Award by the Indian Physics Association and Institute of Physics, UK, in 2016 and the G. N. Ramachandran Award in Physical Sciences by SASTRA University in 2024. He delivered the Sir C. V. Raman Memorial Award Lecture at Calcutta University (2014) and the Prof. Abdus Salam Memorial Award Lecture at Jamia Millia Islamia, New Delhi (2016).

He serves as an Editorial Board Member of the *Scientific Reports* and *Physics (MDPI)* Journals. He served as a Member of the Editorial Board of *Physical Review C* from January 1, 2010 to December 31, 2012 and *Pramana* during 2011–2020. He has been President of the Indian Physical Society since 2014.

Dr. Srivastava has worked at several prestigious labs and universities abroad, including KfK (now KIT) Karlsruhe, GSI Darmstadt, Lawrence Berkeley National Laboratory, Brookhaven National Laboratory, University of Minnesota, Duke University, McGill University, University of Cape Town, Bielefeld University, and the University of Frankfurt.

His research interests include electromagnetic probes of quark–gluon plasma (QGP), relativistic hydrodynamics, production and propagation of charm quarks in QGP, and transverse flow. He has made important contributions to the theoretical understanding of elastic, inelastic, and break-up reactions of light nuclei. Prof. Srivastava has published more than 170 papers and delivered more than 400 talks, seminars, and colloquia. He is the author of two collections of short stories and more recently three books of children's literature on science.

Prof. Srivastava co-authored the book *Climate Change and Energy Options for a Sustainable Future* with V. S. Ramamurthy, which was published by World Scientific in early 2021. He then co-authored *Art and Science of Managing Public Risks* with V. S. Ramamurthy and Shailesh Nayak, which was also published by World Scientific in early 2022.

V. S. Ramamurthy (born 1942) is a well-known Indian nuclear scientist with a broad range of contributions from basic research to science administration. He is currently Emeritus Professor at the National Institute of Advanced Studies in Bengaluru.

Prof. Ramamurthy started his career at the Bhabha Atomic Research Centre, Mumbai, in 1963. He made important research contributions in the areas of nuclear fission, medium-energy heavy-ion reactions, statistical and thermodynamic properties of nuclei, and low-energy accelerator applications. During the period 1995–2006, Prof. Ramamurthy was fully involved in science promotion in India as Secretary to the Government of India, Department of Science and Technology (DST), New Delhi. He was also Chairman of the Standing Advisory Group on Nuclear Applications of the International Atomic Energy Agency (IAEA) for nearly a decade. Other important assignments held by him include Director, Institute of Physics, Bhubaneswar (1989–1995), Department of Atomic Energy (DAE); Homi Bhabha Chair Professor at the Inter-University Accelerator Centre, New Delhi (2006–2010); Chairman, Recruitment and Assessment Board, Council of Scientific and Industrial Research, New Delhi (2006–2010); and Director, National Institute of Advanced Studies, Bengaluru (2009–2014). After retirement from service, Prof. Ramamurthy has been actively involved in human resource development in all aspects. In recognition of his services to the growth of Science and Technology in the country, Prof. Ramamurthy was awarded the Padma Bhushan by the Government of India in 2005.

Prof. Ramamurthy co-authored the book *Climate Change and Energy Options for a Sustainable Future* with Dinesh Kumar Srivastava, which was published by World Scientific in early 2021. He then co-authored *Art and Science of Managing Public Risks* with Dinesh Kumar Srivastava and Shailesh Nayak, which was also published by World Scientific in early 2022.

Shailesh Nayak (born 1953) is currently Director of the National Institute of Advanced Studies, Bengaluru; Chancellor of the TERI School of Advanced Studies, Delhi; Editor-in-Chief of the journal *Indian Society of Remote Sensing*; and Life Trustee of India International Centre. He obtained his Ph.D. in Geology from the M.S. University of Baroda in 1980. He was a Distinguished Scientist in the Ministry of Earth Sciences and President of the International Geological Congress from 2015–2017. He served as Secretary of the Ministry of Earth Sciences, Government of India, from 2008–2015, and provided leadership for programmes related to earth system sciences. He has been credited with launching many research programmes related to monsoons, air quality, seismology, triggered earthquakes, coastal vulnerability, solid earth, climate change, polar science, ocean minerals, etc.

Prof. Nayak set up the state-of-the-art tsunami warning system for the Indian Ocean in 2007 in just two years' time, which has been providing tsunami advisories to the Indian Ocean rim countries. He pioneered the development of algorithms and methodologies for the application of remote sensing to the coastal and marine environment. He generated a baseline database of the Indian coast and developed services for fishery and ocean state forecast. The generation of detailed information on the Indian coast has influenced the development of policy for the zoning of coastal zones to regulate coastal activities.

Prof. Nayak is a Fellow of the Indian Academy of Sciences, Bengaluru; the Indian National Science Academy, New Delhi; the National Academy of Sciences, India, Allahabad; and the International Society of Photogrammetry and Remote Sensing; he is also an Academician of the International Academy of Astronautics. He was awarded honorary Doctor of Science degrees by Andhra University in 2011, Assam University in 2013, Amity University in 2015, and Graphic Era University in 2015. He was conferred the prestigious Indian Science Congress Vikram Sarabhai Memorial Award in 2012 and Bhaskara Award in 2009 for his outstanding contributions to remote sensing and the Geographic Information System. He has published about 200 papers in peer-reviewed journals. In recognition of his services to the country, Prof. Shailesh Nayak was recently (2023) awarded the Padma Shri by the Government of India.

Prof. Nayak co-authored the book *Art and Science of Managing Public Risks* with V. S. Ramamurthy and Dinesh Kumar Srivastava, which was published by World Scientific in early 2022.

D. Suba Chandran is Dean of the School of Conflict and Security Studies at NIAS. Within the School, he heads two programmes — Science, Technology, and International Relations (STIR) and Conflict Resolution and Peace Research. The STIR programme looks at Science and Technology through an International Relations prism and vice versa. STIR specially focuses on space, maritime, emerging technologies, AI, and cyberspace.

As a part of STIR activities, Prof. Chandran leads the "Science Diplomacy" initiative within the institute, aimed at bringing S&T and International Relations together. He also coordinates the BRICS-Young Scientists' Forum. He holds exclusive track-two dialogues with a special focus on S&T with specific countries from Europe to East Asia.

Prof. Chandran also coordinates the NIAS Global Politics initiative, which looks at the contemporary world order and focuses on four area studies — Pakistan, Europe, Africa, and Maritime. In the long run, the area studies are expected to branch out into independent disciplines. He also teaches an online course on "Contemporary World Affairs."

Contents

List of Figures

List of Tables

Chapter 1

Celebrating Sharing of Knowledge and Collaboration

> "No single people is an island unless they've been there for a very long time and haven't invented a boat. And that's a good thing: without new relationships between different people exchanging unfamiliar ideas, nothing much would happen at all."
>
> — Josephine Quinn,[1] Historian, University of Oxford, 2024

1.1 Preamble

Cooperation and conflicts among people have left indelible marks on the rich tapestry of the history of mankind. Humanity has always lived in a fragmented world, fragmented by language, culture, religion, geographical formations, political borders, etc. But looking closely, we find that the redeeming feature of humanity has been the cooperation across these boundaries and a celebration of the bounties of knowledge thus available. This idyllic situation is occasionally marred by conflicts though.

The following story about the futility of conflict is well known in Buddhist literature. The King of Kosala was defeated three times by his nephew, Ajatashatru, and was very depressed. When the news reached Lord Buddha, He told His disciples, "Bhikkhus! In one who conquers,

[1] Josephine Quinn, "What the world gets wrong about 'civilisation'." *Financial Times*, February 3, 2024.

enmity and hatred increase; one who is defeated suffers pain and distress." And then He added:

"जयं वेरं पसवति, दुक्खं सेति पराजितो

उपसंतो सुखं सेति, हित्वा जयपराजयं।"

—धम्मपद, सुखवग्गो, 15.201

"Victory breeds hatred. The defeated live in pain. Happily, the peaceful live, giving up victory and defeat."

— Dhammapad, Sukhavaggo, 15.201

Lord Buddha talked about avoidance of war and bloodshed throughout His life. On another occasion, when two neighbouring states of Shakyas and Kolya prepared to go to war because of a dispute over the sharing of water from the River Rohini, during a lean season, He had again intervened and reasoned with them about the futility of war and the importance of cooperation:

"सुसुखं वत जीवाम, वेरिनेसु अवेरिनो

वेरिनेसु मनुस्सेसु, विहराम अवेरिनो।"

—धम्मपद, सुखवग्गो, 15.197

"Indeed, we live very happily, not hating anyone among those who hate; among men who hate we live without hating anyone."

— Dhammapad, Sukhavaggo, 15.197

These teachings have guided people for more than 2,500 years, and they have ignored them to their peril. Even more importantly, these bring out the importance of sane intellectuals in restoring sanity in a crazed world, thirsting for revenge and bloodshed.

Science provides a unique platform to build the edifice of collaboration beyond borders. Scientists have repeatedly transcended the boundaries of their nations in their search for the rules governing the working of Nature. A popular and well-known Sanskrit saying goes on to declare:

"विद्वत्वं च नृपत्वं च न एव तुल्ये कदाचन ।

स्वदेशे पूज्यते राजा विद्वान् सर्वत्र पूज्यते ॥"

> *"Wisdom and kingship can never be compared. While a king is revered only within the boundaries of his kingdom, a wise man is revered everywhere."*

We recall some notable events and initiatives in collaboration, cooperation, and knowledge exchange, sometimes voluntarily and sometimes forced, to emphasise our point and also to set the stage for further discussions.

1.2 Prehistoric Times

1.2.1 *What sets Homo Sapiens apart?*

The story of the evolution of Homo Sapiens — starting from their emergence in East Africa and spreading to all parts of the world — parts of which were even to get isolated, e.g., Americas and Australia for millennia, from the so-called Old World, remains one of the most engaging and fascinating stories mankind has ever known. It started with our ability to walk on two feet — releasing our hands to manipulate things — and scan the surroundings. Next in our favour was the large size of our brain, which helped us devise crude tools, already quite early in our development.

One of the most important turning points in our evolution was the control of fire, which gave us easily digested food and protected us from parasites, infections, predators, darkness, and cold. Fire may have inadvertently led to the discovery of some metals.

However, looking closely, we realise that most animals, birds, dolphins, and whales also have languages. Some animals such as bottlenose dolphins, chimpanzees, elephants, orangutans, sea otters, and gorillas have taught themselves the use of simple tools. River otters make excellent dams and "engineer and alter" the environment to suit themselves. Beaver birds make intricate nests using blades of grass. The fire hawks of Australia are known to start a fire by dropping burning twigs over dry grasslands to gather and kill prey. The bearded vultures carry bones to a great height and drop them on rocks to break them and eat the bone marrow; they are known to repeat the process if the bone does not break on the first attempt.

So what sets Homo Sapiens apart from other animals? It has been suggested that it is rather difficult to be intimate with more than 150 people ("Robert Dunbar's number"). This even applies to chimpanzees, baboons, and macaques. Thus, groups with strong bonds could not be larger than 150. In the initial millennia of Homo Sapiens' history, even a band of 150 people must have been quite formidable and successful in warding off dangers, hunting animals, and collecting food in profusion by a coordinated effort.

We also realise that beyond basic foraging, Homo Sapiens were relentless knowledge seekers. Their survival depended on it. They needed an intricate mental map of their surroundings. To find food efficiently, they studied plant lifecycles and animal behaviours. They learnt which foods sustained them, which caused illness, and even how to use plants for medicine. They tracked the seasons and identified the signs of approaching storms or droughts. Every stream, tree, cave, and flint deposit were carefully examined. Every member of the group needed to know how to craft tools, repair clothing, set traps, and face dangers, such as avalanches, snakes, and lions. Mastering these diverse skills took years of learning and practice.

The development of languages was one of the next important steps. It helped us form groups for hunting, foraging, and protecting "our turfs."

Yuval Noah Harari, in his famous book *Sapiens: A Brief History of Humankind*, has argued (opined?) that a unique feature of human languages is their ability to imagine and transmit information about things that do not even exist. This "cognitive revolution," which perhaps took place about 30,000 years ago, according to him, allowed Homo Sapiens to create legends, myths, spirits, religions, and even nations. These shared myths gave them the unprecedented ability to cooperate flexibly in large numbers. He further argues that even though "ants and bees can also work together in huge numbers, they do so in a very rigid number and only with their close relatives." However, united by legends and myths, he suggests, "Homo Sapiens can cooperate in extremely flexible ways with a countless number of strangers. That is why Homo Sapiens rule the world."

Thus, we see that collaboration and cooperation in large numbers are the most important features that set us apart from animals. It is a different thing that man has also used this ability to wage war, cause destruction, kill, enslave, and exploit other "nations" repeatedly. But then man has also produced great souls like Lord Buddha whose message of peace and universal love brought considerable order to a large part of the world, which continues to be of great relevance, more so, today.

The other major trait of Homo Sapiens is curiosity. It makes us observant and helps us identify patterns, which in turn help us unravel the nature of things and the laws of nature. It also makes us explore newer ideas and newer worlds, and most importantly it adds excitement to our lives. All sciences have their origin in curiosity.

1.2.2 *Special skills and traditional knowledge*

Even the "most primitive tribes" living in deep forests and vast deserts have used this curiosity and observation to develop unique skills, e.g., boomerangs and protection of nature. Thus, simple folk discovered agriculture, magnetic rocks, and coffee. They already had developed rudimentary knowledge of seasons, hints for the approaching weather, so valuable for agriculture and animal husbandry, migration of animals and birds, and perhaps map-making. They even noticed the movement of planets ("wanderers") in the firmament.

In the faraway Pacific, "wayfinding" — the art of navigating using wind, stars, ocean swells, and other environmental cues and a source of pride in the Pacific Island culture and heritage — was how sailors from the Marquesas Islands first discovered Hawaii, about 3,700 km away, more than 1,500 years ago, using canoes. This knowledge is still available to the people there, but it is rapidly vanishing. However, wayfinding was how humans found their way for most of our existence. It enabled the aboriginals, Arab nomads, and Inuits to journey across the monotonous yet shifting landscapes of land, desert, snow, and ice, using the Sun, stars, Moon, waves, currents, clouds, and different animals and birds.

Let us recall that rock honeybees make very large honeycombs, with a large quantity of excellent honey. However, they are extremely ferocious. The tribes living in the forests of the Andaman and Nicobar Islands, who had avoided contact with the outside world till recently, know of a bush, whose leaves they chew, apply the juice over their bodies, and throw on the honeycombs. This drives these bees away, and they easily harvest the honey.

In the deserts bordering Gujarat, there are people called "Paagi" (footprint readers), who are so skilled that seeing the footprints of a camel, they can tell the number of people riding on it, and looking at the footprints of humans, they can guess their weight, their age, and the distance they may have travelled after leaving the footprint, and they have been doing it for centuries. The Indian Army traditionally recruited them as guides. Before radio collars for tigers became common, hunters used to identify individual tigers from their footprints.

For centuries, farmers in Africa have depended on birds such as Glareola, starlings, and ibis to alert them about the impending emergence of locusts and to keep them under control, as they appear at the time of their breeding and also follow them. Birds from Northern Eurasia have been migrating for millennia to escape the harsh winters to warmer regions near equators for feeding and breeding. The observation of their arrival and departure has provided a valuable clue about the impending weather.

1.2.3 *Sharing of special goods and products*

The importance of knowledge has been celebrated in all cultures, and there have been enough exchanges of knowledge and information among people of different regions from prehistorical times. The exchange of goods found in one particular region with goods found in other regions has been prevalent since prehistorical times. The discovery of blades of volcanic obsidian at places far away from their origin (near extinct volcanoes) is one such example. Metals like gold and silver, which are found in their native states only in some places, have been found during archaeological excavations all over the world.

The trade of beads, honey, food grains, cotton, silk, pelts, pearls, copper, tin, bronze, and iron and steel increased contact among different cultures. Horses, bred in Arabia, were famed for their speed and endurance and were sold all over the world. The large caravans, carrying goods over large distances, were one of the early conduits of the exchange of stories, legends, information, knowledge, and culture. These were often accompanied by monks and priests, who spread their religion. Even slaves, both young men and women, were taken and sold all over the world, till slavery was abolished. They left the imprints of their knowledge, cuisine, music, legends, and culture in the countries where they were taken against their will. Several of these poor souls rose to great heights of power and wisdom (recall Aesop and Janissaries of the Ottoman Empire), though, unfortunately, most suffered a miserable life.

The trade of spices and other goods brought Western powers to Asia and Africa, which enriched the Western nations beyond their wildest dreams due to colonisation. It was the quest for a sea route to India, which took Christopher Columbus to the Americas, where Homo Sapiens had not gone since about 20,000 years ago (barring occasional Vikings) when the Bering Strait could still be crossed on foot.

Corn, beans, pepper, potato, tomato, turkey, and countless other things from the Americas have now spread to the entire world. Tea, coffee, rubber, cacao, cotton, silk, cane sugar, etc., which used to be cultivated only in some specific regions of the world are now cultivated in places with suitable climates and soil across the world. One must not forget, though, that some plants and animals in this process have also become invasive and are crowding out native species.

We should not, even for a moment, think that small countries may not have anything to offer. The endemic species of the remote Galápagos Islands allowed Charles Darwin to arrive at his Theory of Evolution. Congo is a relatively poor nation. Yet, it has the world's largest known deposit of cobalt, a key metal in the lithium-ion batteries that power everything from smartphones to electric vehicles. The lithium triangle, covering the border areas of Argentina, Bolivia, and Chile, provides most of the lithium produced in the world. Some of the small countries such as Myanmar, Sri Lanka, Cambodia, Thailand, and Afghanistan produce the bulk of ruby in the world, so valuable for lasers and masers and countless industrial applications. The vast deserts of Chile, Australia, South Africa, and Tibet provide some of the best locations for setting up astronomical observatories. The caravan routes across the Gobi Desert continue to throw up valuable treasures (including ancient manuscripts) and artefacts lost during fierce sandstorms during ancient times.

Guar (cluster beans) has been cultivated in the arid regions of Rajasthan and the neighbouring areas of Pakistan for centuries as a vegetable, feed for animals, and for nitrogen fixation till the use of its gum for thickening ice cream was discovered. The advent of fracking in recent years has increased its demand manyfold and its cultivation is becoming popular in other arid regions of the country and the world. Guar gum is used in oil and gas well stimulation, specifically hydraulic fracturing in which high pressure is used to crack rock and release the oil and gas trapped in it. Guar gum makes the fracturing fluid thicker so that it can carry sand into fractured rock. From a lowly vegetable and crop, guar has moved on to become a major foreign exchange earner (up to 450 million

Fig. 1.1. Aristotle and Alexander (Painting by Jean Leon Gerome Ferris).

USD per year), and India continues to meet 80% of the world's demand. The use of fracking is bound to increase as oil wells dry up or supplies are curtailed due to political situations.

1.3 Historical Times

It is said that when Alexander (356–323 BCE) went to see his tutor, Aristotle (384–322 BCE), before embarking upon his mission to conquer the world, Aristotle (Fig. 1.1) requested him to bring back some wise men (teachers) from India so that he could discuss their world view with them and benefit from the exchange of knowledge.

Archimedes (ca. 287–ca. 212 BCE) was so widely respected for his discoveries and contributions in mathematics, physics, engineering, astronomy, and mechanics that when Romans attacked Syracuse, the Roman general Marcus Claudius Marcellus explicitly instructed his soldiers that Archimedes was not to be harmed and was to be brought to him. Unfortunately, he was killed by a soldier, either because Archimedes (Fig. 1.2) said that he was thinking about a geometrical problem and the soldier should not disturb him or because he wanted to carry the geometrical instruments with him, which the soldier mistook as weapons.

Centuries later, fearing a similar fate for Carl Friedrich Gauss (1777–1855 CE) at the hands of the French Army, which was about to enter

Fig. 1.2. Death of Archimedes (Painting by Thomas Degeorge).

Brunswick (where Gauss lived) during a war against Prussia, Sophie Germain (1776–1831 CE) intervened through a relative, who was a General in the French Army, to ensure that Gauss was not harmed. This was remarkable, as being a woman, Sophie was not allowed to join École Polytechnique in Paris and had learnt and contributed extensively to mathematics by corresponding with several mathematicians, including Gauss and Joseph-Louis Lagrange (1736–1813 CE), by assuming a male pseudonym,[2] M. Le Blanc. Archimedes and Gauss, separated in time by about two millennia, had one thing in common: their thirst for knowledge and their ability to transmit it.

Al-Khwarizmi (Fig. 1.3) or Muhammad ibn Musa al-Khwarizmi (ca. 780–ca. 850 CE) was a Persian mathematician and astronomer, who was possibly born in Khwarezm, in present-day Uzbekistan. Around 820 CE,

[2]But then, the famous Bronte sisters (Charlotte (1816–1855 CE), Emily (1818–1848 CE), and Anne (1820–1849 CE)), well-known poets and novelists, had assumed male pseudonyms for publishing their books, which are now considered classics.

Fig. 1.3. Statue of Al-Khwarizmi in Uzbekistan (Credit: Wikipedia).

he was appointed as the astronomer and head of the House of Wisdom (Dar al-Hikma) library in Baghdad. The House of Wisdom acquired and translated scientific and philosophical treatises from Greek and Sanskrit sources and published original research. This was to prove very valuable for the preservation of Greek philosophy and their mathematical and other scientific work, which were to be lost (obliterated?) during the Dark Ages in Europe and were recovered when these were translated back into Latin, centuries later. However, perhaps his most significant work, with far-reaching consequences for mathematics and science, was the introduction of Indian numerals and decimal systems (finally translated as Al-Khwārizmī: *Concerning the Hindu Art of Reckoning*) for the representation of numbers, and the Indian systems of algebra and geometry (finally translated as *The Compendious Book on Calculation by Completion and Balancing*) to the Arab world and Europe. Numbers in Arabic and Persian are called Hindsa (from India) and are still written from left to right and not from right to left, like the rest of the language. His work on maps and compilation of astronomical tables and table of "sine" from Indian sources also proved valuable [1]. Fibonacci (ca. 1170–ca. 1240–1250 CE), working in the court of Frederick II (1194–1250 CE), was to play a leading role in popularising Indian numerals in Europe.

Al-Bīrūnī or, in full, Abū al-Rayḥān Muḥammad ibn Aḥmad al-Bīrūnī (973–ca. 1052 CE), was a Persian astronomer, mathematician, ethnographer, anthropologist, historian, and geographer, who had become well known for his scholarship while living, studying, and working in Khwarezm. When Maḥmūd (971–1030 CE) became the ruler of Ghazna in 998 CE and expanded the Kingdom of Ghazna, he demanded that all scholars move to Ghazna, including the famous philosopher-scientist Ibn-Sina (Avicenna) and Al-Biruni. Ibn Sina managed to escape, but Al-Biruni (Fig. 1.4) moved to Ghazna around 1017 CE and spent the rest of his life there.

Maḥmūd is believed to have vowed to invade India once a year and led about 17 such expeditions, between 1001 and 1026 CE sacking several temple towns along with enormous wealth, including those of Thanesar, Kanauj, and Somnath. Al-Biruni accompanied Mahmud to India and lived

Fig. 1.4. USSR stamp, Abū Rayhān al-Bīrūnī, 1973 (Credit: Wikipedia).

there for some years. This utter destruction had led to a great dislike for foreigners in India. Al-Biruni writes [2]:

> *"Mahmud utterly ruined the prosperity of the country and performed wonderful exploits, by which the Hindus became like atoms of dust scattered in all directions, and like a tale of old in the mouth of the people. Their scattered remains cherish, of course, the most inveterate aversion towards all Muslims. This is the reason, too, why Hindu sciences have retired far away from those parts of the country conquered by us and have fled to places which our hand cannot yet reach, to Kashmir, Benares, and other places. And there the antagonism between them and all foreigners receives more and more nourishment both from political and religious sources."*

In the face of this antagonism, Al-Biruni faced an uphill task in even talking to people, let alone getting any information. Yet, his scholarship and acquisition of knowledge of Sanskrit slowly won the people and the scholars, and he wrote one of the most authoritative and yet critical books, *aḥqīq mā li-l-Hind min maqūlah maqbūlah fī al-ʿaql aw mardhūlah* (variously translated as *Verifying All That the Indians Recount, the Reasonable and the Unreasonable* or *The Book Confirming What Pertains to India, Whether Rational or Despicable*), in which he explored nearly every aspect of Indian life, including religion, history, geography, geology, science, and mathematics.

We pause here to recall that several Buddhist monks (Fig. 1.5) from India, e.g., Kashyap Matang and Dharmaratna, travelled to China as early as the first century CE, while Bodhidharma, called Putidamo in Chinese and Daruma in Japanese, travelled to China and perhaps beyond, during the fifth or the sixth century CE. All of them preached the philosophy of Buddha. Bodhidharma introduced the dhyana (meditation) aspect of the Mahayana system of Buddhism, which is called Chan in China, Zen in Japan, Sŏn in Korea, and Thien in Vietnam, and guided the monks of Shaolin Monastery — famous for their prowess in the martial arts — in meditation and training.

Faxian (337–ca. 422 CE), also called Fa-Hien, and Xuanzang (602–664 CE), also known as Hiuen Tsang, were famous Chinese Buddhist monks and translators, who travelled by foot from China to India to acquire Buddhist texts. They also left valuable accounts of their travels,

Fig. 1.5. Left to Right: Bodhidharma, Faxian (Fa-Hien), and Xuanzang (Hiuen Tsang) (Credit: Wikipedia).

kings, kingdoms, customs, people, and the flourishing universities in India during their visits. These accounts were to prove valuable in providing accurate historical dates for events in India. The books they carried back were translated into Chinese, and several of those are now available only in translation, as many libraries in India were destroyed.

However, these people travelled in peace and goodwill, while Al-Biruni, as indicated above, must have faced a very hostile reception initially in India because of his association with Mahmud Ghazani, and only his scholarship and intellect must have helped him acquire the pieces of information he sought.

1.4 Medieval Times

A very interesting case is provided by Admiral Zheng He (1371–1433 CE), a Chinese mariner, explorer, and diplomat during China's early Ming dynasty.

With the strong support of the then emperor, Admiral Zheng He (Fig. 1.6) led seven huge armadas from China to the far reaches of the Indian Ocean. The largest of these comprised 300 ships and carried close

Fig. 1.6. Admiral Zheng He (1371–1433 CE) (the Wax Statue in the Quanzhou Maritime Museum [3]).

to 30,000 people. They visited Indonesia, Sri Lanka, India, the Persian Gulf, the Red Sea, and East Africa [4].

The largest of his ships called "treasure ships" were about 127 m long and 52 m wide and had four decks. Other ships carried horses, armoury, and fresh water, and carried navigators, explorers, sailors, doctors, workers, and soldiers, along with translators, diplomats, and diarists. The records provided the sailing time or distance, which took account of the local currents and winds, depths, bays, estuaries, capes and islands, ports and mountains along the coast, and important landmarks, such as pagodas and temples, and shoal rocks.

At some stage, Zheng He is believed to have recorded the following: "We have traversed more than 100,000 li (about 50,000 km) of immense water spaces and have beheld in the ocean huge waves like mountains rising in the sky, and we have set eyes on barbarian regions far away hidden in blue transparency of light vapours, while our sails, loftily unfurled like clouds day and night, continued their course [as rapidly] as a star, traversing those savage waves as if we were treading a public thoroughfare..."

And yet, once the emperor passed away, his immediate successor stopped the expeditions, which briefly resumed during his grandson's rule, and then the operation was abruptly terminated. "The great fleet was dismantled, crucial technical and geographical knowledge was lost, and no explorer of such stature and means ever set out again from a Chinese port" [5].

It is now believed that Zheng He's fleet could have easily rounded the Cape of Good Hope (Fig. 1.7) and reached Europe or even the Americas. Yet, the Chinese rulers restricted their interests and ambitions to the Middle Kingdom's immediate environs. Columbus's fleet of 1492 consisted of three small ships, the biggest of which was 26 m long (Fig. 1.8), manned by 120 sailors, but was "exceptional due to their (European's) unparalleled and insatiable urge to explore and conquer" [5].

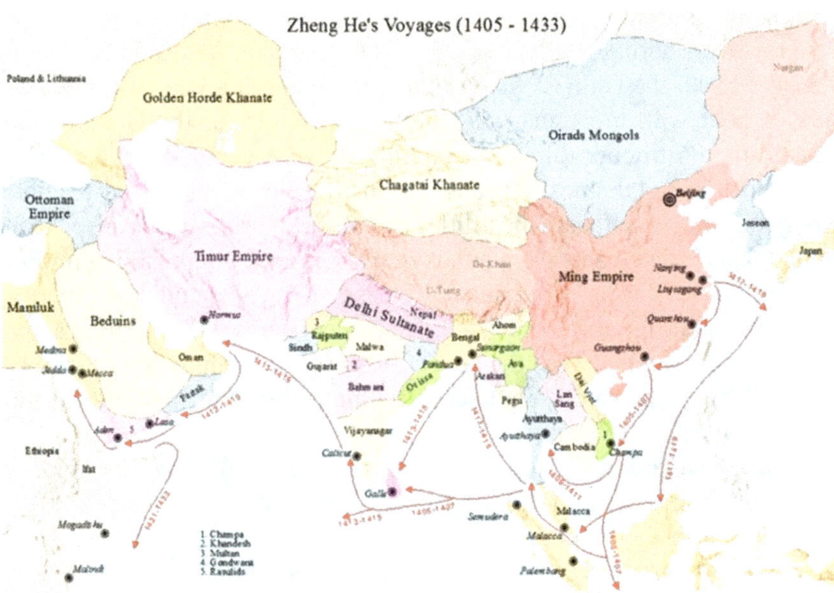

Fig. 1.7. Voyages of Admiral Zheng He [4].

Fig. 1.8. A Comparison of Zheng He's Treasure Ship and Columbus's Ship [6].

We should, however, not forget that the European ships which reached the Americas were guided by a compass, a Chinese invention, and the conquest of the Americas was achieved by using gunpowder, another Chinese invention.

However, during the late 1400s, China retreated into relative isolation. By 1500, the Chinese government had made it a capital offence to build a boat with more than two masts, and in 1525, the Government ordered the destruction of all oceangoing ships. A century earlier, China had a fleet of 3,500 ships. Various reasons have been suggested for this. Firstly, China was facing hostilities from the North. It has also been argued that civil servants were either zealous or afraid of the rising importance of the sea-faring merchants that these voyages entailed and argued that China had everything it needed. We have discussed this point in some detail to point out that even the best efforts at the accumulation of knowledge and its spread can be completely unhinged by an unappreciative and unresponsive society.

While we are discussing this aspect, we recall that the mighty Srivijaya Empire (Fig. 1.9) essentially dominated the entire Malay Peninsula from the seventh century to the twelfth century CE and made no known attempt to reach Australia or New Zealand.

The Srivijaya Empire declined after an invasion in 1025 CE by the Chola Emperor Rajendra I. Even though the Chola forces essentially

Fig. 1.9. Extent of the Srivijaya Empire [7].

controlled the vast and lucrative trade route for several decades (Fig. 1.10), even these did not venture beyond established routes.

It is also surprising to recall that after the Srivijaya fell, it was largely forgotten, until 1918 when the French historian George Cœdès formally postulated its existence.

Returning to our present discussion, we recall that Naṣīr al-Dīn al-Ṭūsī (Fig. 1.11) or Muḥammad ibn Muḥammad ibn al-Ḥasan al-Ṭūsī (1201–1274 CE) was a Persian philosopher, scientist, and mathematician. After his education and early works, he was employed with (or held by) Ismāʿīlītes at Alamut.

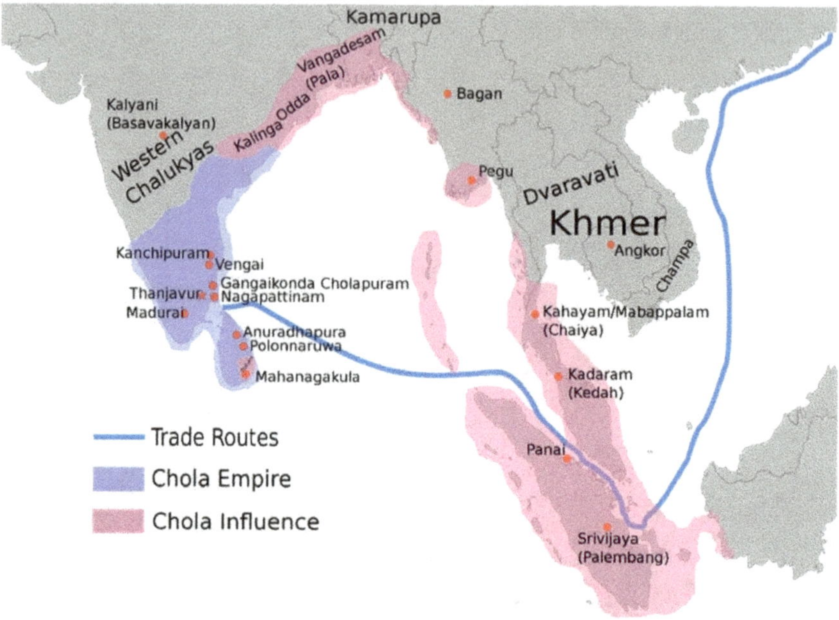

Fig. 1.10. Extent of the Chola Empire during the Reign of Rajendra Chola I (ca. 1030 CE) [8].

When Alamut fell to Hülegü Khan (ca. 1217–1265 CE), the grandson of Genghis Khan, in 1256 CE, al-Ṭūsī accepted a position with the Mongols as a scientific adviser. Later, he persuaded or leveraged the belief of Hülegü Khan in astrology, to build a most advanced astronomical observatory and an excellent scientific library with Indian and Greek treatises, at Marāgheh (now in Azerbaijan) in 1259 CE. He also collected the finest astronomers from the Islamic world and from as far as China. All these activities led to a compilation of the extremely accurate Zīj-i Ilkhānī (1271; "Ilkhan Tables"), tables of planetary movements, which corrected, updated, augmented, and refined the tables of Ptolemy (100–170 CE) for these. There are speculations that these may have helped or inspired Nicolaus Copernicus (1473–1543 CE) to arrive at his heliocentric theory for the solar system. Many consider Al-Tusi as the finest and most illustrious astronomer of the mediaeval Islamic period, with extensive contributions to astronomy, mathematics, architecture, biology, chemistry, philosophy, medicine, physics, science, and theology.

Fig. 1.11. Nasir al-Din al-Tusi at the Observatory in Maragheh (Painting from Persia: 1562 CE, Credit: Wikipedia).

His observatory, unfortunately, survived for only about 50 years and was destroyed by wars, along with the scientific library. However, it inspired the observatory of Ulugh Beg (1394–1449 CE) at Samarkand, the observatory of Taqi ad-Din Muhammad ibn Ma'ruf in Constantinople (Istanbul, 1577 CE), and the 18th-century Sawai Jai Singh II's observatories in India. The Constantinople observatory was destroyed within a few years as religious leaders felt that "it brought the wrath of God," just as the King of Denmark commissioned (ca. 1576–ca. 1580 CE) the observatory of Tycho Brahe (1546–1601 CE), which was to provide data that paved the way for Kepler's Laws.

Maharaja Sawai Jai Singh II (Fig. 1.12) of Jaipur (1681–1743 CE) constructed the observatories called Jantar Mantar (Yantra Mantra or Instruments for Calculations) at Delhi, Jaipur, Mathura, Benares (Varanasi), and Ujjain, during 1724–1734 CE or a few years earlier. The main purpose

Fig. 1.12. Maharaja Sawai Jai Singh II (ca. 1725, Jaipur, Credit: British Museum).

of these was to correct the table of astronomical observations compiled earlier by Ulugh Beg, which were being used to provide times for festivals, prayers, etc. He is believed to have sent emissaries to Europe and Central Asia to collect their tables as well and to have held wide consultations with Indian, French, and Portuguese astronomers before building the construction. These had masonry and metallic instruments, which were to be used with the naked eye during the day or night.

Several speculations had been made about why he opted for naked-eye instruments when observatories using telescopes were already in use in Europe for close to a hundred years. One of the arguments suggests that he was strongly influenced by Indian astronomers, who were still using the naked eye and often handheld instruments, which were subject to the movement of hands. The other argument suggests that while a telescope could not be accessible to all, his large but accurate instruments could be used by all.

Pathani Samanta (1835–1904 CE) living in an isolated hilly area of Odisha used wood and bamboo to make handheld instruments to make very accurate astronomical measurements. He was not even aware of the developments in the West. We shall return to this point in Chapter 13 on astronomy.

Jai Singh's edifices of astronomy did not survive for long. The Mathura observatory was destroyed. The Delhi observatory was severely vandalised when Jawahar Singh, son of Suraj Mal, the Jat King of Bharatpur, plundered Delhi. The Jaipur observatory was converted into a gun factory by Jai Singh's grandson, and he used his ancestral 400 kg brass astrolabe for target practice. Some sloppy efforts at repairing the cracks in the masonry affected the accuracy of several instruments. There have been some efforts to restore the surviving structures.

Monir Uddin Ahmed *et al.* have discussed the possible scientific collaboration among the nations of South Asia, namely Afghanistan, Bangladesh, Bhutan, India, Nepal, the Maldives, Pakistan, and Sri Lanka [9]. In 1801 CE, a Peace and Friendship Treaty between Nepal and the East India Company was signed. It allowed the Scottish explorer Francis Buchanan-Hamilton to collect 2,500 plant and animal species from Nepal, out of which 1,100 were new to Western science. This led to the publication of the first book on the flora of Nepal, in 1825 [10]. This is believed to be the first systematic scientific work in Nepal to be supported by a treaty, a pillar of science diplomacy.

1.5 Modern Times

Consider the case of Alexander von Humboldt (1769–1859 CE), a German naturalist and explorer (Fig. 1.13), who became a leading figure during the classical period of biogeography and physical geography, penning his detailed observations about the Americas in a set of 30 books, and who played a major role in the popularisation of science through his series of five books (he expired before completing the fifth book), called Kosmos.

Earlier, because of his scientific reputation in documenting the geographic and geological details of parts of Germany, he was invited by various groups to join their scientific expeditions. However, he was unable to

Fig. 1.13. Alexander von Humboldt (Portrait by Joseph Karl Stieler [Credit: Humboldt Foundation]).

join these due to the upheavals caused by the Napoleonic Wars (1803–1815 CE). He secured permission to visit Spanish colonies in Central and South America (Venezuela, Cuba, the Andes, and Mexico), which were accessible only to the Roman Catholic Mission and Spanish officials, as well as the United States of America. Even the start of this journey was fraught with danger, as Spain and England were at war and their ship risked being attacked by English warships.

He, along with a French botanist Aimé Bonpland, covered more than 9,600 km by canoe, horseback, and foot from 1799 to 1804. They crossed mighty rivers, climbed some of the highest peaks including many volcanoes, took measurements of longitude, latitude, rain, temperature, barometric pressure, and geomagnetic field, and collected vast amounts of notes about the flora and fauna, as well as a collection of plants. In the process, he discovered the existence of an ocean current off the west coast of Latin America (Humboldt Current, now renamed as Peru Current) and the fact that "mountain sickness" was caused by a lack of oxygen and rarefied atmosphere at greater heights. He also made several observations about the interdependence of flora and fauna, which were later to be taken to their natural height by Charles Darwin, and recorded the effects of deforestation. His *Political Essay on the Kingdom of New Spain* provided valuable information about geology, geography, political, social, and economic conditions, as well as a description of the silver mines of Mexico.

He spent the period 1804–1827 in Paris, getting his 30 books complete with the maps and details of atmospheric and geographic data as well as flora and fauna and published using his personal funds. His trip had also been undertaken using his personal funds. In his writings, Humboldt made passionate pleas against the inhumanities of slavery and colonialism, which became an impediment later as the British East India Company did not allow him to visit India, which he had very much wanted, with a special desire to explore the Himalayas. During this period, he was highly respected and admired for his generosity, humanity, and vision of what science could do. Moreover, he was always willing and eager to help young scientists with guidance, financial assistance, and introductions to the scientific community.

Later, as his funds dwindled, he took up a job as a tutor to the crown prince in Berlin as well as some court duties. He lectured at the University of Berlin and to the common public in one of the first instances of science outreach. He also organised one of the first international scientific conferences in 1828, which was not exactly liked by the governments due to the Napoleonic Wars. He was also to organise perhaps one of the first international collaborations to simultaneously measure the sudden fluctuations of geomagnetic fields across the world (Germany, Sweden, England, Canada, South Africa, Australia, and New Zealand). This wealth of data was used by Sir Edward Sabine to correlate these "magnetic storms" to the periodic movement of sunspots.

During this period, he was invited by the Czar of Russia to visit the Russian Empire and explore its platinum mines in the Ural Mountains. The trip was financed by the Czar, and he travelled comfortably and explored deeply, right up to the border of China. However, Humboldt had to give an undertaking to refrain from commenting on the political situation of the country whose despotism he abhorred. As mentioned above, the British East India Company officials denied his repeated requests to travel to India.

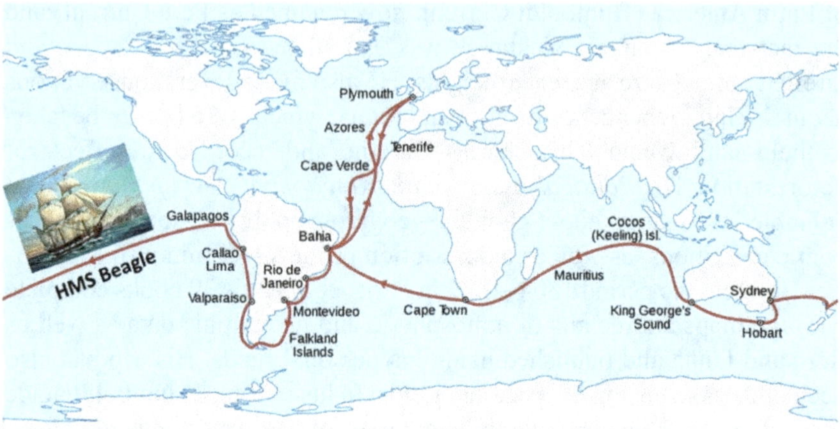

Fig. 1.14. Round-the-World Voyage of HMS Beagle (1831–1836 CE) (Credit: Wikipedia).

Upon his return, Humboldt busied himself with writing the series of books, Kosmos, describing the structure of the universe, as known at the time in a simple and riveting literary style. These books are remembered for conveying the excitement and aesthetic pleasure of scientists in their discoveries.

Several decades later, Rabindranath Tagore wrote a wonderful short book *Vishwa Parichay* (Introduction to Universe) in Bangla (1937), which captured some of the most outstanding developments in sciences, which shaped our understanding.

Charles Darwin (1809–1882 CE) continued the legacy of Alexander von Humboldt, when he joined a voyage around the world (Fig. 1.14) on HMS Beagle (1831–1836 CE) undertaken by Captain Robert FitzRoy as a "gentleman companion" using funds provided by his father, and collected samples of birds, turtles, fossils, flowers, animals, plants, fish, ants, bugs, etc., made careful observations, and took extensive, careful notes. He mulled over these and produced arguably one of the most important and influential books of all times, *On the Origin of Species* (1859), which firmly established the law of natural selection:

> *"As many more individuals of each species are born than can possibly survive; and as, consequently, there is a frequently recurring struggle for existence, it follows that any being, if it vary however slightly in any manner profitable to itself, under the complex and sometimes varying*

conditions of life, will have a better chance of surviving, and thus be naturally selected. From the strong principle of inheritance, any selected variety will tend to propagate its new and modified form" [11].

Dr. Dwarkanath Kotnis (1910–1942) responded to a call by Chinese leaders during the Second Sino-Japanese War in 1938 and went to China along with four others to treat wounded Chinese soldiers. It was a very strenuous job with limited medical facilities; they treated up to 800 soldiers every day, sometimes working for 72 hours continuously without sleep or rest. Dr. Kotnis (Fig. 1.15) became extremely popular with the people of China and learnt to read and write Chinese and sing Chinese songs. While his colleagues returned after a few years, Dr. Kotnis stayed back to serve the Chinese and married there. Unfortunately, the strenuous working conditions took their toll and he succumbed at the very young age of 32.

His memory is kept alive by the Chinese, and Madame Sun Yat-Sen famously remarked, "His memory belongs not only to your people and ours but to the noble roll-call of fighters for the freedom and progress of all mankind. The future will honour him even more than the present because it was for the future that he struggled."

Mao Zedong emphasised the international spirit behind his contribution by adding, "The army has lost a helping hand, the nation has lost a friend. Let us always bear in mind his internationalist spirit."

Fig. 1.15. Dr. Dwarkanath Kotnis (1910–1942).

Fig. 1.16. Tu Youyou (1930–) (Credit: Wikipedia).

The south side of Martyr's Memorial Park and a museum in the Shijiazhuang city of the Chinese province of Hebei are devoted to his memory.

Tu Youyou (1930–) was appointed the head of a secret Project 523 (as it started on May 23) in 1967, when during the Vietnam War, Ho Chi Minh, the then President of North Vietnam, asked the Chinese to help with a cure for malaria, which was taking a large number of lives. The malaria parasite had become resistant to chloroquine, the standard medicine for malaria. She and her colleagues searched ancient Chinese texts to understand how the ancient Chinese handled malaria. By then, about 240,000 compounds had already been tested for use against malaria. The team found a reference to sweet wormwood (*Artemisia annua*) which had been used in China in 400 CE to treat "intermittent fevers." After some initial missteps, Tu Youyou (Fig. 1.16) and colleagues isolated artemisinin (qinghaosu), which has saved millions of lives in the developing world. She received the Nobel Prize in Medicine in 2015.

1.6 The Landing on the Moon

On July 20, 1969, Neil Armstrong and Edwin (Buzz) Aldrin landed on the surface of the Moon (Fig. 1.17). Even though it had its origin in the race for space between the USSR and the USA, it remains one of the most outstanding achievements of mankind. As Yuval Noah Harari famously remarked, it was for the first time in about 4 billion years of the

Fig. 1.17. Neil Armstrong on the Moon (Credit: NASA).

appearance of life on Earth, that any species had overcome its gravitational field and travelled to another cosmic body [5].

The entire event was broadcast live on television across the world. Still, there was a fear that the Lunar Module could fail, and the two astronauts could meet a painful death, being witnessed by all. It was decided that if this happened, the President of the USA was to call the families of the astronauts and then deliver the following speech [12]:

> *"Fate has ordained that the men who went to the moon to explore in peace will stay on the moon to rest in peace.*
>
> *These brave men, Neil Armstrong and Edwin Aldrin, know that there is no hope for their recovery. But they also know that there is hope for mankind in their sacrifice.*
>
> *These two men are laying down their lives for mankind's most noble goal: the search for truth and understanding.*
>
> *They will be mourned by their families and friends; they will be mourned by their nation; they will be mourned by the people of the world; they will be mourned by a Mother Earth that dared send two of her sons into the unknown.*
>
> *In their exploration, they stirred the people of the world to feel as one; in their sacrifice, they bind more tightly the brotherhood of man.*
>
> *In ancient days, men looked at stars and saw their heroes in the constellations.*
>
> *In modern times, we do much the same, but our heroes are epic men of flesh and blood.*

Others will follow and surely find their way home. Man's search will not be denied. But these men were the first, and they will remain the foremost in our hearts.

For every human being who looks up at the moon in the nights to come will know that there is some corner of another world that is forever mankind."

Fortunately, things moved smoothly, and there was no need to deliver this speech. However, we do not think that there could be a better expression of the role of science in bringing out the best in mankind and in nurturing universal brotherhood.

In the pages to follow, we celebrate several international collaborations in sciences in all their glory.

References

[1] B. Kamble, "The Imperishable Seed: How Hindu Mathematics Changed the World" (Gurugram: Garuda Publications, 2023).

[2] "*Al-Beruni's India: An Account of the Religion, Philosophy, Literature, Geography, Chronology, Astronomy, Customs, Laws, and Astrology of India about 1030 AD*" (English Language edition with Notes and Indices by Dr E. C. Sachau).

[3] By jonjanego. https://www.flickr.com/photos/jonjanego/2777527481/, CC BY 2.0, https://commons.wikimedia.org/w/index.php?curid=12114789.

[4] By SY — Own Work, CC BY-SA 4.0, https://commons.wikimedia.org/w/index.php?curid=65379957.

[5] Y. N. Harari, "*Sapiens; A Brief History of Humankind*" (London: Vintage 2011).

[6] https://www.digitalhistory.uh.edu/active_learning/explorations/1492/1492_zhenghe.cfm.

[7] G. Kartapranata, https://commons.wikimedia.org/wiki/File:Srivijaya_Empire.svg.

[8] https://en.wikipedia.org/wiki/Chola_invasion_of_Srivijaya#/media/File:Rajendra_map_new.svg.

[9] M. U. Ahmed *et al.*, "An Overview of Science Diplomacy in South Asia", *Science and Diplomacy* (February 2021), https://www.sciencediplomacy.org/article/2021/overview-science-diplomacy-in-south-asia.

[10] D. Don and F. Hamilton, "*Prodromus Florae Nepalensis*" (London: J. Gale, 1825).
[11] C. Darwin, "*On the Origin of Species by Means of Natural Selection or the Preservation of Favoured Races in the Struggle for Life*", 1st edn. (London: John Murray, 1859).
[12] https://www.archives.gov/files/presidential-libraries/events/centennials/nixon/images/exhibit/rn100-6-1-2.pdf.

Chapter 2

Migration of Birds Across the Nations

> Migration is one the most fascinating natural phenomena, and it is still a highly mysterious one.
>
> — Klaus Schmidt-Koinig, *Avian Orientation and Navigation*, 1979

2.1 Introduction: Cranes

The Krauncha pair on the bank of the Tamasa River was completely obvious to its surroundings and was performing the most beautiful ritual dance of courtship, complete with coordinated bows, running, jumping, and tossing of tufts of grass in the air, as the hunter aimed his arrow and killed one of them. The bereaved mate grieved and gave out a cry of anguish. Sage Valmiki, a witness to this painful episode was so moved that he cursed the hunter:

> मा निषाद प्रतिष्ठां त्वमगमः शाश्वतीः समाः।
>
> यत्क्रौञ्चमिथुनादेकमवधीः काममोहितम्।।
>
> — वाल्मीकि रामायण, 1-2-15

> O hunter! May you not attain fame for many long years as you have killed one of the unsuspecting, devoted, and loving Krauncha birds.
>
> — Valmiki Ramayana, 1-2-15

This became a turning point in Indian tradition, as till then all the literature (in Vedas) was believed to have been revealed and not composed.

Fig. 2.1. Krauncha (Demoiselle Cranes) (Credit: Wikipedia).

Valmiki is revered as Adi Kavi (First Poet), who went on to compose the epic Ramayana, which had the most profound effect on the traditions of India and Southeast Asia.

Krauncha birds (Fig. 2.1) are cranes. They were named Demoiselle by Queen Marie Antoinette because of their delicate, maiden-like appearance, complete with deep orange eyes, long, luscious neck feathers, and slender legs. Demoiselle cranes breed in Central Eurasia, from the Black Sea to Northeast China and Mongolia, and spend their winter in the Indian subcontinent. They are symbols of steadiness, fidelity, mutual help, happiness, longevity, marital and maternal bliss, and good luck, and have found expression through art, legend, religion, mythology, and paintings of many cultures.

Demoiselle cranes are known for enduring one of the most difficult migrations in the world, as they cover up to 5000 km and cross many countries as well as the Himalayas, for which they must fly at a height of 5000–8000 m. Many of them die on the way due to exhaustion,

Fig. 2.2. Sarus Crane (Credit: Sumankalyan Saha).

hunger, and predation by people as well as by the Golden Eagle. However, it is also a matter of great happiness that a particular village, Kheechan, on the edge of the Thar Desert has become a very favourite visiting place for them, where they arrive in thousands by November and stay till March, giving a festive look to the village. The villagers feed them a variety of grains and provide them with protected ponds, dunes, and shades of trees.

While most other species of cranes migrate long distances, the elegant and delightful Sarus Crane (Antigone, Fig. 2.2) does not migrate and lives in wetlands, marshes, and rice fields of the Indian subcontinent, Southeast Asia, and Australia. In India, they are considered symbols of marital fidelity and are believed to mate for life and pine the loss of their mates, even to the point of starving to death. During colonial rule, they, especially the young ones, were hunted for meat.

While in many other parts of the world, they are facing extinction, in India they are revered and live near humans, though the numbers are not very large. In Nepal, Sarus Cranes are revered due to their legendary association with young Buddha, who had saved one and nursed it back to health after it was shot by his cousin Devadatta. Nepal has established a thriving crane sanctuary near Lumbini, the place of birth of Lord Buddha.

These have been extirpated in Malaysia and the Philippines, and are being reintroduced by importing the birds from Thailand. There are fears that the plans for the development of floodplains in Queensland Australia may put their survival there in jeopardy. This is a case where a collaboration between nations can save this marvellous bird from becoming extinct.

2.2 Arctic Terns, Albatrosses, and Penguins

The Arctic Tern (Fig. 2.3) holds the record for the longest migration of the entire animal kingdom. They breed on the coasts and tundra of the Arctic and subarctic regions of Europe, Asia, and North America during summer, and then they follow the Sun and fair weather on their yearly journey to the Antarctic Circle (reaching again during the summer there!), thus enjoying two summers every year. They have been found to cover 70,000–90,000 km in the process.

They often take a somewhat convoluted course to take advantage of prevailing winds and meander across oceans and continents rather than fly

Fig. 2.3. Attacking Arctic Tern (Credit: A. Weith, Wikipedia).

directly north or south. Even though they are abundant, about one million in number now, their habitat is shrinking, and it will be strongly impacted by climate change. The rise in sea temperature is driving away their prey, and an increasing number of sea storms, wind turbines, and deep-sea drilling are posing danger to them along their journey.

Climate change is also changing the weather pattern and availability of food, along their transit route. Given this, measures to protect migratory species taken at the national level are not likely to be very effective. The threat to their survival must be handled along the entire route of their journey. Thus, only a strong, determined, and vibrant international collaboration between all the nations can help save them.

An interesting case is provided by the Wandering Albatross (*Diomedea exulans*), which spends most of its life aloft, circling the Earth over the oceans of the Southern Hemisphere (about 30 degrees south to 60 degrees south) and may cover up to 30,000 km. They alight only to eat and breed. Some of them are known to circle the globe three times within a year.

Even flightless birds migrate. The Adélie penguins (*Pygoscelis adeliae*), living in the Ross Sea region in Antarctica migrate an average of about 13,000 km each year as they follow the Sun from their breeding colonies to winter foraging grounds and back again, keeping close to the edge of the ice.

2.3 Amur Falcons

The Amur falcon (Falco amurensis, Fig. 2.4) is a fascinating migratory raptor. They breed in Southeastern Siberia and Northern China. And then they migrate to South Africa, Kenya, and Somalia in large numbers, with a stopover in India and flying over the Arabian Sea. This arduous return journey of up to 20,000 km includes a non-stop flight over the Arabian Sea after passing across India. They congregate in large numbers in Nagaland, Manipur, and other Northeastern states of India, arriving in late October and staying for about 45 days before flying on to Africa. On their return journey, they take a route lying further north. Their numbers, congregating at Doyang Reservoir constructed near Pangati village in Nagaland, now literally run to hundreds of thousands, lending a great festive look to the skies. They eat mostly termites and wonderer dragonflies, which follow similar routes for migration between India and South Africa. In South Africa, farmers await their arrival to feed on locusts and help get farmlands ready for the next crop.

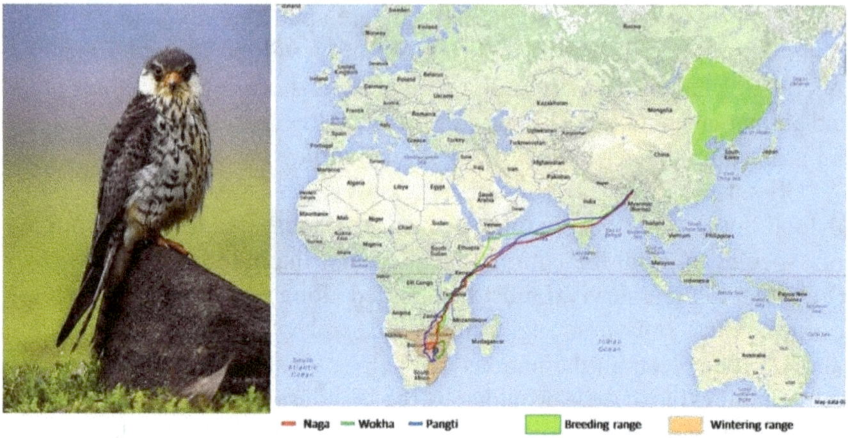

Fig. 2.4. Left: Amur Falcon (Credit: Wikipedia); Right: Journey of Three Amur Falcons; Naga, Wokha, and Pangti [1].

Yet, until 2012, 10,000–12,000 of them were killed every day in Nagaland for their meat. A documentary maker and conservation enthusiast from Nagaland, Bano Haralu, a former TV journalist and founding trustee of Nagaland Wildlife and Biodiversity Conservation Trust produced a documentary (*The Pangati Story* [2]) about their gruesome slaughter and brought it to the notice of authorities. The Amur falcon is protected under the Indian Wildlife Protection Act of 1972. Amid national and global outcry, the dialogue to protect the falcons bore fruits as India had joined the Convention of Conservation of Migratory Species of Wild Animals (1983) of the United Nations, under which it is "duty-bound" to provide safe passage and ensure the conservation of migratory species.

As a result, their hunting was banned, the ban was strictly implemented and an educational programme involving students and communities was initiated with the help of the Nagaland Forest Department and Wildlife Trust of India. Extreme interest and social awakening were generated by radio-tagging a few birds with the help of the Wildlife Institute of India, Dehradun, and tracking their journey and sharing it via social media. The study of the route, timing, and cause would require collaboration by scientists and even more importantly "citizen scientists" reporting from across the world about their movements and sightings, as some of them do drift from the most popular routes [1] due to the alteration of habitat because of climate change or usage of land. A similar programme was enthusiastically implemented in neighbouring Manipur.

2.4 Cuckoos

Most avian visitors to India and other temperate regions of the Earth migrate from North to South. In India, they start arriving around October from Europe and Northern Russia and leave by March to return home. Pied Cuckoo (*Clamator jacobinus*, Chatak, or Papeeha), with its call of “peeu” and its slight variations, arrives in India in late May or early June from Africa and goes back by the beginning of winter. Its arrival heralds the onset of monsoon in the foothills of the Himalayas and plains of North/Central India. Interestingly, the ones found in South India and the Southern African continent are resident birds and do not migrate.

Its call has been lovingly metamorphosed into a call for “the beloved” by poets and writers from ancient times. And its appearance, without fail, just before the rains gave rise to a persistent legend, repeated numerous times in poetry, that it drinks water only from the first rains. Bhartrahari, a famous fifth- or sixth-century Sanskrit poet addressed the Pied Cuckoo:

रे रे चातक! सावधान मनसा मित्र! क्षणं श्रृयताम्।
अम्भांदा बहवो हि सन्ति गगने सर्वेऽपि नैतादृशाः।।
केचिद् वृष्टिभिराद्रयन्ति वसुधां गर्जन्ति केचिद् वृथा।
यं यं पश्यसि तस्य तस्य पुरतो मा ब्रूहि दीनं वचः।।

> O’ Dear Chatak, my friend! Just listen carefully to me. The sky is full of clouds, but all of them are not alike. Only some of them give rain and drench the Earth, while others just thunder aloud. You should not be begging for water from each one of them!
>
> — Bhartrahari, *Neeti Shatak*: 47

It was unjustly named Brain Fever Bird by the British colonial rulers of India. India’s agricultural economy is based on the monsoon, and thus, it is considered one of India’s most auspicious seasons. Farmers of India have traditionally relied on the arrival of the Pied Cuckoo (Fig. 2.5) as a signal to sow seeds, as they know that the monsoon will soon be upon them. This signal is never wrong. No wonder the Pied Cuckoo in North

Fig. 2.5. Pied Cuckoo (*Clamator Jacobinus* or Chatak).

and Central India is such a welcome sight. There are recurring reports that suggest that the arrival of the Pied Cuckoo closely coincides with the variation in the arrival of the monsoon winds. This suggests that they take advantage of these to fly to India. They likely take advantage of "trade winds" blowing in the opposite direction to get back to Africa as the winter sets in North India. Given their importance as a valuable indicator of rain, studies have been initiated by tagging some of these birds with solar-powered transmitters and tracking their journey by the Indian Institute of Remote Sensing in collaboration with the Dehradun-based Wildlife Institute of India and the Department of Biotechnology, Government of India. The study is also expected to help gather data and trends of climate change along its flyway.

A similar study undertaken in China revealed that the Common Cuckoo (*Cuculus canorus*) migrates from Kamchatka and Mongolia to Angola over 12,000 km: the longest migration of any land bird.

The lakes and ponds of India, especially those in West Bengal, and Odisha, provide a haven for migratory birds, which arrive in millions with the onset of winter. Bird sanctuaries of Bharatpur, Chilka, Ranganathittu, and many others are quite famed for their winter guests. However, their habitat, especially ponds and rivers, is facing a severe crisis of pollution and scarcity of water, forcing them to change their migration patterns, and thus researchers around the world have been documenting how various bird species are adapting to climate change and variations.

2.5 International Collaborations

We have taken just a few representative examples to illustrate the migration of birds. It is believed that up to 20% of the birds migrate. Billions of birds around the world migrate in search of breeding and wintering grounds each year. Depending on the season, we have seen that some bird species migrate over huge distances to take advantage of the varying environmental conditions for themselves and their offspring. This allows them to benefit from the high food supply in one area while raising their offspring before migrating to other regions with more favourable living conditions.

We should remember that migratory birds play an essential and indispensable role in the ecosystems they reside in and travel through. When they raise their chicks, they devour a vast number of insects and other organisms that harm the environment and crops, and act as pest control. It is believed that locust attack is caused by the absence of birds. It is also felt that birds disperse seeds along their route and even fish eggs! The droppings of birds, known as guano, are rich in nitrogen and act as organic fertilizers. In the Cappadocia region of Turkey, people make shelters for birds to collect their droppings to be used in their vineyards as fertiliser.

It is estimated that up to one billion birds are lost every year during this North–South (Europe–Africa) migration alone due to the drying up of water bodies, climate change, storms, poaching, hunting, trapping, and changes in the availability of food as more and more land is brought under cultivation, use of pesticides and insecticides, pollution, collision with high rise buildings, power lines, windmills, and exhaustion, as they do not find a safe place to rest and feed during their long journeys. Given their importance for the biodiversity of the Earth, an international collaboration between all nations is imperative to save these magnificent creatures. Many of them show site fidelity and return to it year after year to the delight of observers. They use stars, the Sun, the magnetic field of the Earth, and landmarks to guide them along their long journeys, which are often facilitated by thermals in the atmosphere.

2.6 Animals, Fish, and Whales

Many wild animals too migrate over long distances. In Tanzania, for example, wildebeests (*Connochaetes taurinus*), zebras (*Equus quagga*), and gazelles (*Eudorcas thomsonii*) migrate in huge herds (up to two million of them), which is the largest mammal migration on Earth. They roam the Serengeti (Tanzania and Kenya) looking for fresh grass and water, which are hard to find during the dry season. Their journey is not easy, as they encounter crocodile-infested rivers, predators, and natural disasters such as droughts and flooding in a daily struggle for survival.

Salmon and Hilsa start life in rivers and migrate to the sea to feed and grow. After spending several years in the ocean, they migrate back to the rivers they were born in so that they can spawn. Some studies suggest that possibly the smell of the place of their birth gets imprinted on their brains, and they follow it to locate it years later. Pollution of the river bodies is likely to disorient them.

Humpback whales live in all the oceans of the world. Their numbers had reduced close to extinction by 1985 when almost 95% of them were killed. Finally, a moratorium on their killing since then has led to an increase in their number. They cover great distances every year, their migration being one of the longest of any mammal on the Earth. Some populations swim over 8000 km, from tropical breeding grounds to colder, more productive feeding grounds in the Polar regions. Many still get killed by entanglement in fishing nets. Their protection became possible only by an international collaboration.

Getting killed by entanglement in fishing nets was also a fate that awaited the gentle whale sharks when they came to the coast of Gujarat to deliver babies. A joint movement of social awareness involving students, religious leaders, NGOs, and government officials has put an end to it. They succeeded in persuading the fishermen to release them by cutting their fishing nets, for which they are partially compensated. In a very moving emotional appeal, the whale sharks were compared to married daughters, who come to the home of their parents to deliver their babies, in an age-old Indian tradition. Till about 2003, up to 200 or more of them used to be killed every year. This killing has stopped. Their movement has been extensively studied in collaboration with scientists from Australia by tagging and tracking them. The study also revealed that they spend most of their time in the Arabian Sea between India and Africa, and do not travel to Australia, as originally believed.

2.7 Butterflies

The monarch butterfly (*Danaus plexippus*) has possibly the most highly evolved migration pattern of any insect. They migrate to avoid the freezing temperatures of Canada and the Northern USA during winter, and they fly all the way to Mexico. Their return journey is completed over many generations, as they stop to lay eggs (on milkweed plants) along the way. The caterpillars eat the milkweed and then finish the journey as butterflies! Climate change is likely to disrupt their annual migration pattern by affecting weather conditions in summer breeding grounds as well as their wintering grounds. A hotter, drier summer could shift suitable habitats north, and colder, wetter winters could be lethal to these creatures. Habitat loss, both in Mexico and the USA, is also creating unprecedented hurdles for them, as mountain forests, their winter habitat in Mexico, are being destroyed by human activity, while the use of herbicides is decreasing the availability of their primary food source, the milkweed plant (Asclepias).

Many birds and mammals inhabiting the mountain peaks migrate to lowlands during winter in what is called altitudinal migration.

2.8 Conclusions: Migratory Bird Day

World Migratory Bird Day, observed on the Second Saturday of October, has been proving to be very valuable in raising global awareness about ecological importance and the need for international cooperation to conserve migratory birds and minimise the dangers these visitors, great indicators of the climate along their flyways spanning scores of countries, face. Thus, the theme for the year 2022 was "Light Pollution," which "is known to adversely affect many bird species. Light pollution is a significant threat to migratory birds, causing disorientation when they fly at night, leading to collisions with buildings, perturbing their internal clocks, or interfering with their ability to undertake long-distance migrations [3]." It was "Water and its importance for migratory birds" for the year 2023 and "Protect Insects, Protect Birds" for the year 2024.

There are fears that ongoing climate change may put a large number of species of birds at risk of extinction. Climate change is also leading to a growing mismatch between the date when birds arrive in their breeding territories and the peak dates for blooming plants, emerging insects, spiders, seeds, and other resources that breeding birds and their chicks depend on.

For the conservation of migratory birds to succeed, different countries need to cooperate to ensure that birds are conserved across their range. The world has already lost many magnificent migratory birds with their wonderful and resplendent plumage, captivating calls, delightful chirpings, melodious songs, and graceful flights. It has been found that birds in their wondrous variety, are a fascinating subject for creating an interest in their scientific study and just "birding," with groups across the world, especially students, sharing information about their appearance and their photographs, which should delight anyone engaged in scientific outreach.

While one may think of occasions when wild animals get into conflict with humans, e.g., when wild elephants, wild boars, or *neel gais* destroy their crops, especially at the edge of forests, it is hard to recall any instance when birds cause them any problem (they may carry viruses, which needs to be studied). We have seen that they help mankind by consuming vast quantities of insects and pests.

The Bonn Convention mentioned earlier suggested that the signatory parties (countries)

(a) should promote, cooperate in, and support research relating to migratory species;
(b) shall endeavour to provide immediate protection for migratory species.

The contribution of the birds to man's desire and success in flying transformed the world forever, and now, man aspires to populate other planets! Let us not forget that the fastest fighter jets are modelled on the Peregrine Falcon and the fastest (bullet) train engines are modelled on Kingfishers. The fascinating study of an extremely rapid murmuration of starlings has given rise to the algorithm of swarm intelligence used extensively in computerised optimisations, acquisition of library items, communications, categorisation of medical datasets, dynamic controls, planning of heating systems, and tracking and prediction of moving objects, etc.

One of the most beautiful sights in the autumn (September/October) in India is the sky full of scattered clouds and flocks of migratory birds arriving from the northern hemisphere, in a typical V-formation. The sight is repeated in April when they fly back. A study of their flight and formation, and its adoption by air force pilots, led to a saving of up to 20% of fuel.

We need to act together to protect these wonders of creation and marvels of nature to protect our biodiversity.

References

[1] R. Suresh Kumar, "Flight for Freedom", *Saveus* 3 (2014) 24–31, https://www.deccanherald.com/science/satellite-tagged-amur-falcons-fly-back-to-manipur-village-after-29000-km-migratory-route-908651.html.

[2] https://www.youtube.com/watch?v=kJrPg2rWav0.

[3] https://www.worldmigratorybirdday.org/news/2021/world-migratory-bird-day-2022-theme-focus-light-pollution.

Chapter 3

Coffee and Tea: The Beverages Over Which Knowledge is Exchanged

"*A mathematician is a device for turning coffee into theorems.*"

— Alfred Renyi (1921–1970), Hungarian mathematician

"*There is something in the nature of tea that leads us into a world of quiet contemplation of life.*"

— Lin Yutang (1895–1976), *Importance of Living*

3.1 Introduction

When one of the most famous scientists of his time, Prof. Paul A. M. Dirac (1902–1984 CE), visited India during 1954–1955, he paid a visit to the other intellectual giant of that period, Prof. Satyendra Nath Bose (1894–1974 CE). Prof. Bose, to give Prof. Dirac a flavour of the intellectual atmosphere of Calcutta, took him to the famed Coffee House on College Street in the city (Fig. 3.1).

Indeed, coffee houses or "kahava-khanas" have remained a meeting place for writers, thinkers, poets, singers, performers, painters, scientists, journalists, revolutionaries, and politicians from the early 16th century when these spread from Mecca to Turkey, Persia, England, Europe, and America. In America, the resentment against the high taxes levied on tea by Britain led to the famous "Boston Tea Party" that involved the dumping of tea chests into the sea and a switch to coffee in place of tea which

Fig. 3.1. P. A. M. Dirac and S. N. Bose in College Street Coffee House in Kolkata 1954–1955 (From L to R: Tapen Roy, Prof. S. N. Bose, Prof. P. A. M. Dirac, Prof. Gagan Behari Bandopadhyay, and Purnima Sinha).

played an important role in America's ultimate liberation from the British colonial yoke.

These often-boisterous places, where the consumption of coffee stirred the intellect and the emotions of people (and prompted them to think), were frowned upon by the clergy and the authorities, and they even tried to close them down at different times either in the name of religion or in the name of disturbance to public order. However, good sense always prevailed because of the protests and the inherent good qualities of coffee — in contrast to beer and wine, which made people lethargic, and the bans were lifted — allowing people to have their intellectual pleasure. In the early days, women were not allowed to enter coffee houses except to serve, but this practice was dropped after protests from them. By now, more than 2 billion cups of coffee are consumed every day across the world, filling people with energy and alertness, with the highest consumption seen in the Scandinavian countries (Fig. 3.2).

Several methods for the preparation of coffee have been developed across the world, and connoisseurs strongly defend their preferences. Thus, for example, in South India, coffee is brewed, preferably using freshly roasted and freshly ground beans, a special filter, and fresh milk, and served in traditional brass tumblers, frothing to the brim. People in South India call it "kaapi," and their day starts with it (Fig. 3.3). Charred

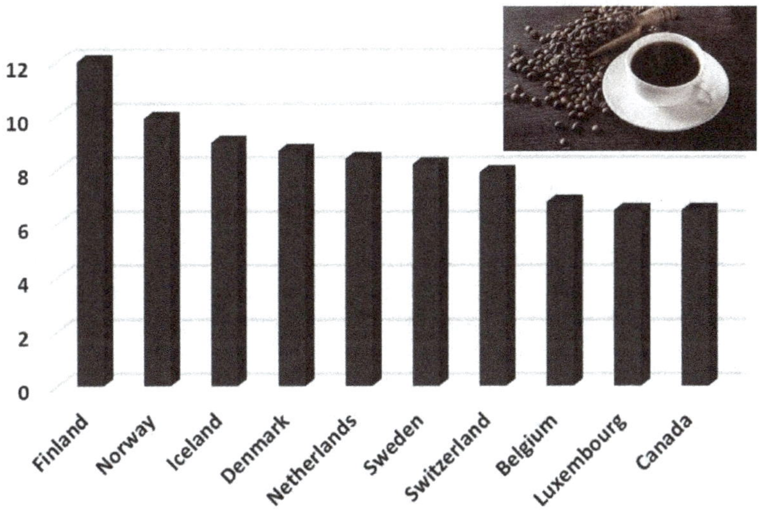

Fig. 3.2. Top 10 Coffee-Consuming Nations of the World.

Fig. 3.3. South Indian Filter "Kaapi."

grain, mostly corn and powdered root of chicory, is sometimes added to coffee to enhance its taste or as a substitute for coffee in poorer regions of the world. A recent survey reported that this "kaapi" from South India is the second-best coffee in the world.

3.2 Discovery of Coffee

While the history of the discovery of coffee is lost (never properly recorded?) in the mists of time, at least three enchanting legends survive. According to perhaps the oldest legend, sometime around 850 CE, a goat-herd named Kaldi, or perhaps Khalid, from Ethiopia noticed that his goats took to jumping and dancing and they became so energetic that they even attacked wolves after they ate some berries from a plant. He chewed the berries himself and felt a similar flush of energy and wakefulness (Fig. 3.4). He took the berries home, and his wife advised him to take them to a neighbouring Sufi monastery. The monks felt that it was a handiwork of Satan (Devil) and threw the berries into the fire. Unknown to them at the time, roasting released the characteristic aroma of the

Fig. 3.4. Kaldi with His Goats After Chewing Coffee Beans. (Credit: www.coffeecross-roads.com/coffee-history/origin-of-coffee-kaldi-and-dancing-goats.)

coffee, which arises from the very complex chemical transformations within the beans. As a result, soon the monastery was filled with the aroma of the roasted coffee. They took the roasted beans out, ground them, and added water to the powder, drank the brew, only to find that it helped them remain alert, improved their concentration, and helped them stay awake during night-long prayers.

There are suggestions that Ethiopians used to consume clarified butter (ghee) laced with coffee powder or coffee beans to stave off hunger and work long hours. It is believed that wild coffee plants, probably from Kaffa, Ethiopia, were taken to southern Arabia, where it was cultivated in the fifteenth century.

The second legend involves the Yemenite Sufi mystic Ghothul Akbar Nooruddin Abu al-Hasan al-Shadhili, who observed birds of unusual vitality, and upon trying the berries that the birds had been eating, experienced the same vitality.

The third legend involves Ali bin Omar al Shadhili, known simply as Omar, who was a Sufi dervish who used to heal people by prayers in the port town of Mocha in Yemen. For some reason, he was exiled into the hills/desert/caves to die of starvation. Very hungry, he noticed red berries on a plant and chewed them. Finding them too bitter, he boiled them with water and drank the resulting brew. It energised him, and he was able to keep his hunger at bay. Finding him still alive after several days, the townspeople welcomed him back, ordained him a saint, and started taking the "miracle brew" to cure themselves of many illnesses.

Pilgrims, merchants, and wars spread the fame of the "miracle drink" across Turkey, India, and Europe. Due to its commercial value, Arabs imposed strict regulations on the export of raw beans, and only roasted beans could be taken out of Arabia.

3.3 Spread of Coffee

It is said that Baba Budan (Fig. 3.5), a sixteenth-century Sufi saint from Chikmagalur, Karnataka, in India, while on a Hajj pilgrimage to Mecca, grew very fond of the coffee and smuggled seven seeds of the raw coffee beans by hiding them in his beard, or taping them to his stomach, or hiding them in his cummerbund, depending upon whom you ask, and planted these seeds on his return at his hermitage at the place now called Baba Budan Giri. The seeds sprouted and grew into lusty coffee plants, and

Fig. 3.5. Baba Budan with Coffee (Credit: Indian Coffee House).

coffee, according to some, spread across the world from there, at least definitely along the Malabar Coast (Fig. 3.6).

It is generally believed that Indian coffee, grown on the Malabar Coast, is one of the highest-quality shade-grown coffee beans in the world. The monsoon conditions, with ample rains and sunshine and intercropping with spices such as cardamom, cinnamon, clove, and nutmeg, lend the beans a distinctive flavour which is liked all over the globe. No wonder such high demand encourages India to export nearly 80% of its beans.

By now, it is generally accepted that Pieter van den Broecke, a Dutch merchant, succeeded in obtaining some of the closely guarded coffee bushes from Mocha, Yemen, in 1616. Alternative suggestions include his obtaining (stealing?) some raw beans from merchants in Venice. These were planted in the botanical gardens of Amsterdam and in the controlled environment of the greenhouse, which thrived and spawned several healthy plants. In 1658, the Dutch introduced coffee plants in Ceylon (Sri Lanka) and southern India. However, they soon abandoned these

Fig. 3.6. A Coffee Plantation in Chikmagalur (Credit: Prashantby).

plantations and concentrated on the rich forests of Java in Asia and Suriname in the Americas.

Gabriel de Clieu, a French naval officer, brought coffee seedlings from Jardin royal des Plantes (which in turn got them from the Dutch) to Martinique in the Caribbean in 1720, where they flourished. Soon, it was also being cultivated in Haiti and French Guiana. In a very romantic story, it has been suggested that around 1727, the King of Portugal sent Francisco de Melo Palheta to French Guiana to obtain coffee seeds. To cut a fascinating story short, Francisco faced severe difficulty in finding the seeds and finally succeeded by charming the wife of the French Governor, who gave him either seeds or saplings concealed in a flower bouquet, and coffee was introduced in Brazil, where it was first cultivated by a massive clearing of the Atlantic Forest and then moved to Sao Paulo. By the 1850s, Brazil became the largest producer of coffee, and it has held that position since then.

The requirement of no frost, plenty of rain, and ample sunshine provides that coffee mostly grows in the tropical "coffee belt" between the Tropic of Capricorn and the Tropic of Cancer (Fig. 3.7).

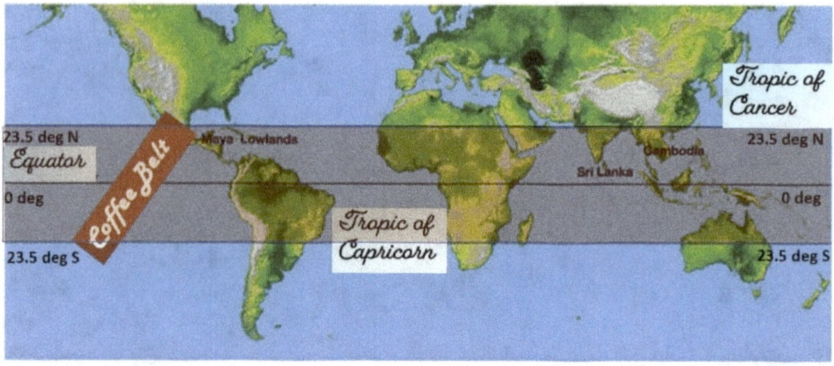

Fig. 3.7. The Coffee Belt.

While now hundreds of distinct coffee flavours are grown across the world, the two main flavours are Coffea arabica and *C. canephora* (or Coffea robusta). Arabica is generally considered a milder and more flavourful and aromatic brew than Robusta, which has a larger content of caffeine, which makes it less vulnerable to pests and has normally a larger yield. In addition, the combination of altitude and temperatures influences the type of coffee the crops will yield and its demand by consumers. Thus, most of the countries in these regions have favourable conditions that make the production of coffee stand out.

The two optimal conditions for growing Arabica coffee are obtained in the equatorial regions at 1200–2100 m and at latitudes lower than 10 degrees, where frequent rainfall causes almost continuous flowering, resulting in two harvesting seasons. Kenya, Columbia, and Ethiopia are examples of countries experiencing this climate. Again, well-defined rainy and dry seasons, and altitudes between 600 and 1200 m in the subtropical regions, result in one coffee-growing season and one maturation season, usually in the coldest part of autumn, in countries, such as Mexico, Jamaica, the Sao Paulo and Minas Gerais regions in Brazil, and Zimbabwe (Fig. 3.8).

As mentioned earlier, Robusta coffee is grown at much lower altitudes (sea level to 1000 m) in areas 10 degrees north and south of the equator. That is because it is much more tolerant to warm conditions than Arabica coffee.

Even though the yield of coffee is generally measured in terms of the number of 60-kg bags, we give it in millions of metric tonnes in Fig. 3.9,

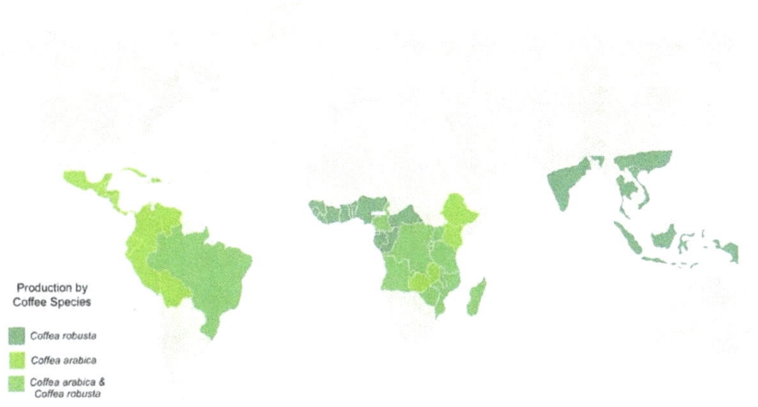

Fig. 3.8. Regions of the World Producing Robusta and Arabica Coffee (Credit: Wikimedia Commons).

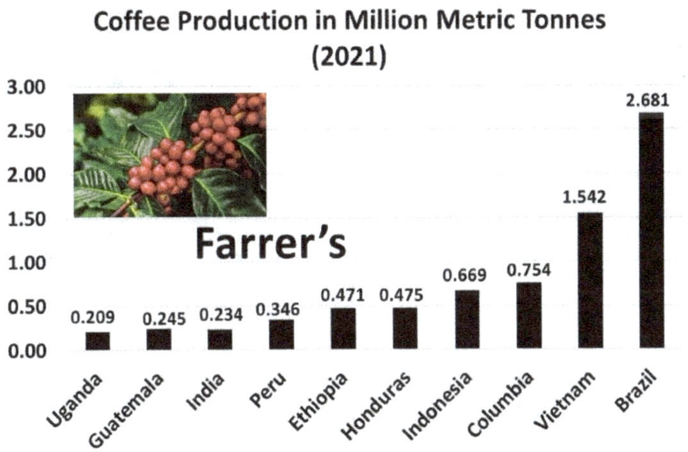

Fig. 3.9. Top 10 Producers of Coffee.

for easy visualisation. Coffee is the second most traded commodity after petrol.

Most coffee fruits have two seeds (beans). However, 5%–15% of the fruits yield only one seed, either because of insufficient pollination or because of exposure to extreme weather conditions. These are called peaberries, and the coffee prepared from them is often sold as a speciality coffee.

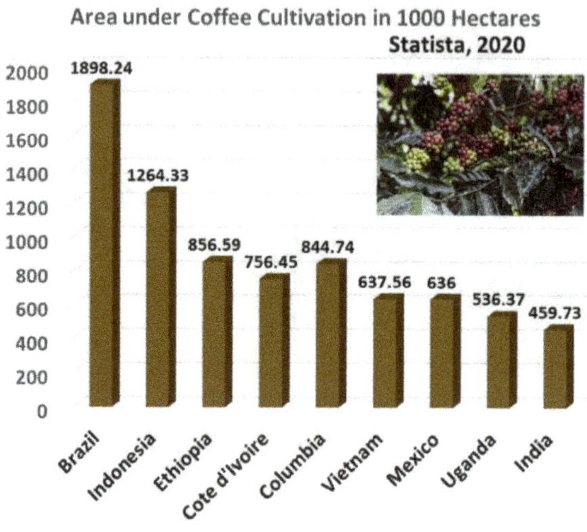

Fig. 3.10. Area Under Coffee Cultivation.

Coffee cultivation (Fig. 3.10) is practised over more than 11 million hectares. At the global level, 25 million people directly make a living from the crop, but it is estimated that around 100 million people are involved in this agricultural sector. Moreover, 70% of the farms are held by smallholders who own less than 10 hectares [1].

Coffee plants left to themselves can reach a height of up to 10 m, but soon it was found that if these are pruned to a height of about 2-3 m, they yield a much more plentiful crop. This is now generally practised across the world.

3.4 A Sad and Cruel History: Slavery

It is well known that by the time of the onset of the Industrial Revolution, British naval ships were sailing across the oceans and "conquering" new lands. This was followed by getting partial or total control over new lands for economic exploitation. There are suggestions that the British (and other) colonisation was "justified" as "scientific racism," as they had superior weapons and were more technologically advanced than Africans. This led to the infamous and cruel Triangular Trade. By the 1600s, occupation in parts of Africa and the Caribbean was well underway. The slave trade was fuelled by the increasing demand for coffee, sugar, cotton, and tobacco, and this helped establish the Triangular Trade. The Triangular Trade consisted of three journeys: ships carried manufactured goods and weapons from Europe to Africa. In Africa, they loaded the captured African slaves who were taken to the Caribbean and the Americas, where they were made to work in coffee, sugar, tobacco, and cotton plantations under appalling conditions. The working and living conditions of these slaves were often so bad that they survived just for about 7 years, and the planters found it cheaper to buy new slaves than to take better care of them. Finally, the third leg of the Trade carried the produce of this slave labour to Europe. Until the abolition of slavery in the early 1800s, Britain was responsible for transporting a third of up to at least 12 million slaves across the Atlantic. Several estimates suggest that up to half of the captured slaves may have perished during the journey across the Atlantic.

3.5 Child Labour

The cultivation of coffee requires intensive manual work, such as picking, sorting, pruning, weeding, spraying, fertilising, and transporting products. Even though coffee is one of the most popular drinks in developed countries (Fig. 3.2), it is mostly grown in developing countries (see Figs. 3.8 and 3.9). Plantation workers are normally paid very low wages, and thus the family of the workers often put their children to work to augment the family income. Children as young as 6 years old often work 8–10 hours a day and are exposed to the many health and safety hazards of coffee harvesting and processing, from dangerous levels of sun exposure and injuries to poisoning from contact with agrochemicals. Children, and women, are mostly hired as temporary workers, especially during the harvesting of coffee, and are therefore paid even less than adult male workers and denied the normal benefits.

When the price of coffee rises, the incentive for families to pull their children out of school increases. At the same time, the fall in the price of coffee leads to rising poverty in these regions, which again prevents children from attending school and forces them to work. Thus, child labour results in maintaining a cycle of poverty over generations by preventing them from having an education and getting a better job. This can only be addressed by providing the workers with a decent living wage, which leaves enough for the education of their children and keeps them in school.

It has also been reported that many coffee workers are effectively enslaved through debt peonage or debt servitude, which is forced labour to repay debts. Thus, the landed elite in coffee-producing regions own large plantations and employ a permanent workforce. Most of the time, on these large plantations, the only source for essential goods is the estate shop run by the landowners since workers are prevented from shopping elsewhere by their long hours of work, lack of transportation, or constraints on travelling out of the estate [2]. Since the workers earn even less than the legal minimum wages and must pay excessively inflated prices at the estate shop, they wind up with very little or nothing, even after their long hours of hard physical labour. As a result, they become indebted to the plantation and are forced to work to pay off their debts. There are reports of families, who work and live on plantations for generations, trapped in debts. It is known that in tea plantations in South India during colonial days, a similar system was in place, and sometimes the workers were paid in currency bills valid only within the estate, effectively blocking their escape forever.

3.6 Exploitation of Animals

When the Dutch started coffee plantations in Indonesia, the workers used to get their wages in coffee beans. As the price of coffee shot up in the European markets, this practice was stopped and they were paid in cash, which effectively pushed coffee beyond their reach, even for their personal use. The workers found that palm civets were eating coffee beans and their droppings contained partially digested beans. They started collecting it and after washing it, started consuming the coffee prepared from these droppings. When the plantation owners came to know that the workers were having coffee, they first assumed that they were stealing the beans. Once they came to know the truth, they started to collect the droppings and marketed the resulting coffee as a speciality coffee, "Kopi Luwak," at a very high price. Kopi Luwak remains one of the most expensive coffees in the world and a cup of this coffee can cost 50–80 USD. Soon, in Thailand, people started to feed raw coffee beans to elephants and the coffee made from their droppings is sold as "Black Ivory Coffee."

Unfortunately, the greed and insensitivity of humans have led to thousands of civets being captured and put in congested and unhygienic cages, where they can hardly move and who are fed nothing but coffee beans to maximise profits. The producers of Black Ivory Coffee claim that the elephants are not harmed; however, the implication of this forced diet, rich in caffeine, has not yet been properly investigated.

This has a parallel in the story of Indian Yellow paint, which was perhaps introduced in India from Persia. It was dissolved in water or mixed with oil to produce a transparent yellow paint which was used in Indian frescoes, oil paintings, and watercolours. It produced a clear, deep, and luminescent orange-yellow colour which, due to its fluorescence, appeared especially vivid and bright in sunlight. Its use in painting fabrics did not take off, however, due to the foul smell that the paint emanated. It was also exported to Europe, where many well-known painters used it in their works. It was prevalent from 1600 to 1900 CE and became commercially unavailable by 1921. The paint was obtained from the concentrated urine of cows which were fed (only) mango leaves. This made them quite weak, and they perhaps contracted jaundice. This cruel practice was ended by legislation in 1908, perhaps under the Acts for the Prevention of Cruelty to Animals [3].

We hope that a worldwide outcry against the cruel practice of keeping civets in unhygienic cages and being fed only coffee cherries can bring their suffering to an end.

3.7 A Ray of Hope: Araku Valley

Amidst all this, let us discuss the remarkable success story of organic coffee from Araku Valley, because of its message of hope and prosperity for the indigenous people across the world, as coffee is often grown in the hilly tracks which have traditionally been inhabited by indigenous people [4]:

> "Araku Valley is a small hill station and a valley region in the Eastern Ghats near the Andhra-Odisha border, at a height of about 1000 metres above the sea level and about 110 kilometres from the city of Vishakhapatnam. It is a notified tribal area with a population of 0.6 million tribal or indigenous people, who are mainly forest dwellers. When the pressure on forest land and foraging increased, they were forced to adapt to agriculture, mainly shifting cultivation in the 1940s. Till the 1980s there was a barter economy, and the region was characterised by loss of biodiversity, rapid erosion of soil, and the felling of trees. The

valley was also plagued by social problems, till the 2000s. Literacy was low, infant and maternal mortalities were high, and the poor returns from the poor-quality farm produce were pushing debt-ridden farmers towards alcohol and drugs.

Coffee was first introduced in this tribal belt in the 1920s by British Revenue officials. However, most tribals engaged in agriculture were using primitive agricultural techniques and working as daily wage earners, if at all they could find jobs. Around the year 2000, the Coffee Board and Integrated Tribal Development Agency provided coffee saplings to tribals. But they got meagre yields of only about 85 kilogrammes of coffee fruits per acre against the possible yield of up to 400 kilogrammes, and the nexus of local traders and middlemen forced them to sell their produce at a ridiculously low rate of Rs. 2 to 3 per kilogramme.

A social sector organisation, Naandi Foundation, stepped in, formed Small and Marginal Tribal Farmers Mutually Aided Cooperative Society, and trained them in best practices of production of organic coffee. By 2010 they started getting bumper crops per acre and sold the coffee fruits at Rs. 18/kilogramme, ushering in prosperity and happiness to the tribal belt.

By now, the organisation has created a centralised coffee processing unit in Araku, which enables farmers to process their products efficiently and consequently fetch higher prices; expanded the cultivation to include pepper, planted 29 million trees and by direct marketing raised the Araku Coffee to the world stage, winning the Gold Medal for the best coffee pod in the Prix Epicures OR 2018 Award in Paris, France. In the process, the earnings of the farmers have gone up by as much as a factor of sixty.

Naandi's brand of terroir-based Araku Coffee is now grown on 25,000 acres of land in the valley which is the world's largest certified organic plantation. The effort has benefitted about 25,000 families, encouraging others to follow suit. Farmers are no longer at the mercy of middlemen and loan sharks. The Foundation supplies coffee across the world through its online portal. The indigenous peoples living in the Araku Valley are an example of how a lack of knowledge and support has seen them struggle over the years, but also of how cooperation and proper training can drastically overhaul operations and consolidate thriving, self-sustained industry, transforming lives. This has been termed The Blue Revolution in the literature."

It is our firm belief that this approach can be adopted by cooperatives of small farms, and it can benefit small farmers everywhere in the world.

3.8 Climate Change and Deforestation: Effects on Coffee

We saw earlier that coffee was initially planted in Brazil by clearing vast tracks of the Atlantic Forest. As these tracks lost their fertility, the growers moved to areas near Sao Paulo. We have also noted that the best coffee is grown in shade, which also helps arrest the run-off of the topsoil. However, even though this coffee has an excellent and unique flavour, its yield per hectare is lower than coffee grown in the sun, using chemical fertilisers. The latter process, however, requires massive deforestation and leads to degeneration of the land, forcing growers to look for newer areas for plantation, for which forests are cleared again.

Moreover, the onset of climate change [5] is likely to reduce the area most suitable for growing coffee by half by 2050, mostly because of rising temperatures in the main producing areas, namely Brazil, Vietnam, Colombia, and Indonesia. This will require the exploration of newer areas for coffee plantations and the development of newer hardier seeds. Will this retain the unique flavours of the variety of coffees, which have given us so much pleasure and satisfaction over centuries?

3.9 Tea: Discovery and Spread

Tea is the second most popular beverage after water. More than 4.5 billion cups of tea are consumed every day. It has continued to impart freshness and relief from fatigue to people across the world for more than 400 years. The people of China, of course, have been enjoying the drink since 2732 BCE when, according to the most prevalent legend, Emperor Shen Nung, who had introduced agriculture in China and advised his people to boil water before drinking for better health, discovered tea. It is said that when the water was being boiled for his use, a gust of wind blew leaves from a tree into his pot. He was attracted by the pleasant smell emanating from the brew and tasted it. It is said that he experienced a warm and pleasant feeling over his body as if the liquid was "investigating every part of his

Fig. 3.11. Chinese Character for Ch'a (Tea).

body." Emperor Shen Nung named the brew "ch'a," which is the Chinese character implying "to check" or "to investigate."

The following Chinese character (Fig. 3.11) which stands for ch'a was adopted in 200 BCE, when the then Emperor belonging to the Han Dynasty ruled that the character referring to tea must include wooden branches, grass, and a man between the two, to symbolise the "way tea brought humankind into balance with nature for the Chinese culture."

By the 8th century CE, the cultivation of tea had picked up considerably in China and even some interesting practices were associated with it, like employing only young girls to pick the leaves to maintain their "purity." These young girls were not allowed to eat garlic, onions, or strong spices to "ensure" that the odour from their fingertips might not contaminate the precious tea leaves.

By the 9th century, tea was introduced in Tibet, where caravans of yaks carried it across rugged terrain, travelling for over a year. It has been reported that 200–300 yaks entered Tibet every day carrying tea. The Tibetans developed their unique style of preparing tea, where the tea leaves were boiled for almost half an hour and then mixed with yak butter and salt.

Tea was introduced in Japan around the same time by Japanese monks returning from China, who carried the seeds of tea with them. Soon, Japan was to develop its famed, sacred tea ceremony "Chanoyu" which honours the "harmony, purity, respect, and tranquillity" of Zen Buddhism, and continues to this day.

The Chinese made a gift of tea to Czar Alexis of Russia in 1618. The tea had covered 18,000 km and taken 18 months to reach there. Six thousand strong camel caravans carried tea to Russia every year, till 1903, when the trans-Siberian railways started operations. At some stage, maybe around the 1840s, the Russian custom of drinking tea by sucking it through a lump of sugar became popular, as regularly shaped sugar cubes became available.

The Portuguese and the Dutch introduced tea in Europe. In 1662, King Charles II married the Portuguese princess Catherine of Braganza, who introduced him and England to tea. As a part of a dowry, King Charles II also received the island of Bombay (now Mumbai). He transferred its control to the East India Company, and it soon developed into a centre for the trade of tea. The English used to buy tea, silk, and porcelain from China for sale in England. However, China did not want to buy one commodity they could supply, cotton, from them and thus the East India Company had to pay for the tea in silver. By then, the East India Company had become a very powerful monopoly, with tea as its primary commodity. They were given the right to acquire territory, mint coins, keep armies and forts, punish lawbreakers, form foreign alliances, and even declare war!

To circumvent the drain of silver to the Chinese, the English started growing opium in the newly acquired areas of Northern India ("Bengal, Madras, Bombay, Patna, Benares, and Malwa") and started forcing opium into China, which they converted to silver, which was then used to buy tea. In 1839, an upright Chinese Commissioner, Lin Zexu (1785–1850 CE), destroyed over 20,000 chests of opium by dumping them into the sea. This led to the First Opium War (1839–1842 CE) which the Chinese lost due to the modern technology employed by the English, and they were forced to grant favourable tariffs, trade concessions, reparations, and territory (including Hong Kong) to them. During the Second Opium War (1856–1860 CE), many Western nations fought against China, which resulted in further grants and concessions. We must mention the remarkable letter that Commissioner Lin Zexu wrote to Queen Victoria of Britain, which must stand out as a cry of exasperation and despair, against the most immoral opium trade carried out by her subjects, which was destroying the Chinese [7] but was forbidden in the Great Britain.

Tea was considered the property of the Chinese Empire, and it was not possible to take its cuttings or seeds out of China. Around this time, the British colonists spotted tea plants growing wild in Assam, which were traditionally used by people there. Robert Fortune succeeded in getting Chinese tree plants by a variety of means, and some Chinese tea manufacturers introduced the techniques of fermenting tea in India. However, the Chinese tea plants did not flourish well in India, and ultimately only the plants found in Assam were planted in all the tea gardens. Only the

Darjeeling area has some strains inherited from a few Chinese plants which survived. By now, it is established that Chinese and Assam tea plants belong to the same species, *Camellia sinensis*, though the names C. Sinensis var. Sinensis and C. Sinensis var. Assamica are sometimes used to distinguish between them.

The British colonists soon introduced the cultivation of tea in South India, Ceylon (Sri Lanka), and Kenya.

Finding the local population uninterested in the hard work associated with clearing forests (which affected the ecology of the area), planting, cultivating, picking tea leaves and processing them, and taking advantage of famines, people in large numbers were (often forcibly) taken from Bihar to Assam, and their descendants now number several million. Their working and living conditions and wages were, however, very poor and many of them perished due to disease and some were even killed for protesting. The situation remains far from ideal, even now. We have already indicated the strategies adopted by the plantation owners to keep workers from escaping and to keep them in perpetual debt and poverty.

Workers were also taken from Tamil Nādu to Sri Lanka, and their descendants still form the backbone of the vast tea industry of that country and yet continue to live in extreme poverty.

3.10 Consumption and Production of Tea and Area under Tea Cultivation

Two leaves and a bud (Fig. 3.12) stand for the most delicate part of the tea plant, which is processed in different ways to provide white, green, black, oolong, and other varieties of tea. A speciality tea called "pu-erh" or "pu-er" from China is pressed into bricks and left to "age" for several years. It is believed that tea leaves must be plucked very expertly and delicately to retain the flavour. In India, 5 to 6-day-old leaves are plucked in Darjeeling and are processed in an orthodox manner to retain the very unique flavour of the tea. On the other hand, 7 to 10-day-old leaves are harvested in Assam tea estates, which are processed using the Crush, Tear, and Curl (CTC) procedure and employing machines to produce tea with robust colour and its distinctive bitter taste. There are indications that the CTC procedure was invented by Sir William McKercher in the 1930s to utilise more abundant mature leaves, and the tea prepared by boiling it with milk and sugar was especially popularised by a variety of means in India and

Fig. 3.12. Two Leaves and a Bud of Tea, to be Plucked Expertly.

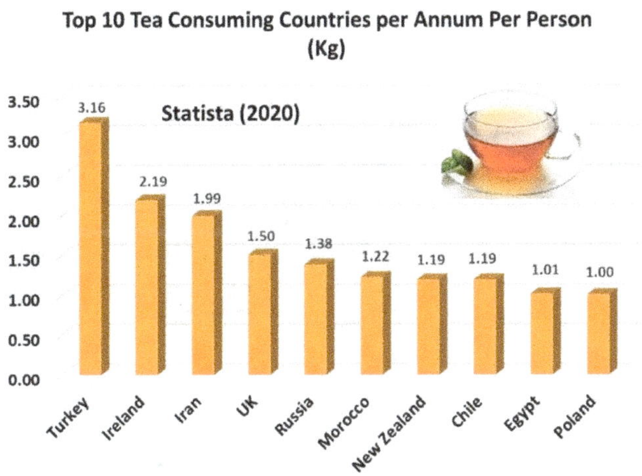

Fig. 3.13. Top 10 Consumers of Tea.

Africa, where people started adding spices and herbs to it, to give rise to "masala chai."

Starting from a novelty drink, available only to a few wealthy persons, tea has now reached a stage where more than four and a half billion cups of tea, prepared in various ways, are consumed across the world every day, and in fact, Chinese are no longer the biggest consumers of tea (Fig. 3.13).

Again, starting from the chest loads of tea produced in the 1820s in China, its production has reached millions of tonnes per year (Fig. 3.14).

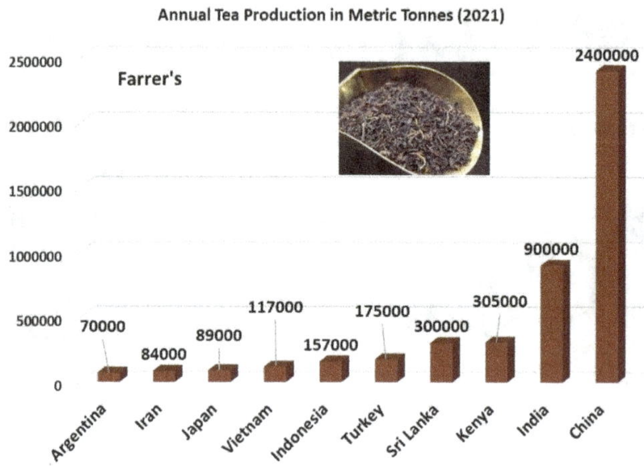

Fig. 3.14. Top Tea Producers in the World.

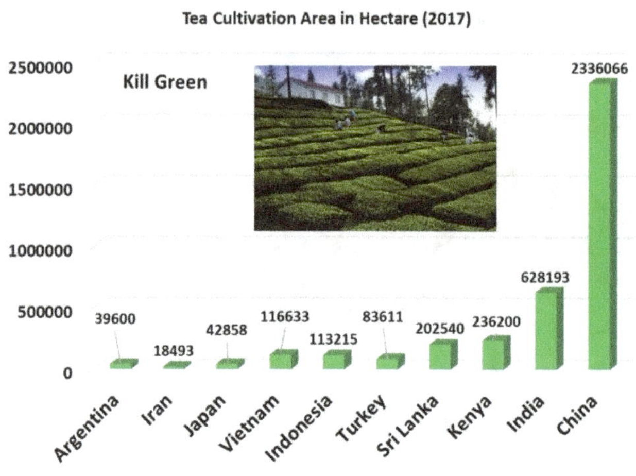

Fig. 3.15. Area Under Tea Cultivation.

By now, tea cultivation employs more than 13 million people and covers about 4 million hectares of area (Fig. 3.15).

3.11 Climate Change and Tea

Ideally, tea belongs to monsoon lands where high temperatures, long growing seasons, and heavy rainfalls help the growth of tea plants. An average temperature of 21°C during the growing season of not less than 8 months is considered ideal as warm summers and frequent rains promote rapid leaf reproduction and increase the number of annual pickings. Even though tea plants can handle light snow or frost, they cannot handle prolonged cold winters or heavy freezes. In fact, during the cold season, e.g., in China and Japan, the tea bushes lie dormant and there is no picking. In India and Bangladesh, the highest yields are obtained from June to September when the weather is hot and rainy, but the best-quality tea is derived from the earlier and later pickings when the climate is cooler and drier, whereas, in Sri Lanka, picking can be carried out all the year-round, as the growth is not restricted by a distinct season.

Tea cannot tolerate stagnant water. Waterlogged lowlands are least suited to tea cultivation. Thus, mountain slopes have been adopted for tea gardens all over the tea regions of the lands served by monsoons. Undulating and well-watered tracts, where the water escapes freely without serious soil erosion, offer the best conditions for tea cultivation. There are indications that tree grows well in shades, which can protect tea plants as well as workers from strong sunlight.

These requirements immediately indicate that tea production is likely to be greatly hit by the onset of climate change and rising temperatures. We know that it may involve less rainfall, with rising incidences of heavy rainfall, and prolonged summers. The heavy rainfalls can remove the topsoil and higher temperatures enable insects that attack tea plants to survive winter, giving them more time to reproduce and multiply [8].

Careful and detailed studies are needed to face these challenges which if left unaddressed will lead to a drop in the production of tea and even a reduction in its flavour. Some of the mitigation strategies involve intercropping tea with legumes, which can help nitrogen fixation, and growing tea from seeds, which gives sturdier plants and a more extensive root system, instead of cuttings.

In an exciting development, Chinese farmers have found that when some insects attack tea plants, the plants react by releasing some chemicals which impart a unique flavour to the tea; this tea is now being marketed as a speciality tea! The sharing of knowledge from over 60 countries where tea is grown is essential to meet these challenges.

3.12 Conclusions

Coffee and tea are just two of the many examples (e.g., cocoa) that we shall discuss in this book. They started as beautiful discoveries of unique plants, which spread across the world in regions suitable for their cultivation, and whose products have given people across the world immense pleasure and led to the intellectual awakening of mankind and enormous progress in painting, music, literature, science, and politics through stimulating discussions. By now, coffee is cultivated in over 11 million hectares, and tea is cultivated in over 4 million hectares, giving jobs to tens of millions of people. Yet, only an active international collaboration in the study of their production can save their delights for future generations, as the onset of climate change and global warming is likely to seriously impact their production.

We would like to re-emphasise that the cultivation and harvest of coffee and tea are labour-intensive processes and employ humans, and in the case of the tea mostly women, to pick the leaves. Their wages, living conditions, and welfare need to be addressed urgently to eliminate their sufferings perpetuated over generations. The world which has enjoyed the fruits of their labour for centuries owes it to them.

The role of coffee and tea in the spread of knowledge and ideas via discussions and open sharing cannot be overemphasised.

The Chinese had already elevated the drinking of tea to a high level of intellectual and cultural activity by the eighth century CE, best exemplified by the famed pagoda poem (Fig. 3.16) by Yuan Zhen (779–831 CE).

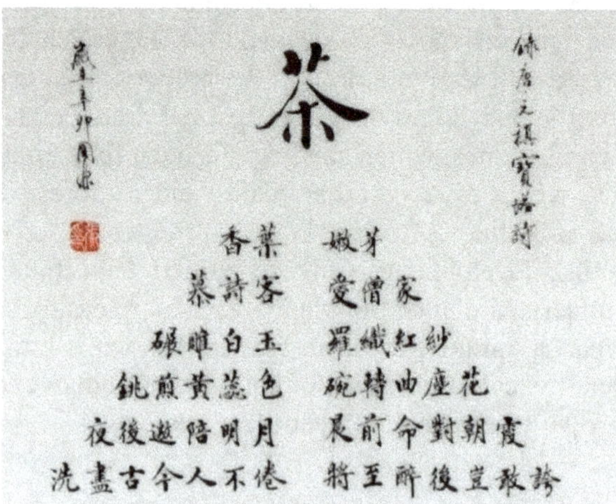

Fig. 3.16. Ch'a, a Pagoda Poem by Yuan Zhen.

Several attempts have been made to translate it, and one of them [9] is given below, which however struggles to retain the grammar of the Chinese metre for distributing 28 characters: one character in the first line, two in the second, three in the third, four in the fourth, five in the fifth, six in the sixth, and seven characters in the seventh line.

"Tea
Fragrant leaves, delicate buds.
Admired by poets, loved by monks.
Crushed with white jade, filtered through red gauze.
Cauldron brewed to the colour of gold, served in cups aswirl in bubbles.
Inviting the moon at night for company, facing alone the twilight before the sunrise.
Feeling energised in past or present, praise fully rinsing away drunkenness."

The sharing of this knowledge and expertise, even if for commercial benefit, has transformed the world.

References

[1] "From the Seeds to Your Cup: Global Coffee Production" (December 2021), https://livelihoods.eu/from-the-seeds-to-your-cupglobal-coffee-production-in-5-key-facts-figures-2/.

[2] https://foodispower.org/our-food-choices/coffee/.

[3] T. N. Mukharji, "Piuri or "Indian Yellow", *Journal of the Society of Arts* 32 (1884) 16–17; R. R. Feller (ed.), "*Artists' Pigments: A Handbook of Their History and Characteristics*" (Washington: National Gallery of Art, 1985); J. Bailkin, "Indian Yellow. Making and Breaking the Imperial Palette", *Journal of Material Culture* 10 (2005) 197–214; D. Tamburini *et al.*, "New Insights into the Composition of Indian Yellow and Its Use in a Rajasthani Wall Painting", *Microchemical Journal* 137 (2018) 238–249.

[4] V. S. Ramamurthy and D. K. Srivastava, "A Multidimensional Approach to Sustainable Rural Development", *Science and Culture* 87 (2021) 349–358.

[5] D. K. Srivastava and V. S. Ramamurthy, "*Climate Change and Energy Options for a Sustainable Future*" (Singapore: World Scientific, 2021).

[6] http://www.coffeeteawarehouse.com/tea-history.html.
[7] https://cyber.harvard.edu/ChinaDragon/lin_xexu.html.
[8] A. Nowogrodzki, "How Climate Change Might Affect Tea", *Nature* 566 (2019) S10–S11.
[9] https://www.teasenz.com/chinese-tea/pagoda-poem-tea-yuan-zhen.html.

Chapter 4

Paper: The Vehicle for Sharing Knowledge

"The visionary starts with a clean sheet of paper, and re-imagines the world."

— Malcolm Gladwell (1963–)

"When you read a book, you hold another's mind in your hands."

— James Burke (1939–)

4.1 Early Mediums for Writing and Communication

As one stands at one of the more than 700 rock shelters in Bhimbetka ("the lounge of Bhima") in the foothills of Vindhya Mountain Range in Central India, a UNESCO World Heritage Site, and looks around, and as one's eyes adjust to the dim light in the dark part of the shelter or cave, the outlines of the cave paintings, which are up to 10,000–30,000 years old, emerge (Fig. 4.1).

Similar paintings of comparable antiquity have been found in Africa, Europe, Australia, and the Americas. These have remained a subject of much scholarly study. Here, we shall discuss the urge of man to use this and other mediums to communicate to others, to "spirits" or "gods," and perhaps to the "after-life." Did these suggest or hold some magic before proceeding on to a hunt or a skirmish? It is believed that some of the rock arts in such cases (Fig. 4.2) could be of even much greater antiquity [2].

Fig. 4.1. Cave Paintings of Bhimbetka Rock Shelter [1].

Fig. 4.2. One of the Oldest Rock Arts in the World, in the Auditorium Cave, Bhimbetka [2].

All animals, including humans, have developed and perfected a system of communication with others of their species, which is perhaps also understood to some extent by some of the other species, e.g., the domesticated animals, at least as a signal. In the case of humans, this system of communication started perhaps as a sign language, which then culminated in a spoken language, and later into a written language, using some symbols or notations, which evolved into scripts among the more advanced societies.

Humans are the only species capable of spoken and written language. It has been argued [3] that between 70,000 and 30,000 years ago, the Homo Sapiens underwent a Cognitive Revolution for some unknown reason, which "enabled them to think in unprecedented ways, …, which ultimately helped them conquer the world, …, and as far as we know only (Homo) Sapiens can talk about entire kinds of entities that they have never seen, touched, or smelled, ..., and legends, myths, gods, and religions appeared for the first time with the Cognitive Revolution."

It has been further argued that humankind's success was due to its ability to create and sustain grand, collaborative myths. We can easily imagine that scripts must have facilitated this Revolution to spread rapidly and become deeply rooted. This is a fascinating and vast sphere of study and much remains to be revealed as some of the scripts, e.g., those of the Indus or Indus-Sarasvati Valley Civilisation (Fig. 4.3), remain undeciphered.

Returning to our discussion about communication, we note that it is generally believed to be a fundamental instrument to inform, motivate, and perhaps establish authority and control, as well as allow for the expression of emotions. And yet spoken words depend solely on the memory of the listener for their accurate onward transmission.

Al Biruni (973–ca. 1050 CE), while discussing the writing by Indians, waxed eloquent [5] about the importance of writing:

> *"The tongue communicates the thought of the speaker to the hearer. Its action has, therefore, as it were, a momentary life only, and it would have been impossible to deliver by oral tradition the accounts of the events of the past to later generations, more particularly if they are separated from them by long periods. This has become possible only by a new discovery of the human mind, by the art of writing, which spreads news over space as the winds spread, and over time as the spirits of the deceased spread."*

Fig. 4.3. The Longest-Known Indus Script Inscription [4].

In the case of Vedic Mantras, for example, words must be spoken in the same order, with the same correct accent, to have the "magic" or the "divine effect." These were orally transmitted over several millennia from father to son or from teacher to disciple, to retain the pristine correctness and the purity of the accent and the words. This was achieved by several ways of recitations (paths) by repeating the words in a very different order so that the words of a mantra are strung together in different patterns, such as "vakya," "pada," "krama," "jata," "mala," "sikha," "rekha," "dhvaja," "danda," "ratha," and "ghana." This unique tradition was proclaimed as a "Masterpiece of the Oral and Intangible Heritage of Humanity by UNESCO" on 7 November 2008 [6].

Man has used a vast variety of materials to write, inscribe, paint, and leave instructions for history, or magic. Wood, bone, clay, leather, metals (mostly copper but also silver and gold), the bark of trees, bamboo, silk, and leaves of some trees have all been used at different times. In fact, "writing on the winds" or "writing on water" is often used as a euphemism for statements likely to be lost, almost immediately.

Still, one must mention that in a remote mountain village, Kuskoy, high above Turkey's Black Sea coast, there are villagers who still communicate across valleys by whistling, using what they call "bird language" to communicate! Similar methods are used by an ancient tribe in the Canary Islands. With the advent of mobile phones, these unique languages are dying.

Examples of writing on bone have been found across the world. Oracle bones from 2000 to 1000 BCE used in China are examples of this medium (Fig. 4.4).

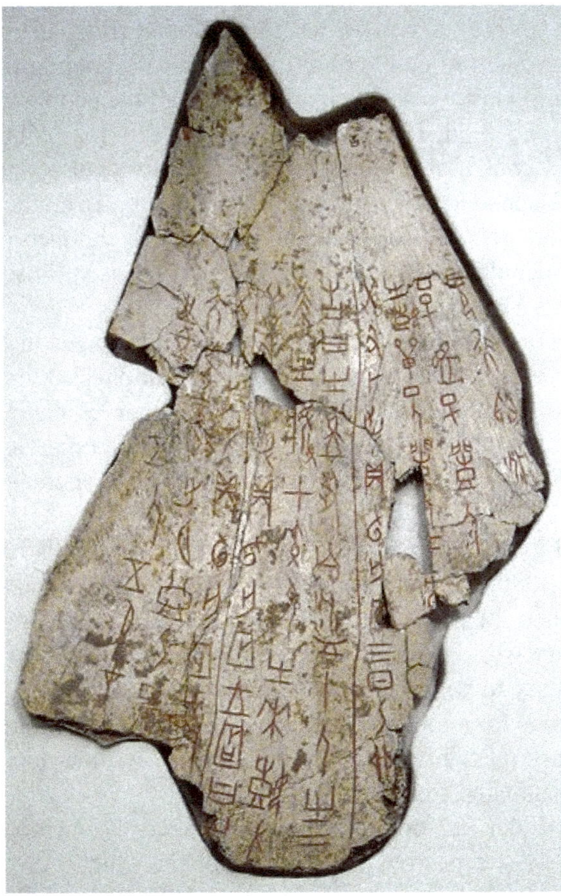

Fig. 4.4. Oracle Bones with Ancient Chinese Symbols (About 2000–1000 BCE) (Credit: Wikipedia).

Writing on clay tablets (cuneiforms) was more prevalent in the Ancient Near East (Mesopotamia) by 3000 BCE and was continued for several millennia. A stylus (mostly a pen of reed, with a sharp triangular tip) was used to write on wet clay tablets, which were then dried in the sun or wind.

If the writing was for a temporary purpose, the clay tablet was wetted in water, reshaped, and reused. For more permanent records of myths, fables, essays, hymns, proverbs, epic poetry, business records, laws, plants, and animals, these clay tablets were also fired. One example is the *Epic of Gilgamesh* written on a tablet. Libraries were established with books on clay tablets, and thousands of complete tablets, as well as several thousands of broken tablets, have been found (Fig. 4.5).

A unique system of coloured strings (Fig. 4.6) hanging from a rod or a string tied into knots was used by the Incas and the people of the Andes, called Quipu or Khipu, for several thousand years (ca. 2600 BCE–17th century CE). It was used to keep records of crops, cattle, and taxes, and may have also encoded stories, histories, and poetry. Even though most of the Quipus were actively destroyed by the Spanish conquerors, out of the fear that those could be records of Incan history, it is still used by people to keep records of things like telephone numbers.

As an interesting aside, we add that many languages in India have a phrase, "tie it into a knot and hold on to it" to emphasise remembering a piece of advice or an event, while some Arab tribes have a charming and quaint expression, "will make a ring of it and wear it on my ears," for the same, which is a testimony to the ingenuity of humans to create symbols.

Ancient Egyptians were already using rolls or scrolls of papyrus to write, around 3000 BCE, and once adopted by the Greeks and the Romans, its use continued till almost 1100 CE because of its ease. This revolutionised the literary world as it provided a smooth and flexible writing material that was easy to write upon using ink and which could easily survive in dry conditions for a long time. Due to this, papyrus remained in use for longer than any other material in the history of written documents. The word paper originated from this.

Pliny the Elder (23 or 24–79 CE) described the method of making papyrus rolls or sheets from the papyrus plant, which grew abundantly along the Nile (Fig. 4.7). Even though these plants grow elsewhere too, only in Egypt, these have a thick and tall stem, with abundant pith. Strips of this pith were laid vertically and horizontally, dampened, and then

Fig. 4.5. Sumerian Clay Tablet, with the Text of the Poem *Inanna and Ebih* by the Priestess Enheduanna (now at the Oriental Institute, University of Chicago).

pressed together. The sap from the plant acted like glue and a sheet was produced. Several such sheets could be glued to form a roll. There are indications that the sheets could be optionally covered with glue and smoothened using pebbles or shells. The ink was prepared using carbon from vegetable matter (black) or red ochre (red).

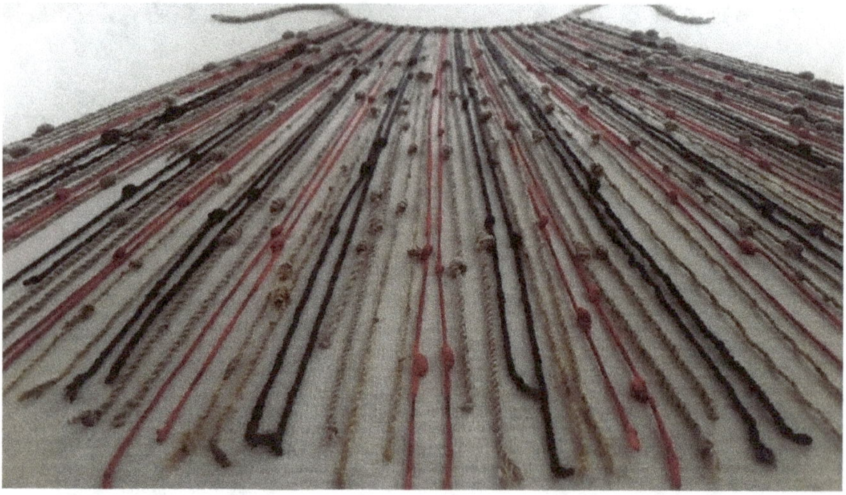

Fig. 4.6. Quipu in the Museo Machu Picchu, Casa Concha, Cusco [7].

Fig. 4.7. Papyrus Plant (left), Peeling the Papyrus Pith (right) (Credit: Wikipedia).

When Ashoka, the Great Mauryan Emperor, who ruled the Indian subcontinent from ca. 268 to 232 BCE, decided to spread his rulings and messages for amicable governance across the country, he used the medium of stone and erected stone edicts (Fig. 4.8, rock and pillar) across present-day Afghanistan in the west to present-day Bangladesh in the east and the entire Indian subcontinent except for parts of present-day Tamil Nadu. He also used several languages, e.g., dialects of Prakrit, as well as Greek and

Fig. 4.8. (Left) A major Pillar Edict of Ashoka, in Lauriya Araraj, Bihar, and (right) a Major Rock Edict of Ashoka, at Dhauli, Odisha (Credit: Wikipedia).

Aramaic, and Brahmi, Kharosthi, Greek, and Aramaic scripts, which helped James Prinsep (1799–1840 CE) decipher the Brahmi and the Kharosthi scripts. These scripts were no longer understood already by the time Firoz Shah Tughlaq (1309–1388 CE) shifted two Ashokan pillars from Yamuna Nagar and Meerut to Delhi.

Copper, mostly in the form of plaques (Fig. 4.9), was used extensively for the grants of lands to officials, poets, and temples. There is a famous eulogy addressed to King Bhoja of the Paramara dynasty, who ruled from ca. 1010 to 1055 CE, "lamenting" the scarcity of iron (due to the making of shackles to bind his enemies) and copper (due to the grant of lands to temples and wise men)! We have already seen its use in the Indus Valley Civilisation (Fig. 4.2).

Al Biruni [5] has described other mediums used by Indians, like Bhurjapatra (bark of Bhurj trees [Fig. 4.10], i.e., *Betula utilis*, Himalayan birch, now called Bhojpatra growing in the Himalayas) used in North India, and palm leaves in South India, with the exclusion of vellum (hide) commonly used in Europe and Arabia. He further writes:

> *"In Central and Northern India people use the bark of the tuz tree, one kind of which is used as a cover for bows. It is called bhurja. They take a piece one yard long and as broad as the outstretched fingers of the hand, or somewhat less, and prepare it in various ways. They oil and*

Fig. 4.9. The Sohagaura (Uttar Pradesh, India) Copper Plate Inscription in Prakrit in Brahmi Script (third century BCE) [8].

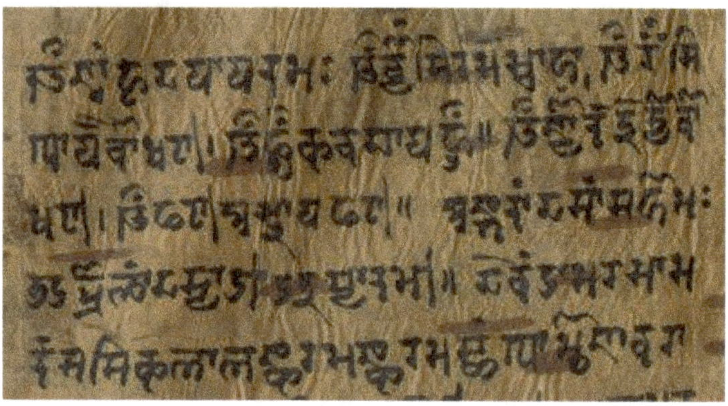

Fig. 4.10. A Manuscript from Kashmir on Bhurjapatra (Sharda script, Seventeenth Century CE) (Credit: Wikipedia).

polish it to make it hard and smooth, and then they write on it. The proper order of the single leaves is marked by numbers. The whole book is wrapped up in a piece of cloth and fastened between two tablets of the same size. Such a book is called puthi (cf. pusta, pustaka, pothi). Their letters, and whatever else they have to write, they write on the bark of the tuz tree."

He also noted that "The Hindus have in the south of their country a slender tree like the date and cocoa-nut palms, bearing edible fruits and leaves of the length of one yard, and as broad as three fingers one put beside the other. They call these leaves tari (taal or taar = *Borassus flabellifer*) and write on them. They bind a book of these leaves together by a cord on which they are arranged, the cord going through all the leaves by a hole in the middle of each."

Palm leaves were used to write (using a stylus) manuscripts (Fig. 4.11) in the Indian subcontinent and Southeast Asia from about the fifth century BCE till the nineteenth century CE. Vellum (parchment) was not used in the regions of the East of Arabia (Turkic-Persian areas of influence), due to the spread of Buddhism, which abhorred the killing of animals.

Chinese had used bamboo strips (Fig. 4.12) as well as silk to write upon. However, bamboo books tended to be heavy, and silk was prohibitively expensive.

In India, cards of silver or even gold in the shape of peepal (Bo tree) leaves, bhurjapatra, and silk scarfs (rolled over a decorated rod of silver or sandalwood) have been used for centuries to send ceremonial invitation cards by the rich. Traditional families still get the horoscopes of babies inscribed on bhurjapatra or palm leaves.

In the Americas, a kind of bark called amate was used to write folded books (codices), before the arrival of Columbus. The amate paper was prepared by peeling the outer bark of fig or mulberry trees, boiling the inner bark and soaking it in water overnight, and then beating with a smooth flat stone until it became pulp, which was spread and dried.

Fig. 4.11. A Palm Leaf from the "Bower Manuscript" (Sixth or Seventh Century CE) Discovered Near Kucha, China (Credit: Wikipedia).

Fig. 4.12. *The Art of War* by Sun Tzu with the University of California, Riverside, Written on Slips/Strips of Bamboo.

The Spanish rulers banned its manufacture; however, the process survived in remote and rugged mountainous areas.

4.2 Paper: The Ultimate Writing Material

Mobility of knowledge across the continents and millennia is a fascinating subject matter of intense study, and several valuable volumes are available about it [9]. Knowledge spread rapidly once (first) cheap paper and (then)

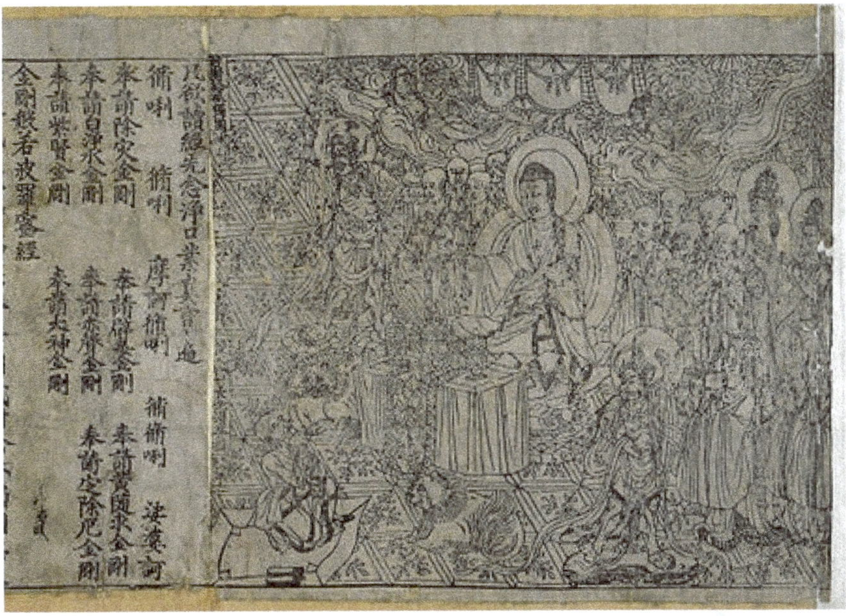

Fig. 4.13. *Diamond Sutra*, the Earliest Known Printed Book (Dated 11 May 868 CE) (Credit: British Library).

printing became available, though the widespread use of printing in the West had to wait for its reinvention (or reintroduction?) by Johannes Gutenberg (ca. 1400–1468 CE), in Germany.

Thus, for example, the Mahayana text of Buddhist sutras, the *Diamond Sutra*, was widely read and translated in China and elsewhere. A copy printed using wooden blocks and dated 11 May 868 is stored in the British Library.

Chinese records (prepared at a later date) suggest that paper was first made in Lei-Yang, China, by Ts'ai Lun (or Cai Lun), a Chinese court official, in 105 CE, though there is evidence of the use of paper from a few hundred years earlier. It is believed that Ts'ai mixed mulberry bast (Fig. 4.14), hemp and rags, linen and old ropes, damaged fishing nets, and other textile waste with water, beat it into pulp, pressed out the liquid, thus isolating the cellulose part, and hung the thin mat to dry in the sun.

China closely guarded this technology because of its immense economic value. However, "Chinese Buddhist monks and missionaries carried paper and papermaking at an early date to Korea, Japan, and Vietnam, and they must also have simultaneously brought that knowledge to India,

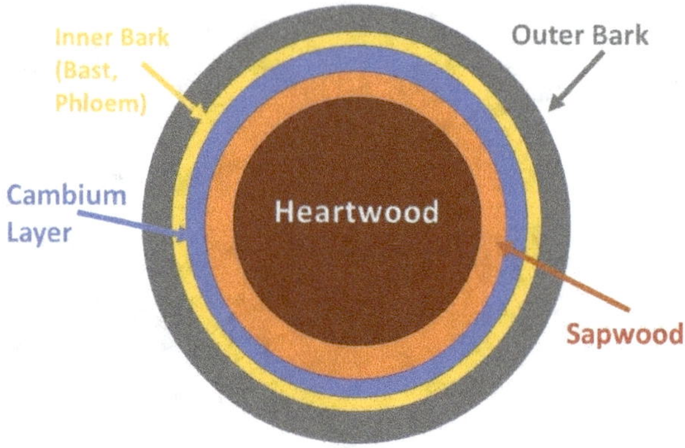

Fig. 4.14. Cross-Section of the Trunk of a Tree and Bast.

Fig. 4.15. Spread of Paper-Making Technology [10] (Design by J. M. Bloom; Cartography by V. Schniepp).

for they went there to collect and copy original Buddhist texts. Yet paper does not seem to have been used or manufactured in India until well after the region was conquered by Muslims in the late twelfth century, presumably because Indian writers were perfectly happy with using palm leaves, the traditional support for writing on the subcontinent" [10].

Muslims themselves learnt the art of papermaking from Chinese prisoners of war, after the Battle of Talas (751 CE), on the border between present-day Kazakhstan and Kyrgyzstan, with the forces of the Tang Dynasty, which they won. They introduced the technology in Baghdad and then all over the Middle East, including in Greater Syria, Iraq, Egypt, and the North African countries, as well as Persia and eventually Spain (Fig. 4.15).

Bloom [10] has noted that in an amusing turn of events, Europeans not only forgot their debt to Muslim papermakers but also remained ignorant of paper's origins in China so that when they first encountered Chinese paper in the sixteenth century, they thought (assumed?) that the Chinese must have learnt the art of papermaking from the ancient Egyptians! Europeans then carried paper and papermaking, along with printing, throughout the globe.

It is well established that the easy availability of paper encouraged an extraordinary culture of book-learning throughout the Muslim lands that was unparalleled in the contemporary Christendom, which continued to rely on relatively expensive parchment, thereby restricting the number of writers and readers. It is also likely that the Christian world was opposed to paper because of its association with Islam, and it continued to use vellum (parchment) for writing. In any case, literacy in that part of the world was quite low, till the start of the Renaissance.

Let us pause here for a moment and add that unfortunately, palm leaves, bhurjapatra, and paper manuscripts can also be burnt easily, and several invading armies have indulged in the ruthless burning of libraries from ancient times as an assertion of their cultural or religious superiority, and disdain for the vanquished. Thus, the great and world-famous libraries of Nalanda (the Ratnasagara: Ocean of Jewels, the Ratnodadhi: Sea of Jewels, and the Ratnaranjaka: Jewel-adorned), Vikramshila, and Odantapuri are believed to have been burnt by Bakhtiyar Khilji around 1193 CE. In 1258 CE, when the forces of the Mongol leader Hulegu Khan laid a siege to Baghdad, he destroyed its great library, Bayt al-Ḥikmah or House of Wisdom, which also housed translations from Sanskrit, Chinese, and Greek writings, and Islamic learning assembled during the period of the golden age of Islam. Sadly, these are not the only examples and even as late as during the 1857 revolt against British Rule, the priceless collection of Sanskrit manuscripts in the library of Rani Lakshmi Bai of Jhansi was burnt by the British forces.

The next great development in paper-making technology was the use of chemically treated wood pulp, which used spruce, pine, fir, larch, and hemlock, as well as eucalyptus, aspen, and birch trees for making the pulp. It has been suggested that the French scientist René-Antoine Ferchault de Réaumur (1683–1757 CE) on a nature walk observed that wasps were making their nests out of a paper-like substance (Fig. 4.16). Paper wasps take wood and chew it down until it is a pulp, mix it with their saliva, and then make their nests from the pieced-together pulp.

Fig. 4.16. Nest of a Paper Wasp (Credit: James Wainscoat, Unsplash.com).

However, mankind took almost 100 years before this observation was converted into technology for making paper.

4.3 Conclusions

The entire civilisation today is a product of knowledge sharing and knowledge dissemination across the world using books, journals, magazines, and newspapers. It has brought literacy and livelihood to people and tremendously improved the quality of their lives and well-being.

The use of paper as a packing material and for insulation, decoration, sanitation, even making shoes, printing money, etc., is only too well known. It is easily recycled, and it is biodegradable.

One may argue that the use of paper may be eliminated as we switch to electronic/digital means of storage of books and records. However, it should not be overlooked that this medium of storage gets outdated quickly, and soon the gadgets needed to retrieve the information get out of circulation. Thus, the cards, tapes, disks, floppy disks, and CDs on which the information used to be stored are already difficult to read. The information stored in computers and clouds is not only vulnerable to the ever-changing technologies, but it also remains vulnerable to power outages, natural events like solar flares, and the ever-present attacks by hackers, while those on paper can be stored for all time to come and easily read.

A recurring refrain in the prayers of many traditions envisages using the Earth as paper and all the oceans as ink, and yet being unable to write all the great virtues of God. This could, perhaps with some poetic licence, also be told about the role played by paper in shaping the world as we know it today.

References

[1] "Rock Shelters of Bhimbetka: Continuity through Antiquity, Art & Environment", A proposal for nomination for inclusion in the World Heritage List, Archaeological Survey of India (2003).

[2] R. G. Bednarik, "The Oldest Known Rock Art in the World", *Anthropologie* 39 (2001) 89–98. http://www.jstor.org/stable/26292558.

[3] Y. N. Harari, "*Sapiens: A Brief History of Humankind*" (London: Vintage Books, 2011), pp. 24–27.

[4] V. Shinde and R. J. Willis, "A New Type of Inscribed Copper Plate from Indus Valley (Harappan) Civilisation", *Ancient Asia*, 5 (2014) 1–10, http://doi.org/10.5334/aa.12317.

[5] "*Al Biruni's India: An Account of the Religion, Philosophy, Literature, Geography, Chronology, Astronomy, Customs, Laws and Astrology of India, about AD 1030*", Vol. 1. P170, (Translation) E. C. Sachau (London: Kegan Paul, Trench, Trubner & Co., 1910).

[6] https://www.kamakoti.org/hindudharma/part5/chap10.htm.

[7] https://en.m.wikipedia.org/wiki/File:Quipo_in_the_Museo_Machu_Picchu,_Casa_Concha,_Cusco.jpg.

[8] B. M. Barua, "The Sohagaura Copper Plate Inscription", *Annals of the Bhandarkar Oriental Research Institute* 11 (1930), 32–48.

[9] See e.g., H. Jöns, M. Heffernan, and P. Meusburger (eds.), "*Mobilities of Knowledge*" (Springer, 2017).

[10] J. M. Bloom, "Chapter 3: Papermaking: The Historical Diffusion of an Ancient Technique", pp. 51–66, in Ref. [9].

Chapter 5

Green Revolution: Feeding the Increasing Population

"*If agriculture goes wrong, nothing else will have a chance to go right.*"

— Dr. M. S. Swaminathan

5.1 Introduction

By now, there is enough evidence to suggest that Homo Sapiens originated in Africa and spread to all the corners of the world, over several hundred to tens of thousands of years. Other human-like species, e.g., Neanderthals, could not adapt to the changing climate and (perhaps), in the face of the advancing Homo Sapiens, went extinct, even though there are some suggestions that they may have intermingled with Homo Sapiens. For a large part of this history, humans survived as hunters and gatherers, learning to make crude tools out of stone and use fire (Fig. 5.1).

The use of fire widened the range of foods the humans ate and gave them warmth, to face cold, and light and protection from predators during the darkness of the night. Cooked meals took less time and energy to digest and left them with more energy to explore wider regions. It may have also led to their brain getting larger. Fire also helped them smelt copper, tin, and iron, which helped make better weapons and "conquer" animals. The lifestyle of hunting and gathering could have supported only a small population.

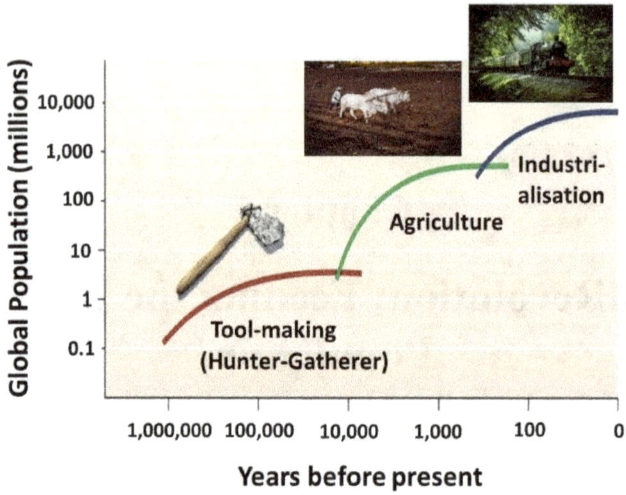

Fig. 5.1. Paradigm Shifts in Development and the Increasing Population [1].

The next revolution in the history of mankind was the discovery of agriculture, which started about 10,000 years ago. These are popularly termed the "hunter/gatherer, shepherd/cowherd, and farmer" avatars of mankind. It led to a quantum jump in the production of food. As a result, the population increased rapidly and gave shape to civilisations.

The Industrial Revolution, which started around the mid-eighteenth century, with the discovery of the steam engine (it existed in primitive forms for quite some time but was not used extensively), and spread rapidly over Europe and America, deriving most of the required energy from coal, was the next and most important development in the history of mankind. It led to a massive increase in the production of food and other goods and could support a much larger population. This rise, propelled by advances in science, technology, and medicine, is continuing. As we shall see, despite several predictions of the impending doom and the population increase exceeding the carrying capacity of the Earth, man has leveraged science, knowledge, and international collaboration to ensure that food production outpaces the rise in population. Man has also become acutely aware of the fact that extensive use of fossil fuels has led to a rapid increase of greenhouse gases in the atmosphere and is fuelling climate change, for which various measures are being adopted, though perhaps not as urgently as required [1].

5.2 Prophets of Doom

The 1960s were a very challenging period for India. It lost its first Prime Minister, Pandit Jawahar Lal Nehru, who had led India on the path of a vibrant democracy and who had presided over the adoption of a scientific outlook for the development of the country. India had a disastrous border clash with China in 1962 and a war with Pakistan in 1965. It lost its next Prime Minister, Lal Bahadur Shastri, who had led India during the war with Pakistan, in January 1966. And then, India suffered severe droughts during 1966 and 1967.

In the 1960s, India was dependent on American aid for food. During 1964–1965, even with an excellent monsoon, India had to import over 7 million tonnes of food, which was more than 10% of the domestic production. Even this production crashed by a further 20% due to the consecutive droughts in 1966 and 1967 mentioned above. Unprecedented food aid had become necessary to save India from mass starvation. Food aid to India and other developing nations used to be provided under the Public Law 480 (PL-480) scheme[1] by the United States of America. However, the supplies were not released on a timely basis, perhaps due to the criticism of the US government for its policies in Vietnam, and this led to what came to be known as a "ship to mouth" existence for the people of India, and the Indian government had to beg American authorities for food repeatedly.

In 1968, Paul Ehrlich from Stanford University published his explosive book, *The Population Bomb* [2]. Right in the beginning, he commented:

> "*I have understood the population explosion intellectually for a long time. I came to understand it emotionally one stinking hot night in Delhi a couple of years ago. My wife and daughter and I were returning to our hotel in an ancient taxi. The seats were hopping with fleas. The only functional gear was third. As we crawled through the city, we entered a crowded slum area. The temperature was well over 100 (sic. degrees Fahrenheit), and the air was a haze of dust and smoke. The streets seemed alive with people. People eating, people washing, people sleeping.*

[1]In a poetic turn of events, the warehouse of the PL-480 scheme was used to house the Department of Science and Technology, Government of India, in 1971, which, among other things, spearheaded the international collaborations in science and technology.

People visiting, arguing, and screaming. People thrusting their hands through the taxi window, begging. People defecating and urinating. People clinging to buses. People herding animals. People, people, people, people. As we moved slowly through the mob, hand horns squawking, the dust, noise, heat, and cooking fires gave the scene a hellish aspect…"

Ehrlich further argued: "The battle to feed all of humanity is over. In the 1970s hundreds of millions of people will starve to death despite any crash programmes embarked upon now. At this late date nothing can prevent a substantial increase in the world death rate although many lives could be saved through dramatic programmes to stretch the carrying capacity of the earth by increasing food production…," and that, "India couldn't possibly feed two hundred million more people by 1980." He went on to add, "I have yet to meet anyone familiar with the situation who thinks that India will be self-sufficient in food by 1971."

Even more gruesome and devastating was the "solution" offered by William Paddock and Paul Paddock, in their book, *Famine, 1975! America's Decision: Who Will Survive?* [3]. They first argued that only the USA could provide food aid. Further, they divided developing nations requiring food aid into three categories:

1. those so hopelessly headed for or in the grip of famine, because of overpopulation, agricultural insufficiency, or political ineptness, and to whom giving any aid would be like throwing sand into the ocean. These nations (like India and Egypt) "could not be saved," and they advised ignoring them and leaving them to their fate.
2. those who are suffering but who will stagger through without US aid, "the walking wounded," and
3. those who can be saved by help.

Ehrlich fully endorsed the views expressed by the Haddock brothers and made some specific predictions:

1. **1969:** "*By 1985 enough millions will have died to reduce the earth's population to some acceptable level, like 1.5 billion people," and "By 1980 the United States would see life expectancy drop to 42 years because of pesticides, and by 1999 its population would plummet to 22.6 million.*"

2. **1970:** "*In ten years all important animal life in the sea will be extinct. Large areas of coastline will have to be evacuated because of the stench of dead fish*" *and* "*Sometime in the next 15 years, the end will come. And by the end I mean an utter breakdown of the capacity of the planet to support humanity.*"
3. **1971:** "*By the year 2000 the United Kingdom will be simply a small group of impoverished islands, inhabited by some 70 million hungry people. .... England will not exist in the year 2000.*"
4. *India is doomed.* "*With luck, it might stagger on to the end of the century, but the train of events leading to the dissolution of India as a viable nation is already in motion*" [4].

They were all echoing the sentiments of Thomas Robert Malthus (1766–1834 CE), an English political economist writing at the turn of the nineteenth century. Malthus is famous for the assertion that in the absence of "preventative checks" to reduce birth rates, the natural tendency for populations to increase — being "so superior to the power of the earth to produce subsistence for man" — ultimately results in "positive checks" that increase the death rate. If all else fails to curb population, "gigantic inevitable famine stalks in the rear, and with one mighty blow levels the population with the food of the world" [5]. Malthus, of course, was alluding to the fact that population increases in geometric progression (1, 2, 4, 8, 16, 32, ...), while food production can at best increase in an arithmetic progression (1, 2, 3, 4, 5, ...) and the increase in population would soon far exceed the food production (Fig. 5.2).

None of these predictions came true. The Green Revolution, started by Norman Borlaug, defied all these prophets of doom, and food production has outpaced the increase in population, which stands at 7 billion now.

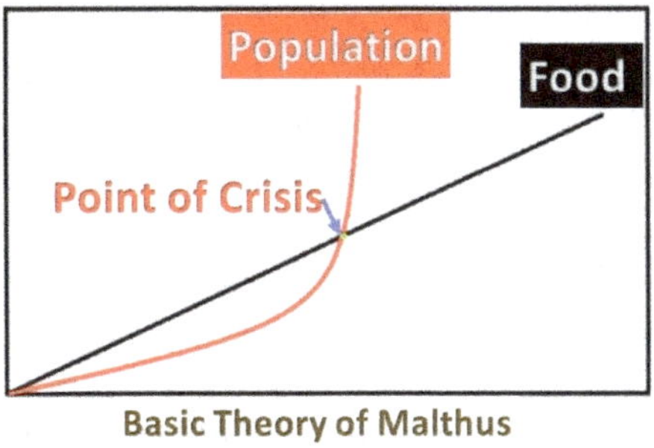

Fig. 5.2. Theory of Malthus.

By most accounts, the world produces enough food to feed all, though an equitable distribution remains a dream.

Yet, we must not forget that in the absence of the Green Revolution, these predictions had a great probability of coming true.

5.3 Famines

The world is not a stranger to famines. Vedic and Puranic literature in India, the Bible, and folk traditions across the world have repeated mentions of famines. By now, it has been realised that famines can arise due to natural disasters such as a long period of drought, floods, extreme cold, cyclones, infestations by pests or diseases of plants, as well as man-made disasters like wars. It is also believed that these can get quite tragic when associated with decisions and policies implemented by government officials, which may be inadequate or even inappropriate.

Kautilya (or Chanakya) (350–275 BCE), writing in his extensive treatise, *Arthashastra* [6], has given detailed instructions for the rulers, in the event of a famine:

> *"During a famine, the king shall show favour to his people by providing them with seeds and provision. He may either do such works as are usually resorted to in calamities; he may show favour by distributing either his own collection of provisions or the hoarded income of the rich among the people or seek help from his friends among kings. Or the policy of thinning the rich by exacting excessive revenue causing them to part with their accumulated wealth may be resorted to.*
>
> *Or the king with his subjects may emigrate to another kingdom with an abundant harvest. Or he may remove himself with his subjects to seashores or the banks of rivers or lakes. He may cause his subjects to grow grains, vegetables, roots, and fruits wherever water is available. He may, by hunting and fishing on a large scale, provide the people with wild beasts, birds, elephants, tigers, or fish."*

Famine has also been studied extensively, and reasonably reliable data on deaths due to famines are now available. In the following, we shall quote extensively from Hasell and Roser's recent study of famines [7]. Some general observations are in order here. Firstly, we know that

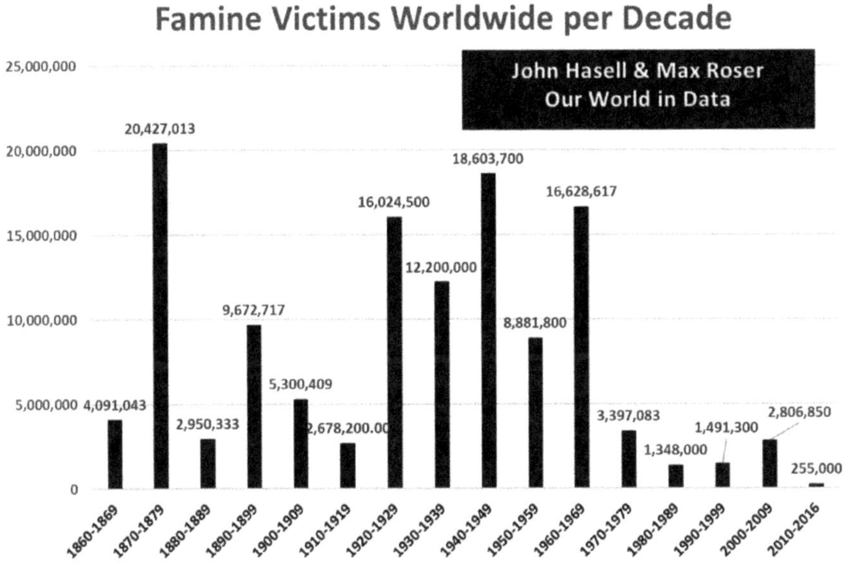

Fig. 5.3. Deaths Due to Famines Per Decade.

famines had ceased mainly in today's developed world by the mid-19th century. Secondly, in the period before the mid-20th century, Asia suffered heavily from major famines. This came to a halt over the mid-20th century. And lastly, since then, famines have almost exclusively been restricted to Africa, with the famine in North Korea being a stark exception. Hasell and Roser have estimated that between 1860 and 2016, a total of 128 million people died in famines, across the world.

Decade-wise deaths due to famines are shown in Fig. 5.3.

Looking more closely, Hasell and Roser found that deaths per 100,000 have reduced drastically in recent times (see Fig. 5.4).

While this comes as a great comfort, this figure must be seen with caution, as famines mostly tend to be confined in some regions. However, this also holds out the hope that the extent of famines has reduced sufficiently and if the world as a whole participates, this ancient suffering can be tackled effectively.

Coming back to the theory of Malthus, following Hasell and Roser, we replot Fig. 5.3 along with the world population (Fig. 5.5).

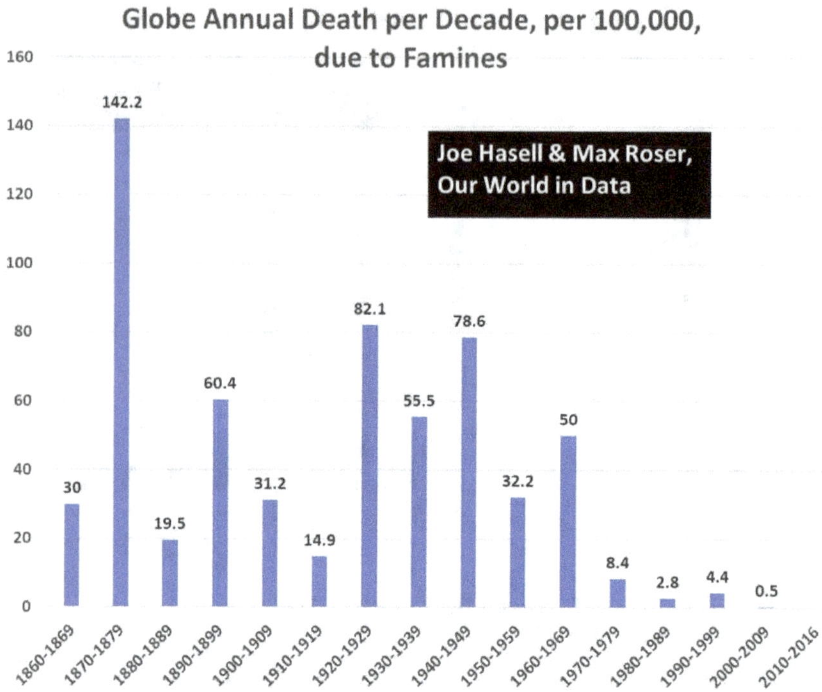

Fig. 5.4. Deaths Per 100,000 Per Decade Due to Famines.

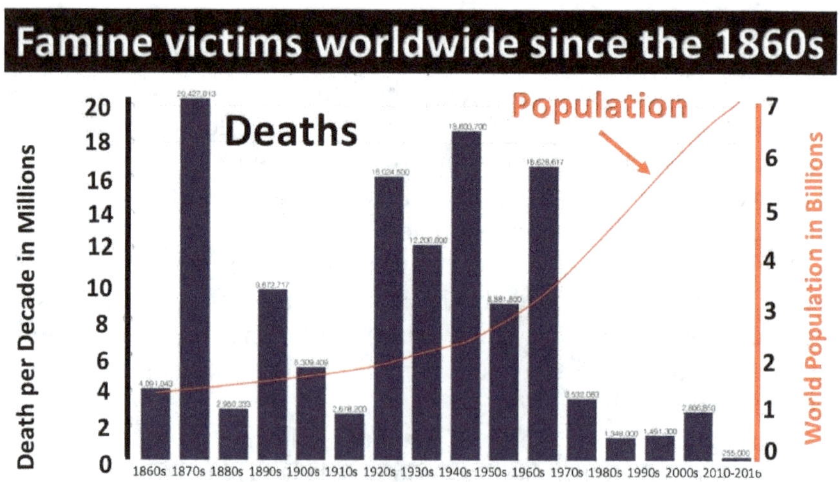

Fig. 5.5. Same as Fig. 5.3, Along with the World Population.

We see immediately that while the world population has continued to rise and crossed the 7 billion mark, the deaths due to famines have reduced drastically, contrary to the beliefs of the prophets of doom quoted in Section 5.2.

Famines are largely expected to ensue when at least 20% of an area's households are facing extreme food shortages with only a limited ability to cope with their current situation. Even before this malnutrition on an extended scale will lead to a number of excess deaths. The 10 deadliest famines in the world are listed below in the reverse order of their severity [8]:

10. Irish Potato Famine;
9. Bengal Famine of 1943, India;
8. Russian Famine of 1921;
7. Great Bengal Famine of 1770, India;
6. Doji Bara Famine ("Skull Famine") of 1789, India;
5. Chalisa Famine of 1783, India;
4. Ukraine Famine of 1932;
3. Northern China Famine of 1876;
2. Chinese Famine of 1907;
1. Great Chinese Famine.

The **Irish Potato Famine** (or Great Famine/Great Hunger, "An Gorta Mór"), as the name suggests, affected Ireland during 1845–1852, where 1.5–2 million people died, and a similar number fled to America. It was caused by a blight of potatoes, which destroyed the entire crop of potatoes. Single-crop cultivation, of potatoes, had been enforced earlier by the landowners, easing out all other crops to maximise profits. There are also suggestions that food transport to Ireland was also pointedly prevented.

We shall discuss the famines of India separately as these were a direct consequence of colonial rule/policies.

The **Russian Famine of 1921**, also known as the "Povolzhye" famine, followed on the heels of the Russian Revolution and the Russian Civil War, which along with droughts killed an estimated 5 million people, affecting the region between the Volga and the Ural Rivers.

The **Ukraine Famine of 1932** (also known as "Holodomor," which means "to starve to kill") is believed to have killed 3–13 million people when the Soviet authorities tried to eliminate farmers throughout Ukraine and eliminate opposition to the soviet rule, policies of collectivisation, and "five-year plans." Reports suggest that house-to-house searches were performed to confiscate every bit of grain and supplies, leaving nothing for the people to survive on, and forcing many to move to cities and still die of starvation.

The **Northern Chinese Famine of 1876** followed 3 consecutive years of drought and may have killed 9–13 million people, while the **Chinese Famine of 1907** was caused by very heavy rains which led to extensive flooding over vast areas and may have killed up to 25 million people. Some reports suggest that the only relief provided by authorities was a suspension of the collection of agricultural taxes.

On the other hand, **the Great Chinese Famine** of the period 1959–1961, which according to some killed anywhere between 10 and 55 million people, was a result of Mao Zedong's "Great Leap Forward" policies, which forced changes which led to a drop in crop production and concentration of millions into "people communes" where they perished in record numbers due to malnutrition, disease, and complete starvation.

5.4 Famines in India during the British Rule

Famines in India under the East India Company and later the Direct British Rule are believed to have killed up to 85 million Indians. The major famines in India, under the British Rule, are listed in Table 5.1 [9].

Indian agriculture is crucially dependent on abundant and regular southwest monsoon (and we return to this again in Chapter 15). A failure of monsoon to bring in rain can lead to severe stress for farmers. The British rulers added to these woes manyfold with their policies aimed at maximising profit. Thus, traditionally Indian rulers used to collect a 10%–15% tax on the produce and used to either waive or defer the collection of taxes in times of distress. The taxes imposed by the British rulers were often as high as 50% and no relief was granted during periods of distress. This quite often left very little for the farmers for their use, and for seeds for the next season — and they often had to sell their cattle to pay the required taxes — under the new Zamindari system which maximised the profits for the rulers, by what is termed rack-renting.

The British also started the practice of collecting levies for wars and exporting food grains like wheat and rice to Britain. They further forced farmers to grow cash crops such as cotton, jute, opium, and indigo, and completely destroyed other domestic industries such as textiles, mining,

Table 5.1. Major famines in India during the British Rule [9].

Year	Name	Region	Deaths	Comment
1769–1770	Great Bengal Famine	Bihar, Western Bengal	2–10 million	About one-third of the then population of Bengal.
1783–1784	Chalisa Famine	Delhi, UP, Punjab, Rajasthan, Kashmir	11 million	Severe famine. Large areas were depopulated.
1791–1792	Doji Bara Famine or Skull Famine	Hyderabad, Central India, Deccan, Gujarat, Southern Rajasthan	11 million	One of the most severe famines known. People died in such numbers that they could not be cremated or buried.
1860–1861	Upper Doab Famine	Rajasthan	2 million	
1865–1867	Orissa Famine	Bihar, Orissa, parts of Southern India	1 million (Orissa) and 4–5 million in the entire region	The British Secretary of State for India, Lord Salisbury, did nothing for 2 months, by which time a million people had died.
1868–1870	Rajputana Famine	Rajasthan	1.5 million	
1876–1878	Great Madras Famine	Mysore and Hyderabad States (Madras Presidency)	6–10 million	
1896–1897	Indian Famine	Rajasthan, parts of Central India, and Hyderabad	5 million	
1943–1944	Bengal Famine	Bengal	3.6 million	1.5 from starvation; 2.1 from epidemics.

metal-working, and ship-building, forcing the artisans to become labourers.

The export of food grains to Britain continued unabated even at the peak of famines. The combination of the above factors led to the **Great Bengal Famine of 1770**, which ravaged Bengal from 1769 to 1773 and at least 10 million people died. It started with a failed monsoon in 1769, which led to a drought which killed two crops of rice. On top of it, the East India Company, which had started collecting taxes, increased them to compensate for wars and forbade storing rice at home. With no food reserves, there was widespread death from starvation. The road and rail transports, which could have brought rapid relief, provided the authorities wanted to implement it, were still decades away.

More than a hundred years later, Bankim Chandra Chattopadhyay wrote his famous novel *Anand Math*, with this famine as the background. The novel contained a song of veneration, "Vande Mataram," to Mother Goddess, who was identified with "Mother India," and Vande Mataram became the battle cry of Indians against British Rule. It is now the National Song of India.

Returning to the present discussion, we recall that the ready availability of road and rail transports was not utilised during the **Bengal Famine** which lasted from 1943 to 1944. Up to 3 million (6 million according to some) people perished in the famine which was entirely man-made (Fig. 5.6).

This has been studied in detail [10]. By now, it is generally believed that the famine was a culmination of several factors, including weather. However, the most important factor was the implementation of the "Scorched Earth" policy, strictly implemented by the British, in preparation for meeting the Japanese Invasion. Thus, all food grains were purchased out of Bengal and all boats, the primary means of transport in Bengal, were destroyed. There is evidence that even though people from other parts of the country were prepared to send food grains to Bengal, they were prevented. Food grains from countries like Australia and America readied for Bengal were diverted to England, with the excuse that they could fall into the hands of the enemies. As a result of all this, whatever little food was available in Bengal went out of reach of people who died in large numbers, a fact which the government of the day tried to suppress using censorship.

Fig. 5.6. The Bengal Famine of 1943: Dead or Dying Children on a Calcutta Street Published in *The Statesman* Newspaper, 22 August 1943.

The complete apathy of colonial rulers is apparent from the fact that when Winston Churchill was informed that famine was raging in Bengal and a vast number of people were dying, he is believed to have quipped, "Why is Gandhi not dead yet?"

Bengal was not the only region of India to suffer massive loss of life due to famines under the British Government.

The **Doji Bara Famine** (or the Skull Famine as bodies could not be cremated or buried and skeletons dotted the countryside) affected subcontinental India from 1789 to 1792 and killed more than 11 million people. It was caused by a major El Niño event that caused a prolonged drought for 4 years in succession. The famine is remembered for attempts at the forced collection of taxes and the absence of any mitigating measures by the British rulers as well as rulers of the princely states, which were affected.

The **Chalisa Famine** (Chalisa or the Forties, as 1783 CE corresponds to the year 1840 of Vikram Era, popular in India) from 1783 to 1784 was again caused by El Niño, which led to extensive drought in large parts of Northern India and killed 11–12 million people.

The **Great Famine of 1876–1878**, also known as the **Southern India Famine** or the Madras Famine, was caused by drought and affected the Deccan Plateau, South and Southwestern India, and claimed 6–10 million lives (Fig. 5.7). "It was part of a larger pattern of drought and crop failure

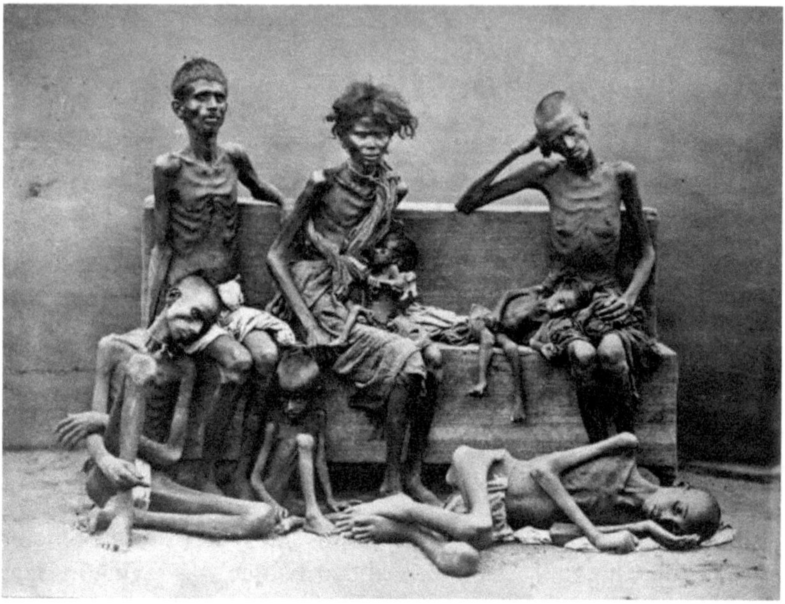

Fig. 5.7. Inmates of a Relief Camp (During the Famine of 1876–1878) in Madras [11].

across India, China, South America and parts of Africa caused by an interplay between a strong El Niño and an active Indian Ocean Dipole that led to between 19 and 50 million deaths" [9].

Once again, it is painful to recall that railways, which could have been used to transport food grains to the affected regions, were instead used to transport food grains to ports for export to Britain, along with cash crops which people had been forced to cultivate.

We know that the Bengal Famine of 1943 was the last substantial famine in India. It occurred only 4 years before the British Empire ended. Famines, which had been a persistent feature of the British Indian Empire, ended abruptly with the establishment of a democracy after independence, as noted by Amartya Sen, the famed economist. Sen further argued that in a functioning multi-party democracy, the government is accountable to the public, particularly through a free press, and uncensored public criticism. Thus, the government has an "excellent incentive to do its best to eradicate famines."

A recent example of this is the distribution of free food grains to 800 million people in India for over 2 years (and continuing as we write this), in the wake of the COVID-19 pandemic which badly affected industry, agriculture, commerce, education, and employment, especially for landless labourers, artisans, and small farmers and traders, which averted large-scale starvation [12].

A recent study has suggested that the population of erstwhile British India (i.e., people of India, Pakistan, and Bangladesh) still carry the scar of the 31 famines which occurred during the British Rule and that they have a tendency to generate and store fat and not burn it off, amassing low lean muscle mass, as they are "starvation-adapted," which may be behind their tendency to be easily affected by Type-II diabetes [13].

5.5 The Haber–Bosch Process

It is known that famines are generally localised and seriously affect 10%–20% of the population. These can, thus, be effectively handled if food aid from other parts of the country or other countries is available and distributed quickly and equitably.

The overall availability of food for the rapidly increasing population is a different and much more complex issue. Production of more food generally requires more land, irrigation, and fertilisers. Land suitable for agriculture is limited unless we cut down more forests. It has been known for a long time that forests play a very important role in shaping our climate, regulating rains and rivers, and maintaining a balance in our climate, biodiversity, and ecosystem. The other possibility is to increase the amount of fertilisers used for the crops. Even though some crops like legumes, e.g., red gram and peas, have roots that host bacteria that replenish nitrogen levels, for thousands of years, fertilisers have been obtained from manures and composting of vegetation. Both are limited and through the 1800s, heavily populated countries in Europe, like Germany and Britain, used guano (the excrement of seabirds and bats) from islands off the coast of Peru and saltpetre from Chile as a source of natural nitrogen fertiliser.

In the early 1900s, Fritz Haber and Carl Bosch developed a process using which atmospheric nitrogen was converted to ammonia by a reaction with hydrogen under high temperatures and pressure in the presence of a catalyst. The ammonia was then used to produce nitrates, which are used as a chemical fertiliser [14]. This is considered one of the most important chemical discoveries in the history of mankind. Today, the world generates about 160 million tonnes of ammonia every year, and it is believed that one-third to half of the world population of more than 7 billion people has food due to this process. This production is likely to double in the coming years. We add that the increased food production also owes a lot to new seeds developed over the last 60 years, which respond well to the increased application of fertilisers, which ushered in the Green Revolution, which we shall discuss in the following.

Before closing this discussion, we add that some serious concerns have also been voiced about the Haber–Bosch process, and methods to circumvent these are badly needed. Firstly, the process is very energy-intensive and uses about 5% of the natural gas production and 1%–2% of the annual energy consumption of the world. There are some developments to produce green ammonia by using hydrogen from water electrolysis and nitrogen separated from the air. These are then fed into the Haber–Bosch process, all powered by sustainable electricity.

It has been argued that the excess chemical nitrogen disturbs the carbon and nitrogen balance in the soil, and the microbes in the soil end up consuming more carbon, which leads to a loss of organic matter in the soil. The weakening of the soil food web leads to the weakening of the processes of fixing nitrogen, building stores of phosphorus, and defending crops against pests and diseases. Farmers tend to tackle this with the increased use of fertilisers, herbicides, and pesticides.

It has also been pointed out that plants take up only 10%–50% of the chemical fertilisers applied and the rest runs off to water bodies like streams and even pollutes wells due to their mobility in soil. It ultimately reaches the oceans, causing "dead zones," as blooms of algae near the surface block out sunlight and kill the fish below. Some of this fertiliser also ends up in the air as nitrous oxide, a powerful greenhouse gas, and leads to acid rain, which makes soils more acidic, disrupts ecosystems, and threatens biodiversity. We shall come back to this point again.

5.6 Food Crisis in India in the 1960s

India became independent in 1947 from British Rule. British India included present-day Pakistan, Bangladesh, and Myanmar. The British gave up Myanmar (Burma), and the rest of India was partitioned. Rice transport from Myanmar was stopped, and well-irrigated and fertile lands,

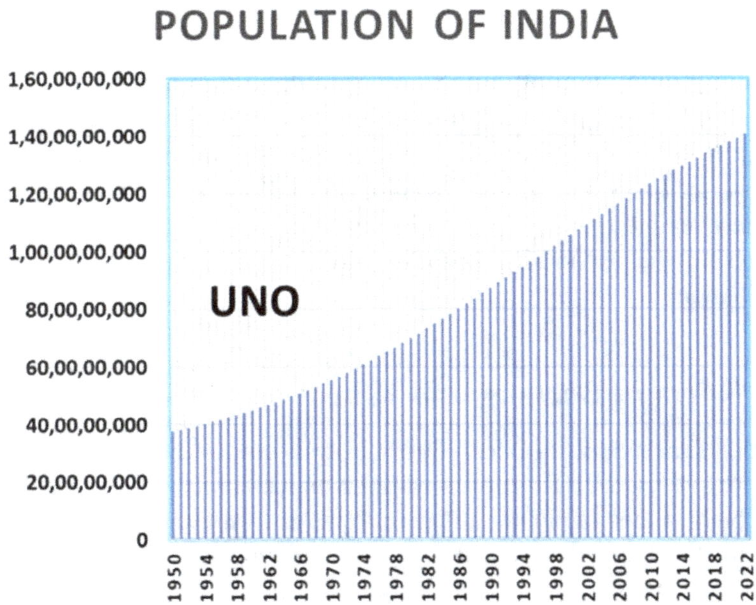

Fig. 5.8. Rising Population of India.

watered by an elaborate network of canals of West Punjab, which used to produce surplus rice and wheat, went to West Pakistan, while the rice-surplus areas of East Bengal went to East Pakistan (now Bangladesh). Only 15% of agricultural land in truncated India was irrigated, and the rest was rainfed. Improved medical facilities, access to modern medicines, and awareness of hygiene had started a rapid increase in the population (Fig. 5.8).

We have already seen how India was pathetically dependent on food aid to stave off starvation on a massive scale due to all the above reasons. We have also noted that this happened during the Vietnam War, to which India used to voice opposition at all forums including the Non-Aligned Movement, to the annoyance of the US government. It is well known that the then American President made Indian politicians and officials beg repeatedly for more food. At least on one occasion, the President's daughter is believed to have pleaded with him to release food as "Christmas was around the corner." This gave rise to the phrase, "ship to mouth" existence. While this averted mass starvation, it also left Indians writhing with humiliation. "I don't ever want us to have to beg for food again," Prime Minister Indira Gandhi is believed to have said after a call with the US President. K. Kamraj, the President of the Congress at the time, had remarked in December 1965 that he "would prefer to starve than receive wheat from the Americans." Lal Bahadur Shastri, the earlier Prime Minister of India, had advised (and himself followed) a fast once a week, to tide over the crisis. The prophecies of doom and a sermon to leave Indians to die as nothing could be done to save them added to the humiliation.

5.7 Norman Borlaug and Wheat

Norman Ernest Borlaug (1914–2009 CE) was born in 1914 in Saude, near Cresco, Iowa, USA. He grew up working on the family farm and left Iowa after high school to pursue a degree at the University of Minnesota. He was greatly inspired by a talk by plant pathologist E. C. Stakman and obtained his master's and Ph.D. degrees, working with him. Borlaug completed his degrees in 1942. As World War II had already broken out, he took a job with DuPont where he conducted research for the United States armed forces. By then, the Rockefeller Foundation and the Mexican government had teamed up to form the Cooperative Mexican Agricultural Programme, in an effort focused on revitalising agriculture in Mexico. Borlaug (Fig. 5.9) joined the programme in 1944 after turning down an offer to stay at DuPont. Borlaug's work in Mexico was to change the world forever. It is also a fascinating story of brilliance, dedication, hard work, and perseverance.

The first major problem which Borlaug faced in Mexico was stem rust. It was caused by a fungus, *Puccinia graminis*, whose spores travelled through the air. When these landed on wheat plants, they caused brown lesions, which soaked up the nutrients meant for the grain and weakened the plants, leading to breakage, desiccation, and shrivelling.

The epidemics of stem rust over 3 consecutive years from 1939 to 1941 had wiped out wheat in the Yaqui Valley of Mexico. Breeding rust-resistant varieties of wheat is a slow process, which can take up to 10 or 12 years. Borlaug, against all advice, decided to take advantage of the two growing seasons of Mexico, employing what is known as "shuttle breeding." Thus, the wheat was selected and bred in the cooler central highlands in the summer and then seeds from the selected plants were bred in

Fig. 5.9. Norman Borlaug Congressional Gold Medal (Front) [15].

the warmer north-western regions during the winter. This halved the time needed for the process to be completed, though it meant much harder work for him and his team. In the process, the selected plants were exposed to varying lengths of daylight and latitudes. Soon, it became abundantly clear that the wheat grown in Mexico could be cultivated in various regions of the world. In the process of this breeding, Borlaug and his colleagues discovered a gene Stem Rust 31, or Sr31, which provided protection against stem rust when present in wheat, and as luck would have it, also led to increased yields.

By this time, more and more chemical fertilisers (nitrate) were becoming available, as World War II had ended (when they were used to make explosives). The increased usage of fertiliser led to better crop growth and higher grain yields. The heavier heads of grain, in turn, created the problem of lodging, as stalks collapsed under the weight of the grain and fell over, ruining the crop and leading to a large reduction in yields.

To prevent lodging, Borlaug bred the tall, thin stalks common in Mexico with shorter wheat stalks of the dwarf variety called Norin 10, originally from Japan. Borlaug began to produce stronger, higher-yielding Mexican varieties, which produced stronger stalks and two to three times more grain than the standard varieties. Next, Borlaug bred the shorter varieties with the stem rust-resistant wheat he had produced earlier, creating semi-dwarf wheat that was resistant to the disease and could be grown in a range of climates, completing his triumphal march. The success of his wheat can be gauged from the fact that by 1963, 95% of the wheat grown in Mexico came from Borlaug's breeding programmes. The wheat harvest that year was six times larger than the harvest just 19 years earlier when Borlaug had arrived in Mexico.

5.8 Ushering in the Green Revolution

As both India and Pakistan were facing food shortages, they looked at Borlaug's success with awe. The Indian programme was led by Monkombu Sambasivan Swaminathan (Fig. 5.10), who had earlier worked with

Fig. 5.10. Prof. M. S. Swaminathan (7 August 1925–28 September 2023).

Borlaug, and was very strongly supported by the Minister of Agriculture (and later the Minister of Finance, who helped by providing subsidies for seeds, fertilisers, irrigation, and pesticides) Chidambaram Subramaniam and the Prime Ministers, Lal Bahadur Shastri and Indira Gandhi. In 1965, India and Pakistan imported 200 and 250 tonnes, respectively, of seeds from Mexico, which arrived after facing several hurdles on the way. By the time the seeds arrived at the ports, a war broke out between India and Pakistan, and Borlaug, who had travelled to the subcontinent to supervise the planting and the cultivation, had to travel between India and Pakistan via Dubai, to cover a small distance between the two experimental farms across the borders. The effort produced the highest-yielding wheat fields those areas of Asia had ever seen (Fig. 5.11), and the next year India imported 18,000 tonnes of seeds, while Pakistan asked for 42,000 tonnes.

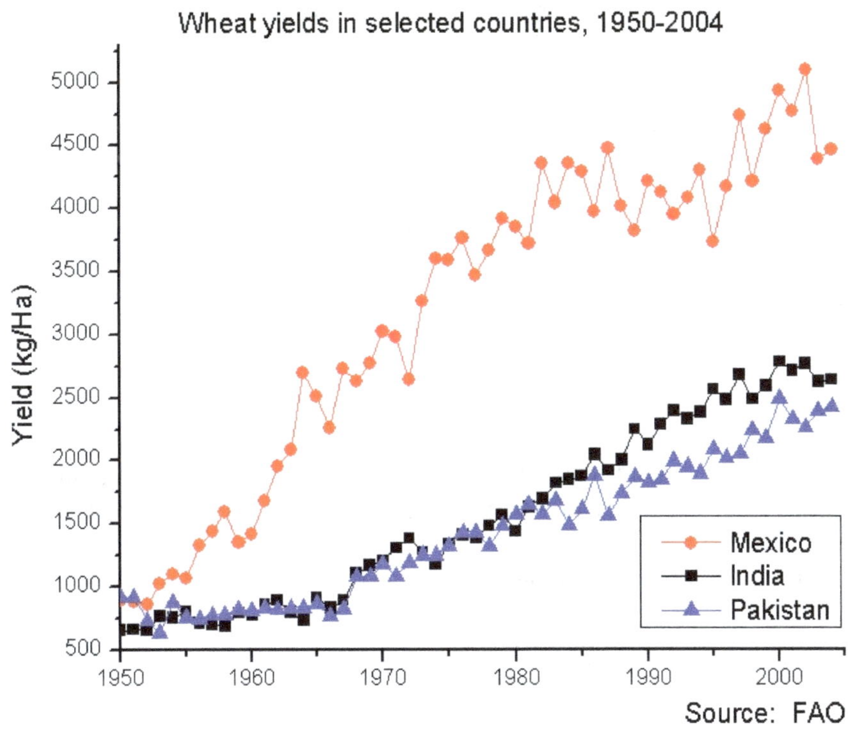

Fig. 5.11. Wheat Yield in India, Pakistan, and Mexico (Credit: Wikipedia).

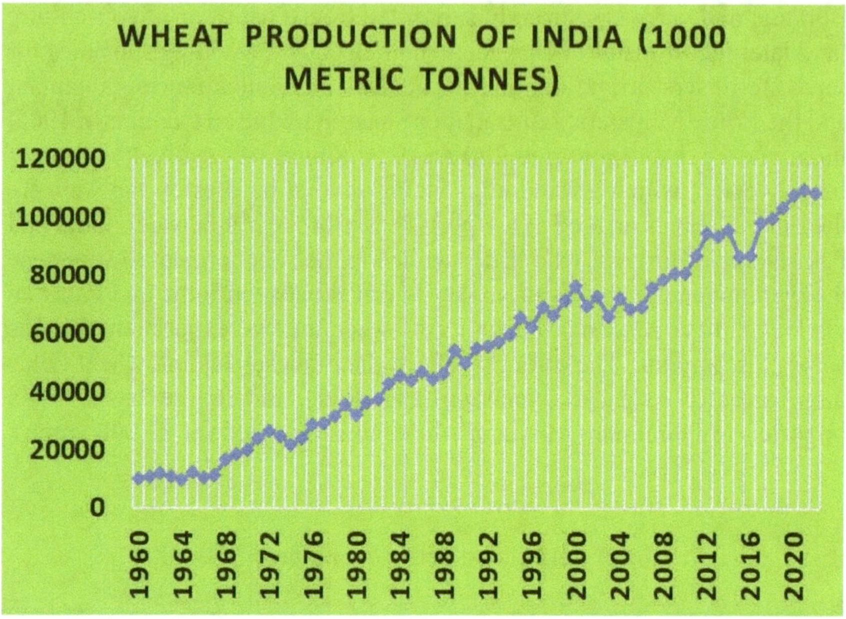

Fig. 5.12. Rise in the Production of Wheat in India.

"By 1968, normal rains had returned, and Mexican wheat varieties had produced so much grain in India and Pakistan that there weren't enough people to harvest the crop. There weren't enough bullock carts to haul the wheat to threshing floors. There weren't enough jute bags, trucks, rail cars, or grain storage facilities. Some towns closed schools temporarily to house the grain. And millions were no longer hungry [16]."

We have seen an enormous rise in the population of India (and the world). However, food production in India has continued to outpace this rise (see Fig. 5. 12).

The Green Revolution had been ushered into India and the world.

5.9 Gurdev Singh Khush and Rice

The same procedure for developing new breeds of rice was used to increase its yield, manyfold, especially at the International Rice Research Institute (IRRI), the Philippines [17].

Rice is a staple food for about half of the global population. It is also the most important cereal crop in the world. It is cultivated in a wide range of climatic conditions spread from 35° north in North Korea to 35° south in Australia, and rice fields cover an area of about 155 million hectares.

In India, it is cultivated from about 2 m above the mean sea level in Kerala to more than 800 m above the mean sea level in the Himalayas, over an area of about 43 million hectares under diverse climatic, soil, and ecological conditions, throughout the year, in one or other part of the country [18]. Special seeds which were developed following the procedures adopted earlier for the breeding of wheat have led to an increase in the production of rice in India by a factor of six, without any significant increase in the area covered by the cultivation of rice (see Figs. 5.13, 5.14).

The IRRI, the Philippines, in November 1966, released IR8, which was called the "miracle rice." It was a cross between a Chinese dwarf paddy Dee-geo-woo-gen and a tall vigorous Indonesian variety. Traditional tall varieties used to have plants that grew to a height of more than 150 cm and had weak stems. When fertilisers were applied, they largely grew still taller and bent over ("lodged"), took 160–180 days to mature, and had a paddy yield of 4–5 metric tonnes/hectare.

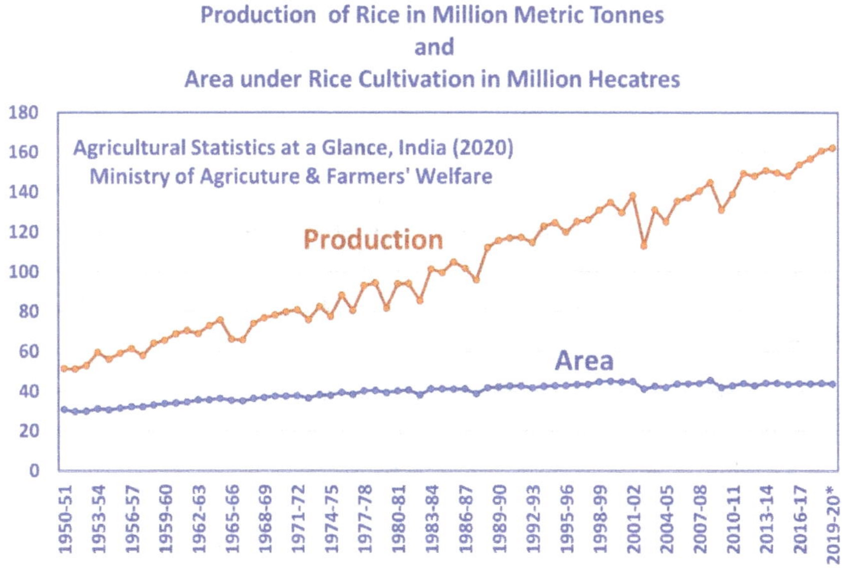

Fig. 5.13. A Six-Fold Increase in the Production of Rice Without any Significant Rise in the Cultivated Area (* Denotes Estimate) [19].

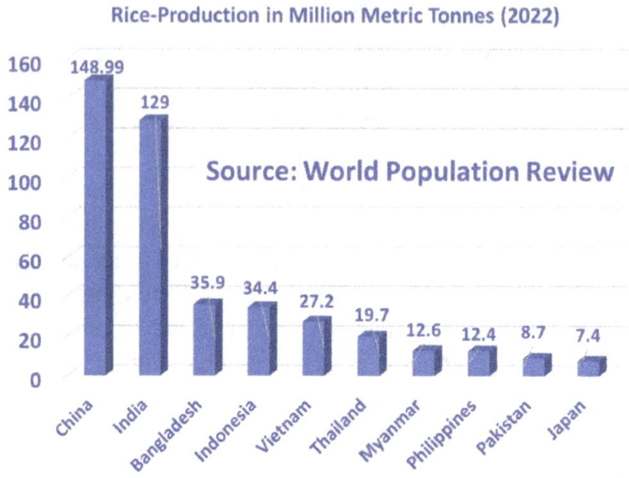

Fig. 5.14. Top 10 Rice-Producing Countries in the world (2020).

The IR8 had plants that were just about 95 cm tall, with sturdy stems that did not lodge when heavily fertilised. The paddy yields were 4.5–5 tonnes/hectare with minimal fertilisers and could go up to 9–10 tonnes with a higher application of fertilisers. However, the IR8 was susceptible to diseases and insect pests. Moreover, its grains were "chalky" which broke during milling and became dry and hard when cooked.

Gurdev Singh Khush (Fig. 5.15) and his team working at IRRI released the high-yielding rice IR36 in 1976, which had genes from 14 indigenous landrace varieties of six countries and one wild rice species, with a resistance to a broad spectrum of pests and diseases. It matured in just 111 days, which enabled farmers to easily grow two crops in a year. Ultimately by the 1980s, it was planted over an area of 11 million hectares annually, which no food crop had ever occupied.

The next miraculous variety of rice to be released by Dr. Khush and his team was IR64, in 1985 [20]. It was planted in an area of over 10 million hectares in the 1990s. It had genes from 20 landraces of eight countries, had a higher yield than both IR8 and IR36, and possessed resistance to multiple diseases and insects, beyond them. The quality of IR64 grains was much better and they had a soft texture when cooked and had better palatability. The PR-106 rice, which had long and slender grains, was developed by Punjab Agricultural University, Ludhiana, using rice seeds provided by IRRI and became quite popular in Punjab, while IR8

Fig. 5.15. Gurdev Singh Khush (1935–): Father of the Green Revolution for Rice.

(initially) and IR36 and IR64 remained popular in eastern, central, and southern India.

It is thus not a wonder that the renowned Indian agricultural scientist and 1996 World Food Prize Laureate Dr. Gurdev S Khush is considered one of the fathers of this significant development, which prevented famine and land degradation mainly in Southeast Asia, where in the 1960s, countries were running out of time because of the increasing pressure of population density. It has been estimated that increased rice production from high-yielding varieties feeds 1 billion additional people who use rice as their staple cereal. It has also been estimated that the price of rice adjusted for inflation is 40% lower in the domestic and international markets than it was in the 1960s. This benefits poor rice consumers who spend 50%–60% of their earnings on purchasing food [21].

The production of rice has risen rapidly in India and the world using the seeds developed using the procedure adopted and perfected by Borlaug for wheat, and it has outpaced the population growth.

Yet, it is felt that in semi-arid regions like Punjab, the area under rice should be "progressively reduced" and the rice area should be replaced with oilseeds and other crops, to avert the excessive use of groundwater. Moreover, the yield of rice (and other food grains) must continue to increase to meet the growing demand for food, under the threats of climate change, temperature fluctuations, low water availability, poor soil quality, and low nutrient availability [18]. There is also an urgent need to continuously develop better seeds which can tolerate various stresses, such as salinity and shortage of water.

We must add that continuing with the development of new seeds, Golden Rice, which makes beta-carotene, has been developed which gives the rice a yellow (golden) colour. It provides additional intervention for vitamin A deficiency, which is the most significant cause of childhood blindness, and also the biggest killer of children under 5 years old, globally. The use of genetic engineering for its development has however made some people wary of it.

5.10 The Road Ahead: Evergreen Revolution

We must add that the Green Revolution in India has been confined to Punjab, Haryana, Western Uttar Pradesh, and regions of Andhra, where irrigation facilities, larger farms, subsidies for fertilisers, pesticides, herbicides, and mechanisation, and irrigation are available. Let us recall the march of the Green Revolution (Table 5.2).

Table 5.2. Major developments in wheat cultivation in India [23].

Period	Steps taken
1967–1968	Green Revolution: Introduction of high-yielding semi-dwarf varieties
1970s	High-yielding semi-dwarf varieties: Wide adoption of varieties, such as Sonara 64, Lerma Roja, and Sonalika
1980s	Mechanisation: Increased mechanisation (use of tractors, etc.) particularly in states like Punjab and Haryana
1990s	Scientific package of practices: Package of region-specific practices identified and recommended to farmers
1990s	Frontline demonstrations: Demonstration of newly released crop production, protection, and management practices
1995	Introduction of 1B/1R translocation, release of PBW-343, popular with farmers, potential to yield 6.3 tonnes/hectare
Early 2000s	Resource conservation technologies such as Zero Tillage, Rotary Till, and Rotavator popularised among farmers
2010s	Release of superior wheat varieties like HD 2967 and HD 3086, and WB 2 as the first bio-fortified variety

There should be no doubt that the Green Revolution brought food security to India, and the confident India that we see today has its foundation in its own food security which generates a surplus to provide food security to other nations. The history of agriculture in the first two decades after independence is not only fascinating but also inspiring for the sagacity of political leaders (Lal Bahadur Shastri, Indira Gandhi, and C. Subramaniam), the commitment of top bureaucracy, the vision of agriculture scientists (M. S. Swaminathan), and the agility of farmers of Punjab and Haryana in adopting new technology.

We also note that since independence, India has observed a considerable increase in the production of oilseeds (through the Yellow Revolution), milk (through the White Revolution), fish (through the Blue Revolution), and fruits and vegetables (through the Golden Revolution) [22].

We also note that irrigation using deep tube wells has encouraged farmers to grow water-guzzling crops like rice and sugar cane, which is rapidly depleting the groundwater. We have already mentioned the ill effects of the excessive use of fertilisers. Even more serious consequences follow the excessive use of pesticides. Urgent corrective measures are needed for these [24].

Given the above, a need is now seen for "A Continuing Evergreen Revolution," which will ensure regular improvements in fertilisers and pesticides, and drought, heat, excess rain, and salt-resistant crops, to feed the ever-increasing population which is projected to reach up to 10 billion in some decades [25]. This would require a serious result-oriented and continuously refined approach (Fig. 5.16), especially because of the ongoing climate change [26]. "An evergreen revolution," according to Dr. M. S. Swaminathan, "implies the enhancement of productivity in perpetuity without associated ecological harm."

We also need to recognise the general reluctance of the masses towards genetically modified crops and address it through a careful, unbiased, evaluation of these crops [12]. Food grain and food (e.g., vegetables, fruits, dairy, and poultry/meat/fish products) preservation is yet another area needing immediate attention as their loss, in the absence of facilities for storage, amounts to a crime. The Green Revolution is also believed to have bypassed the small farmers, and different strategies need to be adopted to bring unparalleled prosperity to them [27].

We note that all this would require a substantial increase in spending on research and development and manpower engaged in research. The expenditure is a very small percentage of our GDP (Fig. 5.17), at the moment.

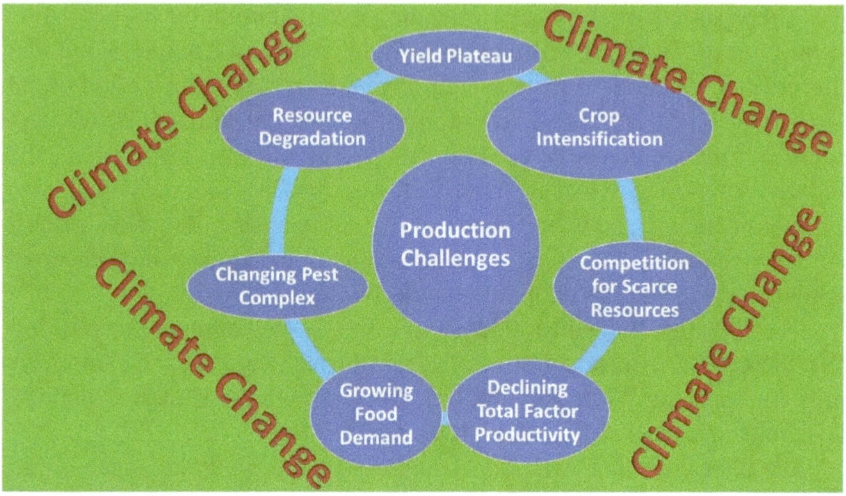

Fig. 5.16. Challenges in Increasing Food Production [26].

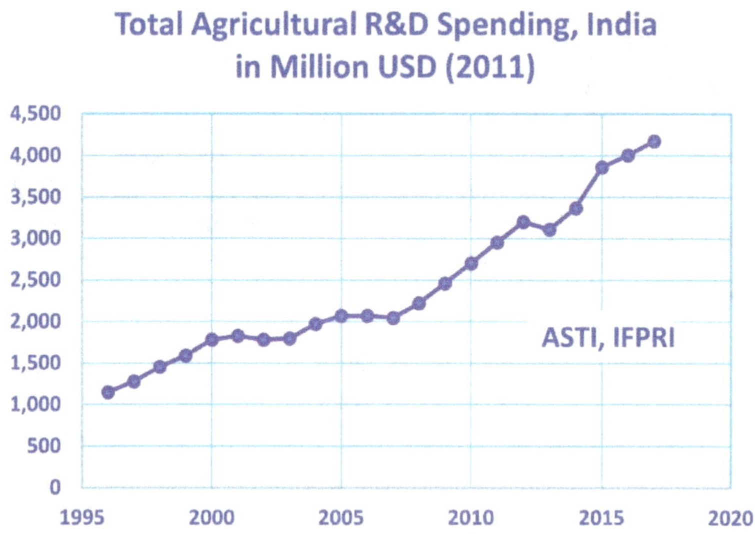

Fig. 5.17. Research and Development Expenditure in India.

The number of (full-time equivalent) agriculture scientists has also remained stagnant for a long period as seen from Agricultural Science and Technology Indicators from the International Food Policy Research Institute (Fig. 5.18).

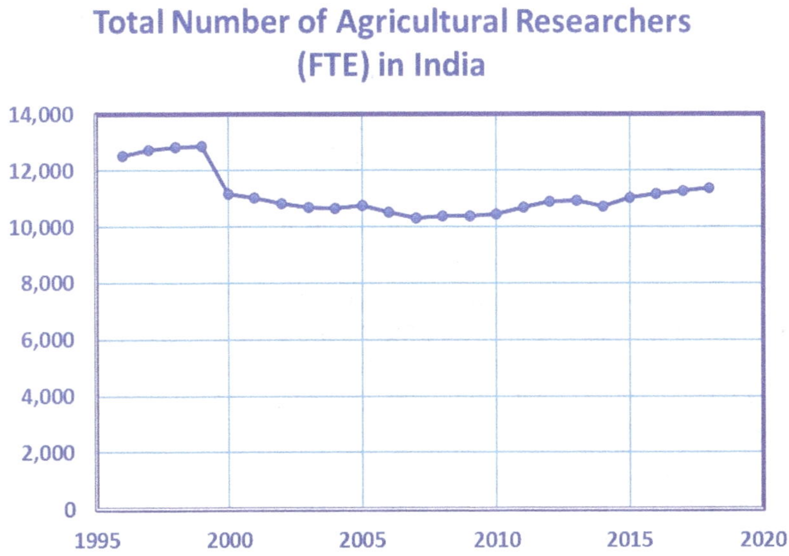

Fig. 5.18. R&D Personnel in Agriculture Research in India.

There are suggestions to go beyond mono-cropping practices, popular for several millennia, because of the ease of managing. We get half of our food from just three crops: rice, wheat, and corn. Mono-cropping is quite vulnerable to losses due to diseases and pests, as these can easily spread over large areas. We need to re-popularise some of the ancient hardy crops such as Amaranth, Fonio, Cowpea, and Taro, which have provided sustenance to communities in South America and Africa for so long. Similarly, hardy crops of millets, which used to be called coarse grains and were eaten only by the poorer sections of the country, have much higher amounts of fibre and essential minerals than wheat, rice, and other cereals. These need to be popularised. Fortunately, this is being pursued. Simple techniques like drip irrigation (in place of flood irrigation) and irrigation during the nighttime also need to be used more extensively to conserve water.

5.11 Conclusions

We close this discussion by recalling the valuable comments of Norman Borlaug, the Father of the Green Revolution, because of which a large number of us are alive today. Writing in 2007, he remarked [28]:

> *"Some 800 million people still experience chronic and transitory hunger each year. Over the next 50 years, we face the daunting job of feeding 3.5 billion additional people, most of whom will begin life in poverty. The battle to alleviate poverty and improve human health and productivity will require dynamic agricultural development."*

We have seen that the breakthroughs in wheat and rice production, known as the Green Revolution, signalled the dawn of applying agricultural science to the Third World's need for modern techniques. "It began in Mexico in the late 1950s, spread to Asia during the 1960s and 1970s, and continued in China in the 1980s and 1990s. Over 40 years, the proportion of hungry people in the world declined from about 60% in 1960 to 17% in 2000." The population of the world increased considerably over the same period.

Counting the environmental benefits of the Green Revolution, he added: "If the global cereal yields of 1950 still prevailed in 2000, we would have needed nearly 1.2 billion more hectares of the same quality, instead of the 660 million hectares to feed the population in 2000. If this land was obtained by converting environmentally fragile land into agricultural land, the soil erosion, loss of forests and grasslands, reduction in biodiversity, and extinction of wildlife species would have been disastrous."

As we have discussed earlier, productivity improvements in crop management can be and have to be made in plant breeding, crop management, tillage, fertilisation, weed and pest control, harvesting, and water use, as well as the recognition of genetically engineered crops. Genetically modified crops may meet the demands of enhanced tolerance to drought, heat, cold, salinity, and waterlogging, which are the consequences of global warming and could prove essential to meet the demands of food, feed, fibre, and even biofuel.

Norman Borlaug presciently ended his remarkable editorial by noting, "The battle to ensure food security for hundreds of millions of miserably poor people is far from won. We must increase world food supplies but also recognise the links between population growth, food production, and environmental sustainability. Without a better balance, efforts to halt global poverty will grind to a halt."

We also note that there are several dimensions of international collaboration which have been identified and discussed [29]. Thus, it has been argued that the technologies relevant to high-yielding variety seeds and climate change have several aspects in common. Firstly, both technologies are affected by multiple causes, scales, and stakeholders or actors. Secondly, the technologies are generally not commercially viable and require a technology push from public institutions to facilitate their deployment. Moreover, it is very well known that international collaborations achieve their fullest potential when there is a minimum absorptive capacity in place. When the domestic capacity does not exist, international collaboration can help build this capacity.

India's success in ushering in the Green Revolution is a classic example of what can be achieved if all the necessary steps are taken. India had an Indian Council of Agricultural Research (ICAR), several small agricultural research centres, and a state agriculture university system across the country. ICAR brought all the pre-independence research institutes under one umbrella and developed a national network of research institutes engaged in basic and adaptive research as well as training at a higher level. Many of the researchers of these institutes were further trained at the institutes of Norman Borlaug and Gurdev Singh Khush. The universities, with adequate land grants, trained scientists, conducted adaptive research, and complemented the ICAR. This combined infrastructure created the robust base for the implementation of thousands of demonstration programmes and the effectiveness of new technologies in communities' fields and generated the demand for high-yielding seeds. The expertise developed in adaptive research went on to play a crucial role in increasing the demand for these seeds, which had to be changed to satisfy the culinary and cultural preferences of Indian consumers as well as to accommodate the varying soil and climatic conditions.

We should also be aware of one more crucial point. As wheat and rice crops are self-pollinating, their characteristics are passed on to the seeds, and thus the private research and development organisations show no interest in their development, as they cannot make money by selling seeds afresh for every sowing. Thus, the world seed market has focused on the seeds of maize (corn), soybean, cotton, and vegetables, for which the second generations do not express the desired traits of the parent. This then provides the seed providers with a safe and regular return on their investment in R&D.

We are convinced that this battle to end global poverty exacerbated by climate change and the increasing population can only be won with global participation and collaboration in sciences.

References

[1] D. K. Srivastava and V. S. Ramamurthy, "*Climate Change and Energy Options for a Sustainable Future*" (Singapore: World Scientific Publishing Company, 2021).

[2] P. R. Ehrlich, "*The Population Bomb*" (New York: Ballantine Books, 1968).

[3] W. Paddock and P. Paddock, "*Famine, 1975! America's Decision: Who Will Survive?*" (Boston: Little, Brown, 1967).

[4] P. R. Ehrlich and A. H. Ehrlich, "*The End of Affluence*" (New York: Ballantine Books, 1976).

[5] T. R. Malthus, "*An Essay on the Principle of Population*" (1798, published anonymously and updated several times), Chapter VII, p. 44.

[6] Kautilya "*Arthshastra*". Translated by R. Shamasastry (Bangalore: Government Press, 1915), Chapter III, p. 296.

[7] J. Hasell and M. Roser, "Famines" (2013). Published online at OurWorldInData.org. Retrieved from: https://ourworldindata.org/famines.

[8] See e.g., https://owlcation.com/humanities/The-Top-10-Worst-Famines-in-History.

[9] https://en.wikipedia.org/wiki/Timeline_of_major_famines_in_India_during_British_rule.

[10] A. Sen "*Poverty and Famines: An Essay on Entitlement and Deprivation*" (New York: Oxford, 1982), Clarendon Press Oxford University Press. by several authors.
M. Mukerjee, "*Churchill's Secret War: The British Empire and the Ravaging of India During World War II*" (New York: Basic Books, 2010), https://en.wikipedia.org/wiki/Bengal_famine_of_1943.

[11] https://commons.wikimedia.org/wiki/File:Inmates_of_a_relief_camp_by_WW_Hooper_2.jpg.

[12] V. S. Ramamurthy, D. K. Srivastava, and S. Nayak, "*Art and Science of Managing Public Risks*" (Singapore: World Scientific Publishing Company, 2022).

[13] https://asianews.network/did-british-colonialism-make-south-asians-prone-to-diabetes-cardiovascular-diseases/; M. Syed, "South Asian Health Crisis, Causes & Solutions" (to be published); See also, J. Li *et al.* "Prenatal Exposure to Famine and the Development of Hyperglycemia and Type 2 Diabetes in Adulthood across Consecutive Generations: A Population-Based Cohort Study of Families in Suihua, China", *American Journal of Clinical Nutrition*, 105 (2017) 221–227.

[14] D. K. Srivastava and V. S. Ramamurthy, "The Incredible Journey of Nitrates and Ammonia", *Science Reporter* (November 2020) 36–39.

[15] https://commons.m.wikimedia.org/wiki/File:2006_Norman_Borlaug_Congressional_Gold_Medal_front.jpg.

[16] B. Ganzel, "The Green Revolution: India and Pakistan", https://livinghistoryfarm.org/farminginthe50s/crops_16.html.

[17] D. S. Athwal, "Semidwarf Rice and Wheat in Global Food Needs", *The Quarterly Review of Biology*, 46 (1971) 1–34.

[18] H. Pathak, M. Kumar, K. A. Molla, and K. Chakraborty, "Abiotic Stresses in Rice Production: Impacts and Management", *Oryza* 58 (2021) 103–125.

[19] "Agricultural Statistics at a Glance 2020" (New Delhi: Ministry of Agriculture & Farmers Welfare, Government of India).

[20] D. J. Mackill and G. S. Khush, "IR64: A High-Quality and High-Yielding Mega Variety", *Rice* 11, 18 (2018).

[21] http://www.goldenrice.org/Content1-Who/who_Gurdev.php.

[22] A. Borthakur and P. Singh, "History of Agricultural Research in India", *Current Science*, 105 (2013) 587–593.

[23] S. Ramadas, T. M. Kiran, and G. Singh, "Wheat Production in India: Trends and Prospects" (2019). DOI: 10.5772/intechopen.86341.

[24] See, e.g., D. A. John and G. R. Babu, "Lessons from Aftermath of Green Revolution on Food System and Earth", *Frontiers in Sustainable Food Systems*, 5 (2021) 1–6.
K. Charyulu and S. Biswas, "*Organic Input Production and Marketing in India — Efficiency, Issues and Policies*" (New Delhi: Allied Publishers, 2011).

[25] P. S. Somvanshi, T. Pandiaraj, and R. P. Singh, "An Unexplored Story of Successful Green Revolution of India and Steps towards Ever Green Revolution", *Journal of Pharmacognosy and Phytochemistry* 9 (2020) 1270–1273.

[26] R. Sendhil, S. J. Balaji, P. Ramasundaram, A. Kumar, S. Singh, R. Chatrath, and G. Singh, "Doubling Farmers Income by 2022: Trends, Challenges, Pathway and Strategies". Research Bulletin No. 40 (Karnal, Haryana: Indian Institute of Wheat and Barley Research, 2018).

[27] V. S. Ramamurthy and D. K. Srivastava, "A Multi-dimensional Approach to Sustainable Rural Development", *Science and Culture* 87 (2021) 349–358.
D. K. Srivastava and V. S. Ramamurthy, "Apiculture: The Sweet Trail to Prosperity", *Science Reporter* (March 2021) 44–48.

[28] N. Borlaug, "Feeding a Hungry World", *Science* 318 (2007) 5849. DOI: 10.1126/science.1151062.

[29] F. Gagnon-Lebrun, International Energy Technology Collaboration and Climate Change, Information Paper, "Case Study 2: Cooperation in Agriculture: R&D on High-Yielding Crop Varieties", Unclassified COM/ENV/EPOC/IEA/SLT (2004) 9, OECD.

Chapter 6

Iron and Steel: The Metals That Built the World

"By preaching the doctrine that nothing is to be admired except steel and concrete, one merely makes it a little surer that human beings will have no outlet for their surplus energy except in hatred and leader worship."

— George Orwell

6.1 Introduction

The efforts to understand the formation of the Earth and the emergence of Life on Earth have been the source of all our knowledge systems. The history of the evolution of Homo Sapiens and human civilisation has continued to fascinate mankind, and newer discoveries are regularly reported from various parts of the world, which keep illuminating this wonderful, rich, and intricate tapestry fashioned by nature, climate, geography, curiosity, and humans.

The division of the prehistory of mankind into the Stone Age, the Bronze Age, and the Iron Age by C. J. Thomsen (1788–1865 CE) from Copenhagen is a rather convenient first step in understanding this evolution. Thomsen had noticed that the artefacts of stone, bronze, and iron, in his collection at the museum, also fell into the three broad categories of time: 3–4 million years ago to about 3000 BCE, 3300 BCE to about 1200 BCE, and 1200 BCE to about 500 BCE (or later), respectively.

Seven metals, gold, silver, copper, tin, lead, iron, and mercury, were discovered and used by early man. Gold, silver, and (occasionally) copper are found in their native state. Gold and silver, but especially gold, were extensively used to make jewellery for several thousand years before the Common Era, due to their lustre, malleability, and ductility, and as they do not corrode. Soon, these were to evolve into an international currency. Copper weapons were not hard enough, but soon man was to discover that its alloy with tin gave rise to a hard metal — bronze which was to rule the world for over 2000 years, till man discovered iron — which was even harder.

The final breakthrough came with the discovery of steel, which was an alloy of iron and a small percentage of carbon. Steel weapons, mostly swords, arrows, and armours, were to rule the world for about 1500–2000 years till swords were made redundant (and reduced to being only ceremonial) by gunpowder, cannons, and finally guns, which of course still used steel.

Almost all ballads that were written during that period talk (or rather boast) of the sharpness of the blades, spears, and arrows of their warriors, their proficiency in swordsmanship and marksmanship, and the swiftness and courage of their mounts or chariots. The might of steel was celebrated by various poets. The famous Russian poet Alexander Pushkin (1799–1837 CE) wrote his poem, *Золото и булат* (*Gold and Steel*), in 1827:

"Всё моё," — сказало злато;
"Всё моё," — сказал булат.
"Всё куплю," — сказало злато;
"Всё возьму," — сказал булат,
which can be translated as follows:
"All is mine," — said Gold;
"All is mine," — said Steel.
"I will buy it all," — said Gold;
"I will take it all," — said Steel.
This firmly established the supremacy of steel over gold! A little later, Rudyard Kipling (1865–1936 CE) went on to write:
"Gold is for the mistress — silver for the maid —
Copper for the craftsman cunning at his trade."
"Good"! said the Baron, sitting in his hall,
"But Iron — Cold Iron — is master of them all."

The story of iron starting from its discovery in meteorites, which were extremely rare, to its smelting from ores to its conversion into steel is a fascinating story of the spread of knowledge across the world. Today, steel dominates all aspects of the life of man across the world, and billions of tonnes of steel are produced every year. As steel can be repeatedly recycled, in principle, all the steel ever produced is available for use and reuse. Some processes used by early man are still not fully understood and remain a subject of intense research.

6.2 Early History of Iron and Steel

“Shooting stars” have fascinated mankind from time immemorial. The Sanskrit word for them “Ulka” has the same connotation. These have also contributed immensely to folklore and fairy tales, including to the rise of that innocent ritual, “wishing on a shooting star”! Most meteorites, for that is what “shooting stars” are, “ignite,” turn into a “ball of fire,” burn up, and explode into minute fiery fragments as they blaze through the atmosphere of the Earth due to friction with the atmosphere and are lost. Most meteorites are asteroids of varying sizes from the asteroid belt between Mars and Jupiter in the solar system, though some could have strayed there from the outer space.

This asteroid belt (Fig. 6.1) is remanent of the cloud of dust and debris of a supernova that collapsed due to another supernova explosion to form the solar system and failed to coalesce to form a planet. Being remnants of a supernova, these are essentially made up of iron and nickel, the heaviest elements which can be produced by the fusion of lighter elements in the stars. One such large asteroid (10–15-km diameter) hit the Earth near Mexico about 66 million years ago and killed all the dinosaurs [2, 3].

Occasionally, a chunk of the meteorite survives the fiery journey through the atmosphere of the Earth and lands on its surface, where it could be found by chance. It is believed that man’s first acquaintance with iron began with this glistening meteorite, which we saw earlier was mostly iron and nickel. “Thus began an obsession that gripped the species. Over the millennia, our ancestors would work the material, discovering better ways to draw iron from the Earth itself and eventually smelt it into steel. We’d fight over it, create, and destroy nations with it, grow global economies by it, and use it to build some of the greatest inventions and structures the world has ever known” [4].

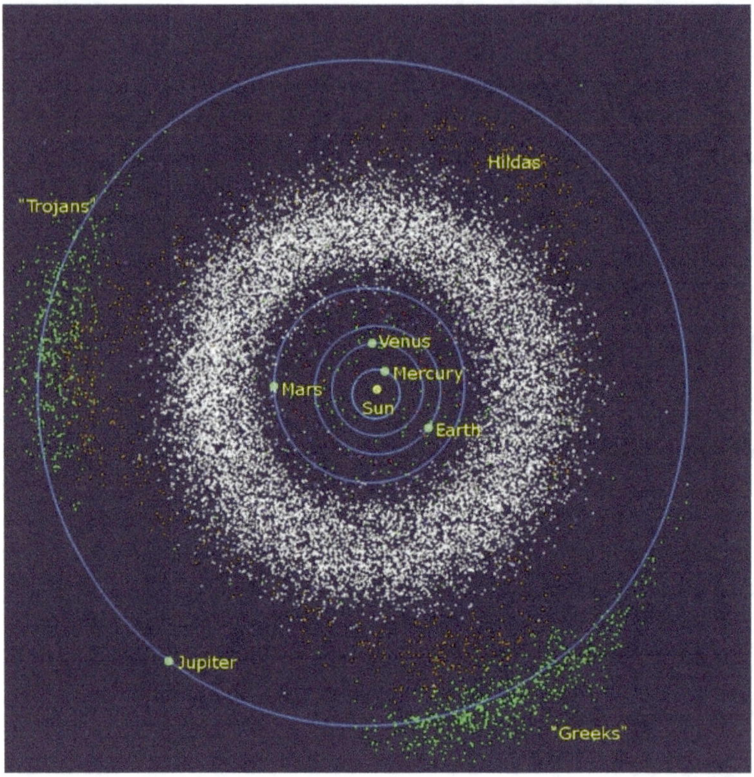

Fig. 6.1. The Asteroid Belt in the Solar System (Dots) [1].

Tutankhamun (c. 1341–c. 1323 BCE), the pharaoh of Egypt, had a dagger made from a meteorite (Fig. 6.2), and it was buried with him!

As indicated above, iron ore is a mineral substance. The most common ores (Fig. 6.3) consist of iron oxides such as magnetite (black iron ore, Fe_3O_4), hematite (red iron ore, Fe_2O_3), limonite or bog-iron (brown iron ore, $2Fe_2O_3 \cdot 3H_2O$), and siderite (pale-brown iron ore, $FeCO_3$), and sulphides like pyrite or "fool's gold" (brass-yellow, FeS_2).

It is by now known that the inner core of the Earth is a hot (more than 5,000°C), dense ball of (mostly) iron having a radius of about 1,220 km, and under tremendous pressure of about 3.6 million atmospheres. The outer core is 2,220 km thick and consists of molten iron and nickel at a temperature of around 4,500°C. Iron ores are abundant and widely distributed in the crust of the Earth, of which they constitute about 5% by weight and are easily

Fig. 6.2. Tutankhamun's Dagger: Made from a Meteorite.

Fig. 6.3. Some Ores of Iron.

mined, mostly using open-cast mines. Iron (compound) is also distributed in low concentrations in soil and groundwater as well as oceans. It is rarely found in its native form except in meteorites, as mentioned earlier.

Iron from the meteorites has a much higher concentration of nickel (up to 7.5% or even higher) than the one extracted from ores, and this fact has been used to conclude that not only the dagger of Tutankhamun but almost all objects made of iron till the end of the Bronze Age were made from iron from the sky! It is not a wonder that the ancient Egyptians called it biz-n-p, while Sumerians called it an-bar, both of which mean "metal from the heaven." But, of course, these finds were extremely rare and thus this "metal of gods" was more valuable than any gems or gold.

Now, of course, iron ore is mined in more than 50 countries. The known iron ore reserve of the world is estimated to be about 175 billion metric tonnes (Fig. 6.4).

Due to its importance in industry and construction, as well as in defence production, iron ore is mined extensively (Fig. 6.5).

6.3 Extracting Iron from Ores

Even though iron ore is quite abundant, it could not be easily extracted as its melting point (1538°C) was beyond the furnaces in use for smelting copper from copper-bearing ores and for producing bronze.

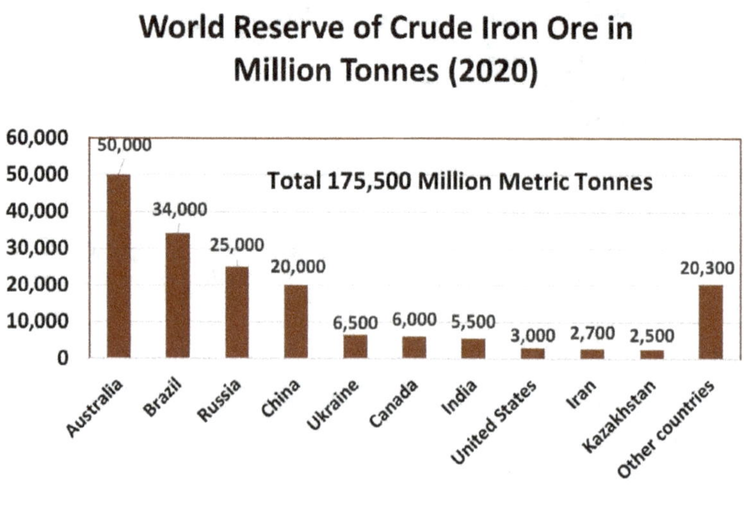

Fig. 6.4. Iron Ore Reserves (*Source*: www.worldsteel.com).

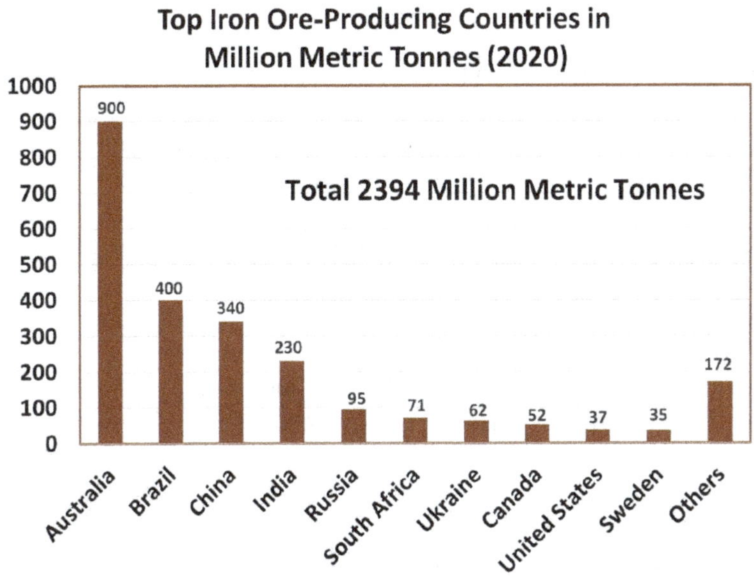

Fig. 6.5. Main Iron Ore-Producing Countries (*Source:* www.worldsteel.org).

According to Strabo (64 BCE–24 CE), one of the early geographers, the Chalybes people living along the Black Sea around 1,800 BCE were the first to produce iron, by heating iron ores into hearths, hammering them, and firing them, to soften repeatedly. In this manner, they could produce sturdy iron weapons. This is what we call wrought iron today. They joined the Hittite people (from Anatolia, modern Turkey) and these weapons along with the use of chariots helped them establish one of the most powerful armies of the ancient world [5].

Cast iron was first produced in China around 500 BCE when they started making a few metre-tall furnaces to burn large quantities of iron ore with wood. This resulted in the smelting of the material into a liquid which was poured into carved moulds, taking the shape of cooking tools and statues.

We now know that the wrought iron produced by the Chalybes was only 0.8% carbon, so it did not have the tensile strength of steel. Chinese cast iron, with 2%–4% carbon, was more brittle than steel. The Chalybes started inserting red-hot weapons of this iron into white-hot charcoal, which gave a thin coat of steel over them as we know it.

6.4 Iron Production in India

Indians were making weapons and surgical instruments of iron by 1000 BCE. The Adivasi tribes of Ashur (Agarias and Birjias) living in the forests of modern Madhya Pradesh, Chhattisgarh, Odisha, Andhra Pradesh, Karnataka, and Tamil Nadu were traditional iron smiths. Often those who produced iron and steel seldom produced implements such as swords, ploughs, chisels, and hammers, a job which was performed by other tribes, mostly Lohars (blacksmiths).

Extracting iron, which was only found in the "metals from the sky" or what was considered the "metal of gods," was considered a very sacred job in all the ancient cultures, and techniques were a closely guarded secret. The craft was practised in utmost secrecy and with sacred rituals, including animal sacrifices [6].

The seminal studies of Agarias, a sub-caste of Asur Adivasis, by Verrier Elwin *et al.* [7–9] have discussed their craft and community in detail. Agarias normally used low-quality ores and preferred it, even if better-quality ore was available. They could identify the ore by its colour, appearance, and density. Ore was normally collected from the surface or at best low pits. Ore-bearing sand washed by rivers and streams was also collected, separated from normal sand/clay using water, and dried before use. This was often the preferred source of iron ore and was believed to yield a better quality of iron. Agarias used to make coal, preferably from dead trees of sal and teak for processing of the iron ore in a low-shaft furnace of clay.

For this, they used to put ore and charcoal in furnaces that were blasted by a pair of bellows worked by the smelters' feet and channelled to the furnace through bamboo tubes, a process that was kept up for hours. Finally, the clay insulation of the kiln was broken up, the molten slag and charcoal were removed, and the ingot was hammered to produce wrought iron.

This ancient craft of Asur Adivasis has almost vanished, due to various reasons. Soon after the British occupied India, they promulgated a series of Forest Acts under which most of the forests were taken over by them and the entry of people into the forests was restricted. Thus, the iron-smelting tribes could not access the source of iron ores and the necessary wood for making charcoal. Even where they were allowed to enter the forests, they had to pay very high taxes/fees for it. By the end of the nineteenth century, cheap, industrially produced steel and iron imported from

Britain flooded the markets and effectively asphyxiated the native tradition, forcing these tribes to abandon their trade which they had practised for millennia. They took to agriculture or worse still, they became daily wage earners, which pushed them into extreme perpetual poverty. The art and science of producing iron and steel (see later) were lost, though several attempts have been made by researchers to try to revive and at least learn and record this craft from the elders of these tribes who may still have a memory of the practices.

Several glorious examples of the height of the iron and steel traditions of India still dot the continent. Prominent among them is the Iron Pillar of Delhi, which stands now in the Qutub Complex (Fig. 6.6).

Fig. 6.6. The Iron Pillar of Delhi [10].

The pillar is 7.2 m high and has a diameter of 41 cm. It is estimated to weigh about 6 metric tonnes. From the inscriptions on the pillar, it is believed to have been constructed by Chandragupta II who ruled from about 375 to 415 CE. It is also believed that it originally stood near Udayagiri rock-cut caves at Vidisha in Madhya Pradesh. It was moved to Delhi during the eleventh century by Anangapal Tomar who ruled from about 1051 to 1081 CE.

It was manufactured by the forge welding of discs of wrought iron. Its surface is famously uncorroded, despite exposure to elements for over 1,600 years. This is now understood due to the formation of a passive protective film at the iron-rust interface, which is rich in phosphorus [11].

It has been suggested (see Ref. [8]) that the presence of a high amount of phosphorus is due to the process of mixing the ore with charcoal in the iron-making process and thus the iron smelted by Asurs contains as much as 1% of phosphorus against less than 0.05% in iron smelted by modern steel mills. Modern steel mills use limestone (as a purifier to eliminate impurities like silica and alumina, but also sulphur and phosphorus) in their blast furnaces instead of charcoal, due to which most of the phosphorus is carried away with the slag.

The iron pillar at Dhar (Fig. 6.7), which was erected by King Bhoja in the eleventh century, whose three pieces have a combined length of

Fig. 6.7. The Iron Pillar at Dhar [12].

about 13.2 m, is still taller. It is understood that one piece is missing, or it was removed to some other place.

The pillar is believed to have been built using the horizontal forge welding technique and was either broken into four pieces or broken by itself.

There are several iron beams in the ancient temple of Konark, probably built in the ninth century. Numerous iron beams exist in the Jagannath temple of Puri in Odisha (which was built in the twelfth century). Most iron cannons of the Mughals and the contemporary rulers were up to 10 m long and weighed as much as 40 metric tonnes. The Jaivana Cannon of Jaipur, manufactured in 1720, weighs 50 metric tonnes, and it was the world's largest cannon on wheels, at the time of its making.

6.5 Wootz Steel

The pride of place in this story belongs to wootz steel, which was produced in South India from at least the mid-1st millennium BCE [13].

The nomenclature wootz popularised by Western travellers and authors seems to be derived either from the mistranscription of "ukku" in Tamil, Kannada, and Telugu or "urukku" in Malayalam. A strong case has also been made recently [14] that it originated from the Sanskrit word "Utsa" (for fountain). In support of this, a photograph of the trading card of James Stodart, who had studied samples of wootz steel along with Michael Faraday during the 1820s and who was a famous surgical instrument maker, cutler, and metallurgist of his time (see Fig. 2, in Ref. [14]), which uses this word for the steel (Fig. 6.8), is often mentioned.

Some historians have also suggested that wootz could be an anglicised word for "uchch (उच्च)," which means superior.

J. Stodart

Surgeons Instruments, Razors & other
Fine Cutlery, made from उत्स (Wootz), a Steel
from India, preferred by Mr. S. to the best
Steel of Europe, after years of comparative trial.

Fig. 6.8. Text of the Trade Card of James Stodart [14].

There are suggestions that Abyssinian traders from the Ethiopian Empire, who sold the ingots of wootz steel of India to Romans, deliberately misinformed them that the steel was from Seres, the Latin word for China, to emphasise that it was from a place too far, and thus Romans called it Seric steel.

We will continue to use the name wootz steel for it, in the following. It was among the finest in the world, and in later years, it was used to fashion the famous Damascus blades of the Middle Ages. There are suggestions that these were called Damascus swords, either because they were produced in Damascus, the capital of Syria, or because of the "damask" or the wavy pattern characteristic of the surface of these swords. The twelfth-century Arab traveller Edrisi [13] is believed to have remarked, "The Hindus excelled in the manufacture of iron, and it is impossible to find anything to surpass the edge from Hinduwani or Indian steel." Some Arab authors have used the name "Teling" for this steel, which could, possibly, indicate its origin in Telangana.

One such sword was believed to have been in the possession of the famous Arab General Saladin (c. 1137–1193 CE). It was so sharp that it could easily pierce European armour and slice a gauze kerchief or a floating feather (or as some say a pillow) into two, effortlessly just as these passed over its blade. Wootz steel was traded all over the ancient and medieval world, including Europe, the Middle East, and China. The qualities of wootz steel were well suited to making weapons, and some of the best specimens of these are in the museums of Turkey. The famous sword of Tipu Sultan (1751–1799 CE) (Fig. 6.9) was made of wootz steel.

One of the earliest mentions of the wootz steel is found in the writings of a Roman historian, Quintus Curtius, who recorded that among the gifts which Alexander of Macedon received from King Puru (Porus) of Takshshila (Taxila) in 326 BCE, there were 100 talents (about three metric tonnes) of wootz steel which was evidently more highly priced than gold or jewels. As mentioned earlier, Arabs started fashioning it into swords and other weapons, and during the Crusades, Europeans were overawed by the superior Damascus swords. This tradition continued till the early twentieth century (Fig. 6.10).

Traditionally, in ancient India, swords used to be forged from the ingots of wootz steel and the red-hot blade was thrust into the green trunks of bananas, and cooled overnight, to increase its hardness. There are also indications of a gruesome procedure for quenching red-hot blades by thrusting them into the stomach of a slave and then leaving them to cool. Arabs, on the other hand, are said to have mounted a horse with the

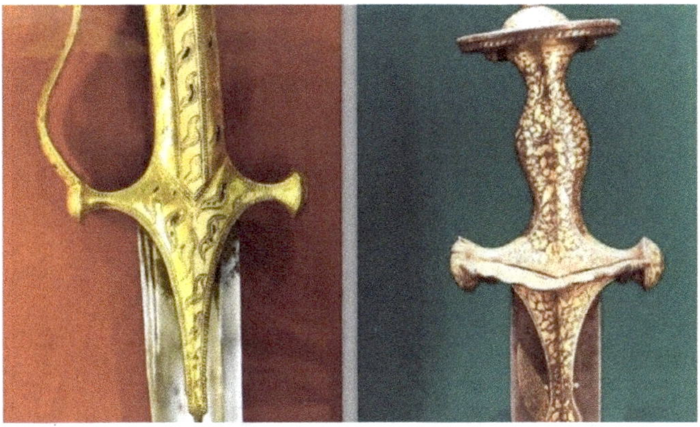

Fig. 6.9. Swords of Tipu Sultan (1751–1799 CE) in the Museums of London Made of Wootz Steel [15].

Fig. 6.10. Sword Makers of Damascus, Syria, in 1900 (Credit: Underwood & Underwood, 1900).

red-hot sword, after quenching its blade by thrusting it into the stomach of a slave and going on a fast gallop to cool it. Moghul, Rajput, Sikh, and Maratha rulers continued to patronise swords, daggers, and armour made of wootz steel.

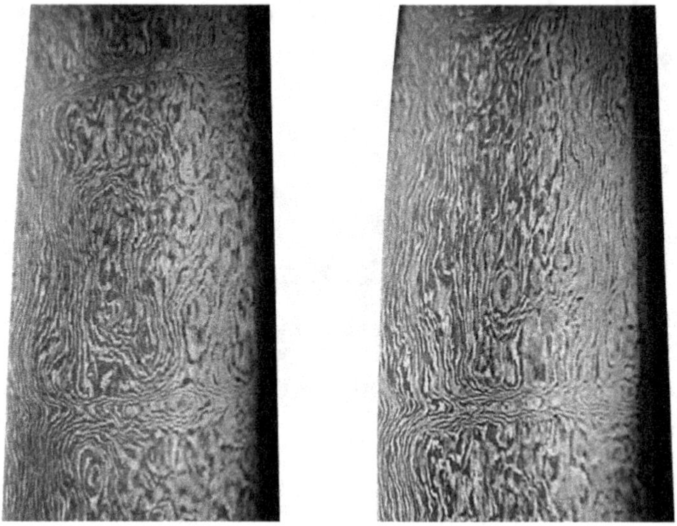

Fig. 6.11. Patterns of Flowing Water on the Swords of Wootz Steel [16].

Wootz steel was generally produced by heating the (preferably black magnetite) ore in the presence of carbon in a sealed clay crucible inside a charcoal furnace to completely remove slag. The source of carbon was bamboo and leaves from plants such as Avārai (*Senna auriculata*). Using buffalo hide bellows, these furnaces could reach a temperature of 1300°C–1400°C which was maintained for several days, during which the iron melted and alloyed with carbon (1%–2%). The crucibles were then taken out and cooled slowly, and finally broken to collect the ingots, which were either traded or made into implements. The famous "flowing water" and other patterns (Fig. 6.11) were created during the process of repeated hammering of the heated ingots to make the blades, which also created intricate microstructures. Similar patterns can also be created by forging alternate pieces of low-carbon and high-carbon steel, but it is often not considered the "true wootz" steel.

Studies suggest that the Chinese and locals in Southwest Sri Lanka adopted the production methods of creating wootz steel from the Chera Tamils by the fifth century BCE. An interesting and unique method to utilise monsoon winds channelled through tunnels, instead of bellows, was also developed in Sri Lanka.

Records suggest that by the seventeenth century, several tens of metric tonnes of wootz steel ingots, produced in over 10,000 iron and steel furnaces, were being exported to Europe, which indicates the presence of a widespread industry for producing it in India.

Wootz steel is credited with the birth of modern metallurgy as Europeans studied it in detail to reproduce it using successively more sophisticated techniques [13]. To produce "true" wootz steel, J. D. Verhoeven, who along with a blacksmith Alfred Pendray [17], performed a series of experiments during the 1990s and identified that the presence of trace amounts of vanadium in the ore used in South India was responsible for producing the patterns, so unique to the swords and knives of wootz steel. The source of this ore may have dried up by the early nineteenth century, which along with the availability of cheaper and industrially produced steel may have led to the "death" of this unique industry in India.

The study of wootz steel production has remained a rich field of archaeometallurgy in India, and the search for abandoned and ancient furnaces along with the study of slag around those places continues to remain very rewarding.

We add that Japanese smiths forged blades for the samurai, which were light and deadly sharp. These were prepared in a very sacred ritualistic manner and involved heating a chunk of wrought iron using charcoal, which was then folded. The process was repeated up to 20 times during which the continued exposure of the wrought iron to carbonaceous charcoal turned it into steel. Knights of Europe at the other end of the world used specially crafted swords, which were forged by twisting rods of iron. This gave unique herringbone and braided patterns in the blades, which were interpreted as dragon coils by the Vikings [5].

Nomads of Russia had developed a technique to prepare bulat (from fulad or faulad in Persian for steel), which was also used by the armies of Genghis Khan for their swords. Unfortunately, the technique is lost, though efforts have been made to recreate it by Pavel Anosov (1796–1851 CE). It is believed that the process involved dipping the finished weapon into a vat containing a special liquid, of which spiny restharrow (*Ononis spinosa*) extract was a part, then holding the sword aloft while galloping on a horse, and allowing it to dry and harden against the wind.

6.6 Modern Times

The modern steel production method is mostly based on using blast furnaces which have been improved and automated considerably since these were invented in the 14th century. However, the process to smelt the ore, inside the blast furnace, where coke, iron ore, and limestone are continuously added and air or oxygen under pressure is blown through it to rapidly remove impurities and produce pig iron (thus named because the cooled ingots look like pigs), remains the same. Pig iron has very high carbon (3.8%–4.7%), silica, and other impurities, and it is processed further to produce iron or steel. Several furnaces now use natural gas, in place of coke, to reduce emissions. The other option is to use an electric arc furnace, where scrap metal or ore is used to produce fresh steel. By now, several thousand varieties of steel have been developed with unique properties, and steel has become the backbone of modern industry, transportation, and civil construction.

Modern steel production in India started in 1874 when the first blast furnace was set up at Kulti near Asansol in 1870 by Bengal Iron Works. Jamsetji Nusserwanji Tata (1839–1904 CE) founded the Tata Iron and Steel Company, to produce iron and steel for the modernisation of Indian industries, after considerable research, exploration, and study, at Sakchi village (around the place which was to grow into Jamshed Nagar).

True to the blatant racism prevalent during the colonial period, he faced ridicule and scepticism, and the British Chief Commissioner of the Indian Railways, Sir Frederick Upcott, famously said, "Do you mean to say that the Tatas propose to make steel rails to British specifications? Why I will undertake to eat every pound of steel rail they succeed in making." The company was formally established by Sir Dorabjee Tata on 26 August 1907. It started pig iron production in 1911 and began producing steel in 1912.

In a poetic justice, the company was to provide extensive support to the World War I efforts of Britain by providing over 2,400 km of rail tracks for Mesopotamia. Dorabjee Tata is reported to have said that if Sir Frederick Upcott had lived up to his word, he would have had "some slight indigestion"! It was also to provide 3,400 metric tonnes of steel for the iconic Howrah Bridge in Kolkata. By 1939, TISCO operated the largest steel plant in the British Empire.

After independence, during the second 5-year plan (1956–1961), three new integrated steel plants were set up with foreign collaboration:

1. Rourkela Steel Plant in Odisha in collaboration with Germany in 1959;
2. Bhilai Steel Plant in Chhattisgarh with Russian collaboration in 1959;

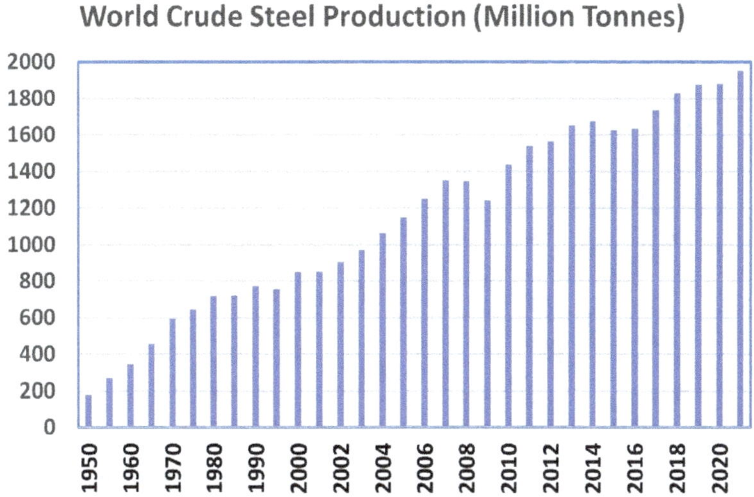

Fig. 6.12. Increasing Production of Steel Over the Years (*Source*: worldsteel.org).

3. Durgapur Steel Plant in West Bengal in collaboration with the United Kingdom in 1962.

In 1964, another integrated steel plant was set up at Bokaro in Jharkhand, with Russian collaboration.

By now, there are more than 50 iron and steel industries in India and several large steel production plants.

We see (Fig. 6.12) that steel production which used to be just about 200 metric tonnes in 1950 has reached 2 billion tonnes/year by now. This is bound to increase to enable rapid industrialisation and development across the world.

Figure 6.13 shows the top steel-producing countries of the world, which is also indicative of the manufacturing capacities of their industries.

The modern era of high-energy accelerators, which revealed the structure of nuclei and elementary particles, as well as the basic constituents of matter, started as more and more powerful electromagnets in various shapes, sizes, and magnetic field configurations and strengths were built by wrapping current-carrying coils around soft iron cores. The beam lines and large detectors in the experiments at CERN in Geneva and the FAIR facility coming up at Darmstadt, and for that matter at the Variable Energy Cyclotron Centre, Kolkata, etc., use massive magnets.

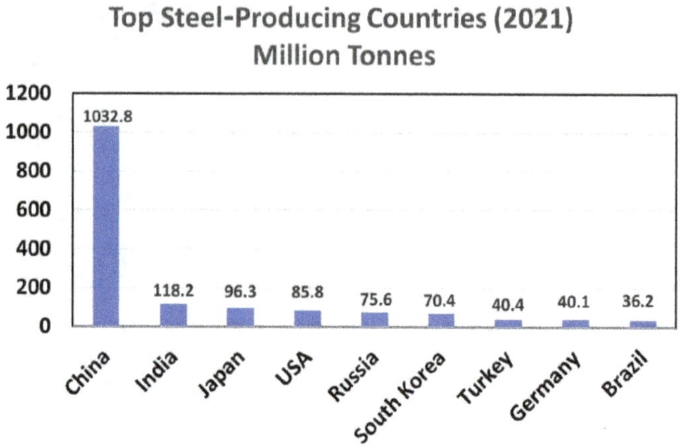

Fig. 6.13. Top Steel-Producing Countries of the world (*Source*: worldsteel.org).

The Industrial Revolution arrived on the tracks of railways, steam engines, machines, and steamships, which used steel extensively. The automobile industry remains one of the major users of steel.

Steel is classified into four main categories: carbon steel, stainless steel, alloy steel, and tool steel, which are based on their composition or method of preparation.

- Carbon steel is unique for a notable absence of other elements in its makeup. It contains only up to 2% carbon or less by weight, and its elemental nature makes it a strong, durable material that is ideal for numerous uses. It is further categorised as low-, medium-, and high-carbon steel. Up to 90% of the steel produced is carbon steel.
- Stainless steel has 10% or more chromium in it, which gives it a lustre and makes it corrosion-resistant and long-lasting by preventing rusting. It is a metal of choice for making most medical equipment and appliances.
- In alloy steel, iron is infused with other elements such as aluminium, copper, manganese, molybdenum, silicon, or vanadium, each of which contributes its unique attributes to the final product.
- Tool steels are used for tool manufacturing machinery. These are produced by tempering, which is the process of heating to a high temperature, cooling quickly, and heating again. It is extremely hard and heat-resistant.

6.7 Steel Industry and Climate Change Mitigation Measures

We have seen that the world now produces close to 2 billion metric tonnes of steel per year. This is expected to rise to 2.5 billion metric tonnes, in years to come, due to the increasing demand for steel across the world. The production of 1 metric tonne of steel produces 1.85 metric tonnes of carbon dioxide on average, and accounts for 7%–9% of the total carbon dioxide emission of the world. This, in turn, exacerbates global warming. The 2015 Paris Climate Agreement stipulated that the steel industry's annual emissions must fall to about 500 million metric tonnes of CO_2 by 2050. Achieving this would require the emission of CO_2 in steel production to be reduced from 1.85 to 0.2 metric tonnes per metric tonne of steel. This would, in turn, require nothing short of a revolution in developing new production technologies and massive investments to implement these.

This can only be done if we lower carbon dioxide emissions from all stages of steel production, including the pelletisation of the iron ore, reduction of iron oxides to iron, and production of crude steel. One such scheme, the Hydrogen Breakthrough Ironmaking Technology [18], aims to replace the coke and other fossil fuels used in traditional, blast furnace-based steelmaking and instead relies on hydrogen created with renewable electricity (Fig. 6.14).

6.8 Conclusions

We have seen that the knowledge of iron and steel has led to the rapid development of human civilisation. The study and recreation of several techniques, which were used by our ancestors to produce unique steels across the world, have led to the development of modern metallurgy, which is now giving us designer metals for unique applications. Sharing this knowledge has been extremely rewarding.

Iron has served the purposes of creation, and it has been used to destroy. The importance of iron in shaping human civilisation has been discussed by several authors since the time it was discovered. We recall the words of Pliny the Elder (23–79 CE), a Roman naturalist and writer, who wrote about iron: "Iron mines bring man a most splendid and most harmful tool. It is through this tool that we cut into the ground, plant

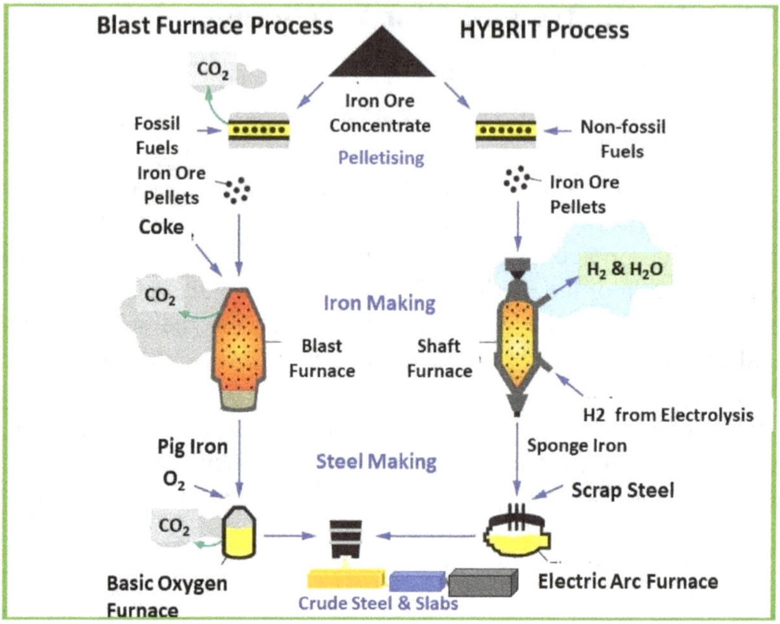

Fig. 6.14. The Hydrogen Breakthrough Ironmaking Technology: A Plan to Produce Green Steel (Adapted from Ref. [18]).

bushes, cultivate flourishing orchards and make grape vines with grapes. By this same tool, we build houses, break stones and use iron for all such purposes. But it is also with the help of iron that we fight and battle and rob. And we do it not only at close quarters; but by giving it wings, we hurl it far into the distance, now from embrasures, now from powerful human hands, and now from bows in the form of feathered darts. This, I think, is the most infamous invention of the human brain. For to enable death to catch up with man faster, it has given it wings and armed iron with feathers. For that many, the blame rests with man and not with nature."

References

[1] https://commons.wikimedia.org/wiki/File:InnerSolarSystem-en.png.

[2] D. K. Srivastava and V. S. Ramamurthy, "*Climate Change and Energy Options for a Sustainable Future*" (Singapore: World Scientific Publishing Company, 2021).

[3] V. S. Ramamurthy, D. K. Srivastava, and S. Nayak, "*Art and Science of Managing Public Risks*" (World Scientific Publishing Company, 2022).

[4] J. Schifman, https://www.popularmechanics.com/technology/infrastructure/a20722505/history-of-steel/.

[5] https://www.popularmechanics.com/technology/infrastructure/a20722505/history-of-steel/.

[6] M. Eliade, "*The Forge and the Crucible*". Translated from Original in French by S. Corrin (Chicago: University of Chicago Press, 1962).

[7] V. Elwin, "*The Agaria*" (Oxford University Press, Humphrey Milford, Bombay, 1942).

[8] A. K. Sinha, "Asur — An Ancient Iron Smelter, Can Get Global Recognition", *The Researchers'* 1 (2015) 1–22.

[9] A. K. Vaish *et al.*, "Iron Making in Ancient India — A Critical Assessment". In: *Traditional and Innovation in the History of Iron Making: An Indo-European Perspective* (Nainital: Pahar, 2002), pp. 237–249.

[10] https://en.wikipedia.org/wiki/Iron_pillar_of_Delhi.

[11] R. Balasubramaniam, "On the Corrosion Resistance of the Delhi Iron Pillar", *Corrosion Science* 42 (2000) 2103–2129.

[12] https://en.wikipedia.org/wiki/Dhar_iron_pillar.

[13] S. Srinivasan and S. Ranganathan, "*India's Legendary Wootz Steel: An Advanced Material of the Ancient World*" (Hyderabad: University Press, 2014); S. Ranganathan and S. Srinivasan, "A Tale of Wootz Steel", *Resonance* 11 (2006) 67–77.

[14] R. K. Dube, "Wootz: Erroneous Transliteration of Sanskrit "Utsa" Used for Indian Crucible Steel", *JOM* 66 (2014) 2390–2396.

[15] https://www.heritagetimes.in/tipu-sultans-swords-and-ring-in-two-museums-in-london/.

[16] https://en.wikipedia.org/wiki/Wootz_steel#/media/File:Watered_pattern_on_sword_blade1.Iran.JPG.

[17] J. Verhoeven and A. Pendray, "The Mystery of the Damascus Sword", *Muse* 2 (1998) 35–43.

[18] M. Pei, M. Petäjäniemi, A. Regnell, and O. Wijk, "Toward a Fossil Free Future with HYBRIT: Development of Iron and Steelmaking Technology in Sweden and Finland", *Metals* 10 (2020) 972.

Chapter 7

Education and Transmission of Knowledge

"आ नो भद्राः क्रतवो यन्तु विश्वतः"

"Let noble thoughts come to us from all sides."

— *Rigveda,* 1.89.1

"I do not want my house to be walled in on all sides and my windows to be stuffed. I want the culture of all lands to be blown about my house as freely as possible. But I refuse to be blown off my feet by any."

— Mahatma Gandhi, *Young India*, 1 June 1921

"Education is the most powerful weapon which you can use to change the world."

— Nelson Mandela

7.1 Introduction

The story is well known. An ageing chieftain decided to take his son on a long hunting trip. When his wife expressed her apprehension that the son was too young to face the dangers of the trip, he replied, "It is the only way our son will learn." He added further, "If he learns it well, he will return, ready to take his place as the next chief. If he does not learn, he will die." No comments are needed to emphasise that education gives us the skills to survive and prosper.

Fig. 7.1. Lord Buddha giving His first Sermon at Sarnath (Credit: Sarnath Museum).

Recall Lord Buddha. When Siddhartha Gautama became Buddha (The Enlightened One) after years of penance and meditation, He did not sit quietly in His newfound Bliss. He shared His understanding and newly acquired knowledge. He went around educating the masses (Fig. 7.1) and explaining His Middle Way, a task which was taken up by his disciples and then their disciples and so on, and which transformed the world.

Education is unique to human beings, and they are the only species on the face of the earth that has education preceding adulthood. The legendary Sanskrit poet, Bhartrahari, bluntly said:

"साहित्यसंगीतकलाविहीनः साक्षात्पशुः पुच्छविषाणहीनः ।
तृणं न खादन्नपि जीवमानस्तद्भागधेयं परमं पशूनाम् ॥"

"*A person who does not know literature, music, and arts is truly an animal without a tail and horns. That he does not eat grass to survive is indeed, a matter of great fortune for the animals! — Bhartrahari, Niti Shatakam, Shloka No. 12.*"

We know that once Homo Sapiens, our ancestors in Africa, became bipedal, this fact set a vast number of changes into motion. It freed their hands for labour and making tools or throwing weapons. It also gave them a wider field of vision. Their legs, knees, ankle joints, spinal vertebrae, toes, and arms changed. Most significantly, their pelvis became shorter and rounded, with a smaller birth canal, making birth more difficult for humans than other primates. This resulted in a shorter gestation period, leading to the birth of babies before their heads became too large, who were also more helpless and not fully developed before birth. The need to care for the babies for years after birth led to strong bonds within families and groups. The continued growth of the human brain after birth helped in social learning and languages.

There have been suggestions that the consumption of meat, by providing a dense source of energy, and cooked food, by using less energy to digest the food and eliminating harmful parasites, helped develop a smaller, more efficient digestive tract, which freed up energy to enable larger brain growth among humans. The development of language led to communication and the formation of larger groups which were so vital for hunting large animals and survival. Every little development, e.g., making stone tools, control of fire, identification of edible foods, and strategies to hunt, perhaps took tens of thousands of years, in the beginning.

Human babies are born with several important instincts. These include sucking and rooting reflexes, which make them suck anything which is put in their mouths and which makes them seek out a nipple when their cheeks are touched; grasping, which makes them close their fingers around anything that is put in their hands; startling, when there is a sudden movement or noise; crying, to communicate and express needs for hunger, discomfort, or tiredness; social interaction, which makes them attracted to human faces; and self-preservation, which makes them blink when exposed to strong light or flinch in response to a loud noise or withdraw from potentially harmful stimuli, e.g., very hot surfaces. These instincts were developed through genetic mutations and variations that enhanced survival and reproductive success over thousands of years and are hard-wired in their genes. Most importantly, these are common to all babies across the world. However, as the child grows, his/her abilities develop further due to the influence of the environment, learning, social interactions, individual experiences, and cultural factors. Additionally, individual variation and cultural factors can impact the expression and significance of certain instincts.

The cognitive development of a child, which encompasses the acquisition of skills such as language, problem-solving abilities, and critical thinking; the acquisition of motor skills enabling them to have physical coordination, dexterity, and the ability to perform various physical tasks as well as hand-eye coordination; the acquisition of social skills; and the acquisition of knowledge, etc., is facilitated by education. Education is responsible for the enormous development of Homo sapiens, and it distinguishes them from other species.

While we can, with some effort, trace the origins of education and schools back to prehistoric times, it is certain that the forms and methods of education were vastly different from what we have today. The first "schools" of humans were likely the fireside where the elders (as well as visitors and guests) shared their experiences. It is believed that in prehistoric times, education for children was primarily informal and centred around practical skills necessary for survival and hunting, gathering, and toolmaking, and for becoming a part of their society.

The discovery of agriculture about 10,000 years ago led to an explosive growth in the availability of food and hence population, and more complex societies emerged. Agriculture released a part of the population to concentrate on arts, culture, literature, architecture, city building, and warfare, and it became necessary to acquire more extended skills. This required a more formal system of education.

We shall first have a brief look at the practice of education in ancient cultures across the world and then discuss it in more detail for India in historical times, before discussing international collaboration and the international nature of education.

7.2 Ancient Cultures

7.2.1 *Sumerians, Akkadians, and ancient Egyptians*

One of the earliest known educational systems developed in ancient Mesopotamia around 3500 BCE. These people, called Sumerians (ca. 5500–ca. 1475 BCE), established schools called "Edubba" (House of Tablets, as by then they had developed cuneiform writing, which used wet clay tablets). It was open to boys from noble families, who were taught subjects like reading, writing (Fig. 7.2), mathematics, religion, and law.

A very insightful excerpt from notes left on clay tablets, which was possibly used as a lesson, as 21 copies of it have been found, points to the

Fig. 7.2. A Sumerian Tablet.

strict discipline enforced in the classroom, reminiscent of more modern times [1].

The relatively short-lived Akkadian Empire (ca. 2334–ca. 2154 BC) followed the Sumerian civilisation, and even though it had speakers of Akkadian as well as Sumerian languages, it adopted the Sumerian way of writing on clay tablets.

The story of deciphering the Akkadian, Sumerian, and other scripts is a fascinating tale of an international effort and scholarship. Hundreds of thousands of clay tablets inscribed in the cuneiform script document the political, social, economic, literary, and scientific history of Sumerians and Akkadians. Yet, most of these documents remain untranslated and inaccessible due to their sheer number and the scarcity of experts able to read them. Recently [2], artificial intelligence has been used by researchers from Tel Aviv and Ariel universities to translate ancient cuneiform texts from Mesopotamia into English. This, if proved feasible on a large scale, would be very valuable.

Egyptians established schools called "Per-ankh" (House of Life) around 2000 BCE, which were associated with temples and offered

Fig. 7.3. Egyptian Hieroglyphs on a Stela in Louvre, ca. 1321 BCE (Credit: Wikipedia).

education in writing (Fig. 7.3), reading, mathematics, religious rituals, and other subjects to the children of the elite, around the same time.

7.2.2 *Indus Valley Civilisation*

The Indus Valley Civilisation or more accurately the Indus-Sarasvati Valley Civilisation (as it is called now) flourished from ca. 3300 to ca. 1300 BCE and extended from Baluchistan in the west to western Uttar Pradesh in the east, and from northeastern Afghanistan in the north to Gujarat state in the south. It succumbed either due to climate change, or prolonged droughts, or the drying up of the Sarasvati River [3], or all of them.

It had standardised weights and measures of length (over the entire civilisation spread over 1.3 million square kilometres, lasting several thousand years); developed several industries and agriculture, including growing of cotton and weaving of cotton clothes; domesticated cattle and fowls; perfected the metallurgy of copper, bronze, iron (towards the end), gold, and silver; produced, used, and exported beads of terracotta, shell, steatite, agate-carnelian, lapis lazuli, turquoise, faience, jasper, onyx, etc., and gold ornaments; made the first wells in the world; had board games like chess and dice; made lovely toys for children and built large and exceptionally well-planned cities with city walls, large public baths, granaries, industrial settings, houses of kiln-fired bricks and mortar, wide paved roads, extensive drainage with interlocked bricks; used bitumen for waterproofing; and built houses with flushed toilets, the world's first dockyard (at Lothal), water storage systems, sewage systems, etc. The most significant trading partners of Sumerians and Akkadians were people from Meluḫḫa or Melukhkha, which is often identified with the region of the Indus Valley Civilisation.

The uniformity of construction (even the sizes of bricks were the same across the civilisation and over thousands of years) over such a vast area, no signs of any warfare, and proofs of flourishing trade across seas with other contemporaneous civilisations — all the way to Egypt — all point to a highly developed and enlightened egalitarian (no palaces for kings or temples have been found so far) society. Unfortunately, the Indus script, mostly seen on fascinating seals (Fig. 7.4), remains undeciphered. Yet, such a magnificent edifice could not have survived for so long, without imparting a rigorous education and training to its young. Alas, we do not have any information about it.

Fig. 7.4. A Unicorn Seal from the Indus Valley Civilisation [4].

7.2.3 *Ancient Greece*

In ancient Greece, education was considered necessary to develop the mind, body, and character. Often, wealthy families employed private tutors. They also sent their children to schools known as "gymnasia" or "academies" (from before about the seventh century BCE), where subjects like reading, writing, mathematics, music, physical education, and philosophy were taught. It is believed that these were significantly shaped by famous philosophers and teachers like Socrates (ca. 470–ca. 399 BCE), Plato (ca. 428–ca. 348 BCE, Fig. 7.5), and Aristotle (384–322 BCE). Aesop (ca. 620–564 BCE), though perhaps a slave, was to become free and famous for his storytelling abilities and fables, which are still popular. We have already seen how Aristotle asked Alexander the Great, whom he had tutored, to bring teachers from India with whom he could discuss the philosophical developments in India. Greek literature, poetry, drama, philosophy, and even history were to scale unparalleled heights, which can all be attributed to a rigorous education and discourse. Greeks put a great emphasis on physical education, e.g., athletics and sports, to build up stamina, strength, and preparedness for war, as well as on music and dance education, with Olympic competitions held at regular intervals.

Fig. 7.5. Marble Copy of the Bust of Plato.

A very rigorous education, aimed at converting children into warriors, was practised in Sparta from about 550 BCE, where they were subjected to extreme hardships, including brutal torture, and were deprived of food, to weed out weaklings, without any pity. This was to give rise to the nomenclature, Spartan, for a rigorously simple, frugal, or austere living.

7.2.4 *Roman Empire*

In the Roman Empire (from around 600 BCE), education was still primarily accessible to the wealthy and privileged. Boys from affluent families received education in subjects like reading, writing, public speaking, law, and physical training. Most of the teachers were often Greek slaves or freedmen. As Romans controlled a vast area, its educational methodology and curriculum were copied in its provinces. This was to become the basis for education systems throughout the later Western civilisation.

Oratory was very important for success in Roman society, and it was a lifelong learning exercise. Cicero (106–43 BCE) was one of the greatest orators and prose stylists of his time. He greatly influenced the Latin language, and his extensive writings have remained standard texts for the education of Latin even today. Quintilian (ca. 35–ca. 96 CE, Fig. 7.6)

Fig. 7.6. Quintilian (35–ca. 96 CE).

taught oratory for over 20 years and has left behind a 12-volume textbook on the theory and practice of rhetoric, *Institutio Oratoria* (*Institutes of Oratory*). The book has detailed instructions on even what to wear and how to move while giving an oration!

Centuries later (in the late 16th century), the Protestant Reformation initiated by Martin Luther (1483–1546 CE) prompted the establishment of compulsory education for boys and girls, first in regions that are now part of Germany and later in Europe and the United States.

The earliest universities in Europe were the University of Bologna (1088), the University of Paris (ca. 1150), and the University of Oxford (1167), followed by many others across Europe. The onset of the Industrial Revolution in England necessitated the requirement of a large number of technicians, scientists, and engineers, and six "red brick" universities, at Manchester, Birmingham, Liverpool, Leeds, Sheffield, and Bristol, the major industrial cities which were established in the early 1900s, to train the students in "real-world skills," sometimes by amalgamating the existing (engineering) colleges.

7.2.5 *Vedic India*

Education played a vital role in ancient Indian society. During the Vedic period (1500–500 BCE), education was primarily transmitted through oral traditions. The oral traditions involved loud rote recitations, which were often hilariously described as the croaking of frogs by poets. The education took place in gurukuls or ashrams (Fig. 7.7), which were residential schools led by a guru (teacher). Students, typically from higher castes, lived with the guru, worked in his fields, looked after his cattle, and did other household jobs. They received instructions in various subjects, including philosophy, religion, literature, mathematics, medicine, and warfare. The intensity of this association is best understood by looking at a prayer from Krishna Yajurveda, *Taittiriya Upanishad* (2.2.2), which used to be chanted at the beginning and the end of classes in these traditions:

"सह नाववतु।
सह नौ भुनक्तु।
सह वीर्यं करवावहै।
तेजस्वि नावधीतमस्तु
मा विद्विषावहै।
शान्तिः शान्तिः शान्तिः ।।"

Fig. 7.7. A gurukul Ashram in Vedic India [5].

It states, "*May we be protected. May we be nourished. May we work together with great energy (perform great deeds together). May our intellect be sharpened. May there be no animosity between us. May this bring peace to us, peace to our nation, and peace to the entire universe.*"

Interestingly, it puts teachers and the taught on almost the same footing, as collaborators and seekers of knowledge. This was to be articulated by Rabindranath Tagore much later in the early twentieth century, as follows: "A teacher can never truly teach unless he is still learning himself. A lamp can never light another lamp unless it continues to burn its flame."

The scriptures such as Vedas and Upanishads ("sitting nearby [to learn]") were taught to pursue knowledge, understand the "self," and gain spiritual wisdom. The education was mostly specific to the caste or the class of the student, e.g., the study of the three Vedas was obligatory to the students from the priestly class, and so on.

The names of legendary teachers and their ashrams survive in the Upanishads and Puranas. They have been immortalised in Indian astronomy in the names of the stars of the star constellation known as Saptarshi Mandal (Ursa Major or Big Dipper), namely Atri, Bharadvaja, Gautama, Jamadagni, Kashyapa, Vasistha, and Vishvamitra, according to *Brihadaranyaka Upanishad* (2.2.4). The names depend upon the source, though. Some other texts have slightly different names (Fig. 7.8).

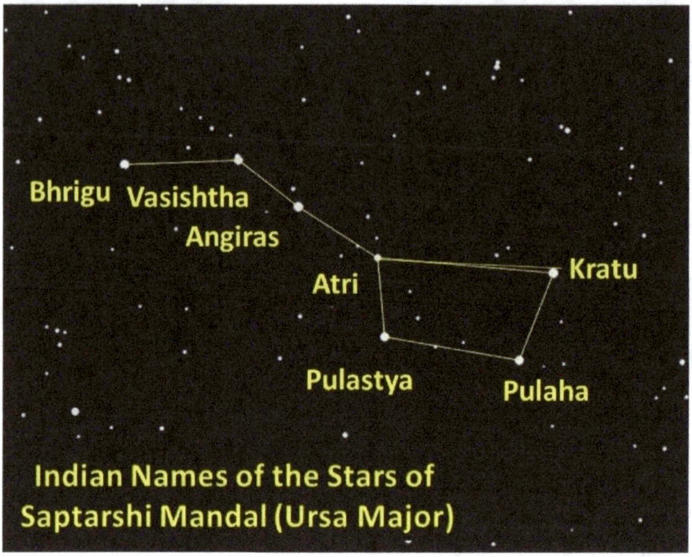

Fig. 7.8. Saptarshi Mandal (Ursa Major) of the Seven Legendary Teachers According to *Vishnu Purana.*

There are strong indications that Boudhayan (ca. 800–ca. 740 BCE) gave several sutras (formulae), including those which gave the approximate value of pi, the Pythagoras theorem, and a very accurate value of the square root of 2. These were useful for making accurate geometrical altars for fire sacrifices or yagnas. One such intricate geometric altar was Shyen (eagle), as shown in Fig. 7.9. It is also the logo of the National Institute of Advanced Studies [6]. Kim Plofker [7] wrote about these "Sulba-sutras" (which were adopted into measuring ropes with appropriately paced knots), making it easy to make intricate altars):

> "*Many of the altars' shapes involved simple symmetrical figures such as squares and rectangles, triangles, trapezia, rhomboids, and circles. Frequently, one such shape was required to be transformed into a different one of the same size. Hence, the Sulba-sutra rules often involve what we would call area-preserving transformations of plane figures, and thus include the earliest known Indian versions of certain geometric formulas and constants.*"

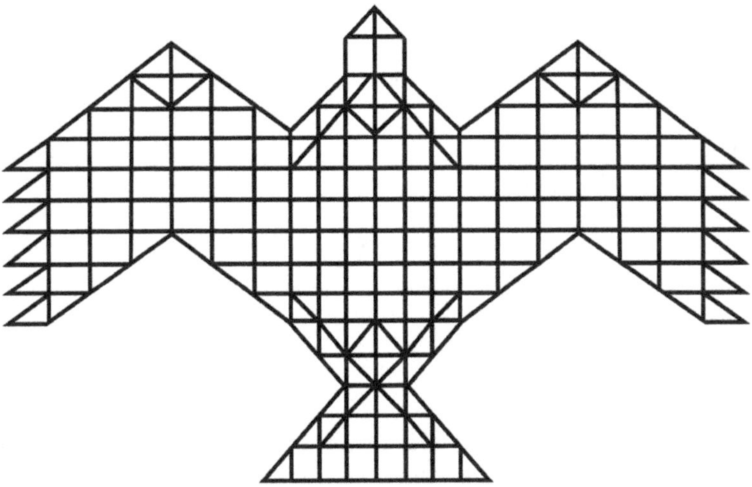

Fig. 7.9. Shyenaciti (Celestial Eagle Fire-Altar) for Agnichayana Sacrifice (Yagya).

7.2.6 *Ancient China and Persia*

In ancient China, initially, education was primarily the responsibility of families and took place in private schools or within the home. Historical records reveal that during the Zhou Dynasty (1046–256 BCE), state-controlled education was established at elementary, secondary, and higher education levels, where students learnt the Confucian classics, history, music, calligraphy, and rituals. A very important development was the introduction of the imperial examination system during the Han Dynasty (206 BCE–220 CE), which selected officials based on their knowledge of Confucian texts and principles. It was a revolutionary step as it selected bureaucrats by merit rather than by birth.

China is believed to have successfully guarded its knowledge of the production of silk, tea, paper, etc., for centuries as these had great economic value.

Persia, during 559–330 BCE, educated its children to prepare them for a rigorous and challenging physical environment, follow the Zoroastrian religion and ethics, and for a militant, expansionist people. As the society had priests, warriors, tillers of the soil, and merchants as the major classes, education was geared to meet their demands. The children were educated at home till the age of 7, and the children of the higher classes received education in law, medicine, arithmetic, geography, music, and astronomy at schools run by the court. There were separate schools to impart military training.

The region north of the Indus River was in regular contact with Persia, through trade, wars, and even occupation, which had profound effects on both societies.

7.2.7 *Mesoamerica*

On the other side of the globe, in Mesoamerica, the Mayan, Aztec, and Olmec civilisations primarily focussed on imparting education on religious and societal roles for the children of elite classes, including noble families, priests, and rulers. It must be emphasised that they were not to come in contact with the Old World till after the conquest by Spain.

The Maya civilisation (200–1697 CE) accorded high prestige to its priests ("ah men" or "ah tzel") because of their extensive knowledge and literary skills. The priests provided religious and moral leadership and served as advisers to the rulers and nobility. The priesthood normally passed from father to son who received a rigorous education in writing, history, medicine, methods of divining, and the calendar system. Mayan cities also had "ch'eenob" or "chooks" where people gathered for learning and discussions. Some Mayan cities had higher centres of learning known as "ixiptla," where specialised training was possibly given to priests, scribes, and medicine men.

The Maya script was still in use when the region was conquered by Spain. Some priests even learnt it. Unfortunately, a sizable number of Maya writings were destroyed, when the Spanish rulers decided to eradicate pagan rites.

The Aztecs (1300–1521 CE) established systems of schools called "Calmecacs" for the children of the nobility where students studied religion, history, mathematics, astronomy, and warfare, as well as how to read and write. It is believed that Calmecacs played a most vital role in the oral transmission of history through oratory, poetry, and music, which were employed to make accurate memorisation of events. The schools for commoners were called "Telpochcalli," where they were taught about religion, history, warfare, agriculture, and crafts.

Not much is known about Olmecs, who lived from 1400 to 400 BCE in what is now Mexico. It has been speculated that they passed on cultural, religious, and artistic knowledge from generation to generation through an oral tradition of storytelling.

7.3 Rise of Early Universities

It is often suggested that by 600 BCE, the class or rather the caste structure in the Indian society had become rigid, Vedic rituals had become highly ritualistic, profiting only the priestly class, and the later emergence of Buddhism and Jainism developed a strong sense of unease among the populace at the large-scale animal sacrifices.

The Buddhists established monasteries, where they gave education to all irrespective of caste, creed, or gender, in the language of common people. These monasteries were often attached to Buddhist temples but educated only those who were its members. The Buddhist doctrines spread from India to China beginning sometime in the first century CE and triggered a profusion of cross-cultural exchanges that had a profound impact on Asian and world history.

It is generally believed that the rise of the Nanda dynasty in about 413 BCE and then the very powerful Mauryan dynasty, some 40 years later, had a profound effect on the earlier Vedic way of life, as the Brahmans gave up their ancient and sole occupation of teaching from their forest ashrams, Kshatriyas gave up their sole occupation of fighting, and lower classes rose and took up all kinds of occupations. Schools were opened in towns and all subjects were taught to the students, to cater to the needs of a growing empire, which needed a vast number of educated employees for all kinds of newly emerging duties and occupations.

7.3.1 *Takshashila*

Takshashila, believed to be one of the earliest universities in the world (Fig. 7.10), had developed an international reputation by the sixth century BCE as a centre of advanced studies and a great centre of learning with several famous teachers, like Panini (the famous grammarian of Sanskrit and author of *Aṣṭādhyāyī*); Charaka (the famous physician and author of *Charak Samhita*); (Sushrut, the author of *Sushrut Samhita* about medicine was perhaps his contemporary and lived in Kashi); Chanakya — the

Fig. 7.10. Ruins of Takshashila University (Credit: Wikipedia).

famed strategist, also known as Kautilya, and famous as an advisor to Chandragupta Maurya — whom he taught there, and author of *Arthashastra*; Jivaka, the famous healer and personal physician of Lord Buddha; Kumaralata, the famed Buddhist philosopher; and Vishnu Sharma — the famed author of *Panchatantra* — one of the most translated books of the world, each having a school of his own. It attracted students from Babylonia, Greece, and China and offered over 60 different courses in various fields, such as science, mathematics, medicine, politics, warfare, astrology, astronomy, music, religion, and philosophy.

Its location at the central junction of the Indian subcontinent and Central Asia (at the intersection of trade routes from eastern India, western Asia, and from Kashmir and Central Asia), on the eastern bank of the Indus River, made it a very important and prosperous city. Kings like Darius I, Alexander the Great, and many others visited it. Alexander is believed to have taken several scholars from there to be brought to Greece with him. The famed Chinese traveller Faxian (ca. 337–ca. 422 CE) visited it around 400 CE when it was fully functional; while by the time the other famed Chinese traveller Xuan Zang (602–664 CE) visited it in 630 CE, it had been laid to waste by the repeated attacks and carnage by White Huns (Hephthalites). Faxian, Xuan Zang, and Yijing (635–713 CE) were among the hundreds of Chinese monks who made pilgrimages and study tours to India during the first millennium CE.

The period of about 500 years between 400 and 800 CE is a golden period of Indian history for education, literature, sciences, and mathematics. During this period, famous universities such as Nalanda, Valabhi, Sharada Peeth, Pushpagiri Vihara, Odantapuri, Vikramashila, and Somapura were established, many of which were also monasteries attached to Buddhist places of worship. Xuan Zang and Yijing visited Nalanda and many other places and wrote extensive descriptions. They also copied and translated many books, to take with them.

7.3.2 *Nalanda and other universities*

Mention must be made here of the most well-known of these, Nalanda University (Fig. 7.11), which is believed to have housed several thousand students and teachers from across the world and imparted education in the Vedas, logic, grammar, Buddhist and Hindu philosophy, astronomy, and medicine. It is also known that more than eight branches of medical science, including surgery and paediatrics, were practised by physicians trained there. Two scholars from that period, namely Aryabhata, who

Fig. 7.11. Ruins of Nalanda University.

introduced the concepts of zero and decimals and authored several treatises including *Aryabhatiya*, and Varahamihira, who was a renowned scholar of all the sciences and arts, from botany to astronomy and from military science to civil engineering, deserve special mention here. Several of these universities, most notably Nalanda, Odantapuri, and Vikramashila, were destroyed along with their students, teachers, and libraries, with their collection of millions of manuscripts.

The (seventeen) most famous teachers associated with Nalanda have been eulogised in a poem by His Holiness the Dalai Lama and portrayed on a historical thangka [8]. Nagarjuna (ca. 150–ca. 250 CE) is widely considered one of the most important Buddhist philosophers, even though there is some confusion about his association with Nalanda, which came into existence quite some time afterwards. Atisa Dipankar Srigyan (982–1052 CE) was associated with Nalanda and Vikramashila and is well known for his work in the Srivijaya Empire (in Sumatra) and later in Tibet (Fig. 7.12).

The Sharada Peeth University in Kashmir is believed to have housed up to 5000 students and teachers and a magnificent library. Xuan Zhang

Fig. 7.12. Atish Dipankar Srigyan (Credit: Wikipedia).

Fig. 7.13. The Buddhist Mantra "Om Mani Padme Hum" in Tibetan Script.

also visited this place. It was also responsible for the popularisation of the Sharda script which was used to write all the Sanskrit texts in Kashmir. Some historians have suggested that the Devanagari script, based on the early Brahmi script, may have been developed here. The famous Buddhist monk Thonmi Sambhota studied here during the seventh century CE and is believed to have devised the Tibetan script based on Sharada and Devanagari scripts (Fig. 7.13).

Several centuries later, Kublai Khan was to ask his Tibetan Imperial Preceptor to devise a script for his empire's Yuan region. The result was the so-called Phagspa script, derived from the Tibetan and Devanagari scripts, which was used to write Chinese, Mongolian, Persian, Sanskrit, Tibetan, and Uyghur for about a hundred years before being discontinued, once the Ming Dynasty prevailed.

The Indian education system had a profound effect on Sri Lanka, China, Tibet, and Central Asia, giving rise to trade, religious, and political influences. Indian scholars were invited to teach (and preach) in China and Tibet, and monks from China and Tibet travelled to India for education and copying/translating Sanskrit and Pali texts. The Buddhist monk Bodhidharma is believed to have introduced Kalaripayattu — the martial art of Kerala, in China — which over time evolved into Shaolin Kung Fu, which in turn gave rise to karate, etc. Indian and Chinese medicines are believed to have had a profound effect on each other as their monks travelled back and forth between the two countries (see, e.g., Ref. [9]); in particular, the works of Nagarjuna on ophthalmology were especially highly valued and venerated in China. The technical knowledge that was necessary for the refinement and production of various types of sugar and bridges was transmitted by monks travelling between India and China [10].

Indian education also profoundly affected a large part of Indochina and the numerous islands of the East Indian archipelago from Sumatra to

New Guinea, from the second century CE till about 1500 CE, when these had Hindu rulers. These influences are quite visible in the vast architectural remains, still extant, and other cultural influences.

7.3.3 *Persia and Greece*

The conquest of the Persian Empire by Alexander from 334 to 323 BCE brought the influence of Greek culture, literature, education, and warfare to Central Asia and India, with profound effects travelling in both directions. Greeks maintained their colonies in the conquered areas, many of which were still flourishing and keeping the Greek traditions alive about a hundred years later when Herodotus visited them to compile his book, *The Histories*.

Ashoka was to leave behind rock edicts in Greek as well as Aramaic, which were spoken in the part of Afghanistan (especially in what is Kandahar now) under Greek influence. Many Greeks were to convert to Buddhism and enrich literature, science, astronomy, drama, as well as astrology in India. Many of these influences survive today in words like "Yavanika" (for curtains in a drama) and arguably in Zodiac signs, as well as perhaps in the mythology of both civilisations.

The most important phase of education, especially higher education in Persia, occurred during the so-called neo-Persian era from 224 to 651 CE when the Zoroastrian system of philosophy and ethics was reemphasised after recovering from the Greek Rule. The greatest achievements of this period were the establishment of the Academy of Gondēshāpūr (Fig. 7.14), where the knowledge about Zoroastrian culture, Indian and Greek sciences, Alexandrian-Syrian thought, medical training, theology, philosophy, and other disciplines was developed to a high degree and imparted to students. The greatest developments were in the field of medicine. The city also gave refuge to various Greek philosophers and Syriac-speaking Nestorian Christians fleeing religious persecution by the Byzantine Empire.

The Academy (with the patronage of the rulers) got Greek and Syriac texts on medicine, astronomy, philosophy, and useful crafts translated into Pahlavi (Middle Persian or Pārsīk or Pārsīg). These works (including the writings of Plato, Aristotle, Hippocrates, Ptolemy, and others) were translated into Arabic when the House of Wisdom (Bayt al-Ḥikmah) was established in Baghdad, under the Islamic Rule during the eighth century,

Fig. 7.14. Ruins of the Academy of Gondēshāpūr in Khuzestan, Iran (Credit: Wikipedia).

closely following its precepts and staffing it with graduates from Gondēshāpūr. Ultimately, these texts were translated into Latin and reached back to Europe when it recovered from the Dark Ages — which had engulfed it during the fifth to the fourteenth centuries — after the fall of the Roman Empire and before the Renaissance. The House of Wisdom was destroyed by Mongols in 1258 CE.

The Sankore Madrasa attached to the Sankore Mosque established around 988 CE in Timbuktu, Mali, was to develop into a great university and a famous library in the Islamic world by the fourteenth century and provided education not only in Islamic studies but also in history, rhetoric, law, science, and, most notably, medicine.

Al-Azhar University or Jāmiʿat al-Azhar was established in 970 CE and was to develop into an important centre of Islamic studies.

7.3.4 *Observatories at Maragheh and in India*

Hulegu Khan, the destroyer of Baghdad, appointed Nasir al-Din al-Tusi as his (scientific) advisor, who went on to establish one of the most advanced astronomical observatories of its time, Maragheh Observatory, in 1259 CE. Nasir al-Din al-Tusi's contributions to science, mathematics, and astronomy are only too well known.

Fig. 7.15. Jantar Mantar: The Naked-Eye Observatory in Jaipur.

The Observatory became a major centre of education where people working on mathematics, science, and astronomy from across the Islamic world and as far away as China gathered. According to texts recovered from the Observatory, the site had a reputation that reached far and wide, and students flocked to it for education and research. However, it decayed due to a shortage of funds and was destroyed by earthquakes. February 24, every year, is celebrated as Engineers' Day to honour al-Tusi's memory in Persia (Iran).

There are also suggestions that the observatories of Ulugh Beg (1425–1450 CE) in Samarkand and those of Sawai Raja Jai Singh (1699–1743 CE), Jantar Mantar (Fig. 7.15) in Delhi, Varanasi, Ujjain, Mathura, and Jaipur, were inspired by it.

7.4 Later Education in Medieval India

The Mauryan Empire, the Gupta Empire, and the Empire of Harshavardhan were also the periods of the rise of large cities. These cities and towns had schools, which also admitted day scholars. Similar arrangements were made by other rulers, including those from the Rastrakuta and Chola Empires.

Higher education was imparted at the following:

- **"Agrahars":** These were situated in rural settings, often supported by land grants, and imparted education in religion, music, medicine, mathematics, astronomy, logic, politics, etc.
- **"Brahmpuris":** These were located often in urban settings and had a settlement of learnt Brahmins, who were well versed in Vedas, Puranas, Smritis, Kavya Bhashyas, etc.
- **"Mathas":** These were essentially monasteries attached to some local temple, where religion and sacred art were taught to monks, ascetics, and other students living on the premises.

These centres of learning in villages, towns, and cities survived the Islamic Rule from about 1200 to the eighteenth century CE, when India came face to face with the British education system. Education in all aspects of life was taken rather seriously during this period in India, and treatises on a vast variety of subjects were produced.

The famous drama *Mrichchhakatika* (*The Little Clay Cart*) attributed to Shudrak, from this period, has a hilarious scene where a thief preparing to cut a hole in the wall of a rich man's house agonises over "how artistic would the hole and the sunlight getting into the house through the hole in the morning look to onlookers so that they would consider him a worthy disciple of the great author of the Treatise on Theft!" Bharat Muni's *Natya Shastra*, a treatise on arts, is also believed to have been composed during this period.

7.5 Education during the Islamic Rule in India

Most Islamic rulers considered it their religious duty to establish madrasas (Fig. 7.16) and to encourage learning, as emphasised by The Prophet. Thus, they established maktabs (elementary religious schools) and madrasas (secondary religious schools), where students were taught theology and ethics, history and politics, accounting and arithmetic, mensuration and agriculture, engineering and astronomy, domestic science and medicine, logic and philosophy, physical and mechanical sciences, grammar, philosophy, mathematics, and law, which was perhaps greatly influenced

Fig. 7.16. The Oldest Madrasa in Delhi, Qutub Complex, Established by Alauddin Khilji (1315 CE).

by the traditions prevalent in Central Asia and Persia. The syllabus was regularly updated. As Persian was the court language, a system of private tutors also emerged to train students in the prevailing law.

The maktabs and madrasas routinely attracted teachers from Central Asia and Persia. Based on these influences, the Dars-i Nizami study curriculum or system was developed by the Farangi Mahal Ulama, with the most famous being Nizamuddin Sihalivi Ansari (March 1677–8 May 1748). This system was followed in madrasas all over South Asia, South Africa, Canada, the United States, the Caribbean, and the UK. One important feature of Dars-i-Nizami was an emphasis on comprehension, rather than rote learning, which was followed for hundreds of years till then.

7.6 Language and Music

One important outcome of the Islamic Rule in India was the emergence of the Urdu (of the camps) language, which had a rich Persian and Arabic vocabulary, in the framework of Hindustani or Hindi. Hindi, in contrast, was rich in Sanskrit and Prakrit vocabulary. Urdu as a literary language reached its zenith towards the end of the Mughal Rule when the most moving and eloquent gazals, nazms, and rubais were composed. Excellent Urdu prose literature was also to emerge soon afterwards. Urdu was adopted as the language of education and administration along with Hindi, in place of Persian in 1837, in India.

Over 70 million people across several countries claim Urdu as their mother tongue, and it has a vibrant literature. The decision to use Urdu as the official language of Pakistan (including East Pakistan) was one of the triggers for the birth of Bangladesh, where most people speak Bangla.

Qawwali, a form of devotional music, also arrived in India with Sufi Islam. Khayal singing was incorporated around the 17th century into the classical music of North India (the so-called Hindustani classical music) which also adopted musical instruments like sitar and sarod. Around 1875, it acquired a remodelled harmonium from its European version, suitable for Indian classical and popular music. This harmonium, by now, has almost completely replaced sarangi as the accompanying instrument.

The classical music of South India (Carnatic music) incorporated the violin as the accompanying instrument towards the late eighteenth century, though the way the violin is held in these performances differs from the way Western classical musicians hold it. Several Indians, e.g., Rabindranath Tagore, were to acquire an in-depth understanding and appreciation of Western classical music and composed famous songs based on these [11]. In more recent times, hugely popular film music as well as popular music has adopted many Western musical instruments and experimented with international genres.

Special mention must be made here about the teachers of classical music and dance of the medieval times and later, who were perhaps more guarded in imparting their knowledge, the intricate details of which they often offered only to their offspring or close relatives. This gave rise to the tradition of "Gharanas" (family or lineage).

The colonisation of Asian, African, and Latin American countries by the British, French, Portuguese, and Spanish (see Table 7.1) played a very historical role in the invasive spread of English, French, Portuguese, and Spanish languages, as well as the education of Europe, at least the part of the education which the colonisers decided to impart to their subjects. The nobility and the wealthy of the subject nations sent their children to the ruling countries for higher education. This gave rise to a new class of

Table 7.1. Main Spanish, French, British, and Portuguese Colonies.

Sl. No.	Colonial power	Colonies
1.	Spain	Argentina, Bolivia, Chile, Colombia, Costa Rica, the Dominican Republic, Ecuador, El Salvador, Equatorial Guinea, Guatemala, Honduras, Mexico, Panama, Paraguay, Peru, Uruguay, Venezuela
2.	France	Benin, Burkina Faso, Cambodia, Cameroon, the Central African Republic, Chad, Comoros, Côte d'Ivoire, Djibouti, Gabon, Guinea, Laos, Lebanon, Madagascar, Mali, Mauritania, Morocco, Niger, the Republic of the Congo, Senegal, Syria, Togo, Tunisia
3.	Britain	Antigua and Barbuda, Bahrain, Bangladesh, Barbados, Belize, Bhutan, Botswana, Brunei Darussalam, Cyprus, Egypt, Fiji, Gambia, Ghana, Grenada, India, Jamaica, Jordan, Kenya, Kuwait, Lesotho, Malawi, Malaysia, Mauritius, Namibia, Oman, Pakistan, South Africa, Sri Lanka, Saint Lucia, Saint Vincent and the Grenadines, Swaziland, Tanzania, Trinidad and Tobago, Uganda, Zambia, Zimbabwe.
4.	Portugal	Madeira, Azores, Cape Verde, São Tomé and Principe, Portuguese Cochin, Portuguese Mozambique, Portuguese Goa, Portuguese Malacca, Portuguese Hormuz, Portuguese Colombo, Portuguese Macao, Portuguese Nagasaki, Portuguese Angola

elites, who were (sometimes derisively) called "Westernised oriental gentlemen" in India, "Ilustrados" in the Philippines, and "Gente de razón" in Latin America. Such people were also derisively called "brown sahibs" or "coconut" (brown outsides and white insides) in India and "banana" (yellow outsides and white insides) in the British colony of Hong Kong.

7.7 Education and the Colonial Period

A vast amount of literature now exists that chronicles the effect of colonisation on the subjects and the confusion created in their minds because of the attempts by the rulers to convince them of their "inferiority as subject races" [12]. By now, it is known that most of the time, the Spanish provided only rudimentary education (reading and writing Spanish) to a selected group of people considered "desirable" or of "pure blood." The French often encouraged centralised French education in government schools, with teachers brought/sent from France/Europe and only a smattering of native teachers. There are suggestions that the colonial rulers may have limited the education of the natives, fearing it would lead to rebellion.

A feeling of how deeply colonial education affected its subjects can be gauged from two examples of autobiographical books from different parts of the world. *L'Enfant noir* (*The Dark Child*) is an autobiographical novel by Camara Laye from Guinea, which was published in 1953 [13]. It became well known for the depiction of the protagonist's family, whose father works as a goldsmith and is also a mystic, and his drifts away from them as he gets more educated and wins a scholarship to study in Paris, as well as its elegant French.

The Autobiography of an Unknown Indian by Nirad C. Chaudhuri (1897–1999 CE) was published in 1951 [14]. It was also praised for its elegant English and was both despised and liked, especially for its dedication which stated:

> "*To the memory of the British Empire in India,*
> *Which conferred subjecthood upon us*
> *But withheld citizenship.*
> *To which yet every one of us threw out the challenge:*
> *'Civis Britannicus sum'*
> *Because all that was good and living within us*
> *Was made, shaped, and quickened,*
> *By the same British rule.*"

Rabindranath Tagore (1861–1941 CE) indicted British Colonialism on India, even more succinctly, when he argued that "India had gained a great deal from its association with Britain, for example, from "discussions centred upon Shakespeare's drama and Byron's poetry and above all … the large-hearted liberalism of 19th-century English politics." However, the tragedy was that "what was truly best in their own civilisation, the upholding of the dignity of human relationships, has no place in the British administration of this country" [15].

Yet, it cannot be denied that the use of English in education and day-to-day affairs of administration and trading is a lot more widespread today in India than it was when India became free. It is also believed that this gives a definite advantage to India in this era of globalisation.

Peggy Mohan has termed English as an *invasive* language [16], which has spread rapidly and is smothering and even killing local languages and dialects. We would like to add that "undergoing a moulting" perhaps most aptly describes the transformation that the students who travel abroad sometimes undergo if their sensitivities to Indian traditions are not sufficiently deeply ingrained. French, Spanish, and Portuguese languages did the same to the indigenous languages of their respective colonies and yet opened them to global influence.

David Arnold [33] has discussed one more aspect of colonialism and concluded, "However, the worst result of colonisation has been the creation of elites which have been so educated that they have no knowledge of their roots, their own culture and history, and greatly share western perceptions of their own societies."

7.7.1 *What if?*

Sometimes, it has been argued that India would not have developed at all without the British education and even may not have become one country. We strongly disagree with the second contention, as India as a country was well established quite early in history. However, the refutation of that point of view is beyond the scope of the present work. What we would like to emphasise is that we have repeatedly seen that "Throughout India's long history, it persistently enjoyed exchanges of ideas as well as of commodities with the outside world. Traders, settlers, and scholars moved between India and further east — China, Indonesia, Malaysia, Cambodia, Vietnam, Thailand and elsewhere — for a great many centuries, beginning more than 2,000 years ago. The far-reaching influence of this movement — especially on language, literature, and architecture — can be seen plentifully even today. There were also huge global influences through India's open-frontier attitude in welcoming fugitives from its early days" [15].

India welcomed and accommodated Jews, Christians, Parsis, Armenians, Baha'is, and others for centuries. Muslim Arabs have had a profound effect on the coastal regions of Kerala as they married local women. Their descendants are still called Moplah (son-in-law). All of them exposed India to international thoughts.

A very important example of getting modernised without coming under foreign rule is provided by Japan. It had extensive trade with the Portuguese, Spanish, and Dutch traders during the 16th and 17th centuries. The efforts of padres, who came along with the traders to convert the Japanese to Christianity and the "unfair" trade practices of the traders, led to their being expelled from Japan in 1639. For two centuries, Japan essentially isolated itself.

Yet, following the "show of strength" by Commodore Matthew Perry of the US Navy, who steamed into the Bay of Edo in Japan in 1853 with four warships in the name of reestablishing contacts, it initiated the reforms of the Meiji Era in 1867.

Thus, Japan moved from being an isolated feudal society at risk of colonisation by Western powers to a modern, industrialised nation and developed into a great power, influenced by Western scientific, technological, philosophical, political, legal, and aesthetic ideas. With the declared intention of seeking knowledge from across the world, Japan "inaugurated a new Western-based education system for all young people, sent thousands of students to the United States and Europe, and hired more than 3,000 Westerners to teach modern science, mathematics, technology, and foreign languages in Japan."

7.8 Education in India during the Colonial Rule

The transatlantic voyages of Christofer Columbus to America from 1492 to 1504, the discovery of the sea route to India by Vasco de Gamma in 1498, and the circumnavigation of the globe by Ferdinand Magellan from 1519 to 1522 started the so-called Age of Discovery or the Age of Exploration. It was also to set the colonisation of Asia, Africa, and the Americas and the gruesome slave trade into motion, which was to continue for centuries and alter the course of civilisation, in addition to causing endless suffering to the hapless slaves.

Vasco de Gama landed at Calicut Port on 17 May 1498. After the Portuguese, the Dutch, the Danish-Norwegians, the Swiss, the Austrians, the British, and the French came to India, they all left their imprints on Indian society, culture, education, and military practices. The Portuguese and the French controlled rather small areas. Austria (and the Habsburg Monarchy) colonised the Nicobar Islands between 1778 and 1886.

The Danish briefly occupied it in 1756, and much later the Japanese occupied it during the Second World War.

7.8.1 *Goa and Portuguese*

Goa became a Portuguese colony in 1510. It was to evolve as a centre for Portuguese rule/control in Malacca, Hormuz, other Portuguese possessions in India, other bases in Indonesia, East Timor, the Persian Gulf, Macau in China, and trade bases in Japan.

Once entrenched, the Portuguese decided to educate the locals so that the natives could serve as government servants since it was difficult to get manpower from Portugal to do clerical jobs. Later, this policy was to be adopted by the British. Initially, the Portuguese introduced secular schools, but over time, parish schools were introduced, which emphasised religious education (see Ref. [17] for details). Goa was also to suffer from the brutal Portuguese Inquisition from 1560 to 1812, with a brief reprieve from 1774 to 1778.

Secondary and higher education was imparted in convents and seminaries founded by the missionaries. The Jesuits, mainly active in Goa, introduced India to both the European college system and the printing of books. Saint Paul's College was established in Goa in 1542 by Saint Francis Xavier, which imparted education in grammar and rhetoric, with lectures on classical authors, and ran a school for local students who were taught reading and writing, as well as a hospital. Later, many more colleges and schools were established. The medium of education was Portuguese. In 1840, the Portuguese started the Medicine and Pharmacy Licenciates, now known as Goa Medical College. Often teachers from Lisbon, Porto, and Coimbra were called to teach. The nobility often sent their children to Portugal for higher studies.

7.8.2 *Pondicherry and French*

India, which was already exposed to the British traders, came strongly under British influence after the Battle of Plassey (23 June 1757) where they defeated the forces of Siraj ud-Daulah. It was further reinforced by their victory over Tipu Sultan on 4 May 1799 at Seringapatam (now called Srirangapatnam).

Recall that Nader Shah of Persia had already delivered a crushing blow to the Mughal Empire and sacked and ravaged Delhi in 1739. He was followed by Ahmad Shah Abdali, who defeated the powerful army of Marathas at the Third Battle of Panipat in January 1761. This eliminated the last possible resistance to invaders/colonisers. Soon after the death of Maharaja Ranjit Singh of Punjab — who had French officers in his employ, in 1839, and the defeat of Sikhs in the successive Anglo-Sikh Wars in 1845–1846 and 1848–1849, mainly due to internal feud and factionalism — British colonialism was quite firmly entrenched in India. It was briefly challenged in 1857 during the Sepoy Mutiny or the First War of Independence (as it is now called) but was brutally crushed. This led to take over of the rule of India by the British government from the East India Company.

As both Siraj ud-Daula and Tipu Sultan were supported by the French forces, the French influence on India was curtailed to disjoint French enclaves at Pondicherry (now Puducherry), Karikal, Yanam (Andhra Pradesh) on the Coromandel Coast, Mahé on the Malabar Coast, and Chandernagor in Bengal. These enclaves were to become a refuge for Indian Freedom Fighters, and dissident writers and poets, the most famous of them being Subramania Bharati (1882–1921 CE), who opposed the British Rule in India. Aurobindo Ghosh (or Sri Aurobindo (1872–1950 CE)) was to move to Pondicherry when he drifted to spiritualism from being a freedom fighter.

The French education system was implemented in Pondicherry, where students from other enclaves often congregated. In 1823, a medical college called Ecole de Medicine de Pondicherry was established. The slogan "Liberté, égalité, fraternité" (liberty, equality, fraternity) from the French Revolution (1789–1799 CE) inscribed at the gate (Fig. 7.17) of many of these enclaves was to act as a beacon of hope for many a revolutionary in their struggle against the British Rule.

7.9 British Colonial Period in India

There are reports that at the beginning of the colonial era, the Bengal Presidency had about 100,000 indigenous elementary schools that had been established and were run by natives themselves, without direct interference from the government [18]. It was also reported that the Madras Presidency had 11,758 schools, and 740 centres for higher education,

Fig. 7.17. The Slogan "Liberté, égalité, fraternité" (Liberty, Equality, Fraternity) from the French Revolution Inscribed on a Gate in Chandernagor.

funded and managed at a community level. Further, the Bombay Presidency had one in eight boys getting an education, while in the areas around Punjab, one in six boys was getting an education.

In 1700 CE, India's population was about twenty times that of Britain. India's share of total world output at that time had been estimated at 24% — nearly a quarter, while Britain's share stood at 3%. Such a high level of economic output and productivity is not possible without the innovative methods and novel techniques of trade and industry imbibed through rigorous education. It is painful to recall that when India became free in 1947, it was reduced from the world's leading manufacturing nation to a symbol of famine and deprivation.

Once the British colonial government felt the need for "native" employees to assist them in governance, it decided to support education. Initially, it decided to establish madrasas in 1781 at Calcutta for Muslim students, Sanskrit College in 1791 at Benares for Hindu students, and Fort William College at Calcutta in 1800 for Europeans to train them as administrators and teach them Bengali, Urdu, Arabic, Persian, Sanskrit, Marathi, and even Chinese. The Asiatic Society was established in 1784 by William Jones. It was to become a meeting and rallying point for Orientalists. James Princep, an assay master in the Calcutta Mint and the founding editor of the *Journal of the Asiatic Society*, went on to decipher Kharosthi and Brahmi scripts and throw light on Ashoka's rock edicts and inscriptions found all over the country, which proclaimed Ashoka's commitment to non-violence, righteousness, and a sophisticated notion of secularism. The institution translated thousands of books from Sanskrit, Arabic, Persian, Bengali, Hindi, and Urdu into English.

Pundit Ishwar Chandra Vidyasagar (1820–1891 CE) studied at Sanskrit College and was later employed at Fort William College. He became one of the most famous educators of his time, a social reformer, a writer, and the "Father of Modern Bengali Prose." He became the principal of the Sanskrit College in 1851, promoted the education of English, and admitted students of the lower classes. He also advocated the education of girls, widow marriage, and the abolition of the Sati system, much to the anger and chagrin of the fundamentalists.

There are even suggestions that the so-called Khadi-Boli (the form of Hindi used for formal writings and speech and different from various dialects of Hindi) was popularised at the suggestion of John Gilchrist (1759–1841 CE), the first principal of the Fort William College, who encouraged (ordered?) another teacher Pundit Lallu Lal (1763–1835 CE) to create (a new) prose of Hindustani by replacing Arabic and Persian words with those of Sanskrit and Prakrit. Thus, Pundit Lallu Lal translated the poetic translation (in Braj Bhasa) of the tenth chapter of *Srimad Bhagwat* by Pundit Chaturbhuj Mishra as *Prem Sagar*. It was one of the first modern Hindi books, composed and typeset between 1804 and 1810 and published in 1810. The College also published textbooks, grammar, and dictionaries for several languages.

Following the announcement of the new education policy by Lord William Bentinck in 1835, which initiated the use of English as the language of administration and of the higher law courts (instead of Persian), Fort William College was dissolved in 1854 (as authorities felt that it was promoting "Indianisation" of the British!). Its magnificent library was divided between the Calcutta Public Library (now the National Library) and the School of Oriental Languages in Paris.

This itself followed the famous (notorious?) minutes of Memorandum on (Indian) Education by Thomas Babington Macaulay, which called native culture and learning inferior and added that Western learning which was superior could only be taught through the medium of English. He further argued that there was a need to produce — by English-language higher education — "a class of persons, Indian in blood and colour, but English in taste, in opinions, in morals and in intellect" who could in their turn develop the tools to transmit Western learning in the vernacular languages of India.

We add here that colonial powers throughout history across the world took pride in their imperial conquests and had an elevated sense of self, with their every fibre nurtured by centuries of predatory accumulation of wealth, knowledge, and privilege. They were completely convinced[1] that their countries fought and defeated particularly immoral, evil, and powerful enemies, and protected civilisation, and all their actions were invariably affected by this conviction.

The introduction of modern Western/English education received strong and vociferous support from the reformist "Father of the Bengal Renaissance," Raja Ram Mohan Roy (1772–1833 CE). Raja Ram Mohan Roy, while vehemently objecting to the education of Sanskrit by opening more Sanskrit pathashalas to Lord Amherst in 1923, is known to have remarked that Sanskrit, "can only be expected to load the minds of youth with grammatical niceties and metaphysical distinctions of little or no practical use to the possessor or the society." He further argued that "the Sanskrit system of education would be best calculated to keep this country in darkness if such had been the policy of the British Legislature" [19]. He helped set up the Hindu College (1817), Anglo-Hindu School (1822), Vedanta College (1826), and (what is now known as) Scottish Church College (1830). Soon thereafter, many educational institutions with Western curricula and medium of instruction as English were opened. The first college for women was Bethune College, Calcutta, in 1879; it was started as a school for girls in 1849. Reports suggest that the Hindu College (now Presidency University) had the Westernised curriculum of a university which had been expressly founded for the "uplift of the natives." Michael Madhusudan Dutt (1824–1873), a brilliant poet of English, one of the pioneers of Bengali literature, and a student of the Hindu College, was a famous example of these conflicting influences.

[1]Recall James Mill, a Scottish historian, who wrote in 1810, "This English government in India, with all its vices, is a blessing of unspeakable magnitude to the Population of Hindostan."

Along with colleges, many of which were opened by philanthropic (and even caste-based and religious) organisations, the University of Bombay (1857), the University of Calcutta (1857), and the University of Madras (1857) were initially opened, followed by the University of Punjab (1882) and the University of Allahabad (1887). By 1947, 18 more universities were opened by the British rulers. Some colleges were also opened to specifically "educate the children of the royalty, nobility, and aristocracy" of India (Fig. 7.18).

In between, Benares Hindu University was opened in 1916 through the personal efforts and donations collected by Pundit Madan Mohan Malaviya (1861–1946 CE). Muhammadan Anglo-Oriental College, founded in 1875 by Sir Syed Ahmad Khan (1817–1898 CE), was the first modern institution of higher education for Muslims in India, which was to become the Aligarh Muslim University in 1920. Vishwa Bharati, the university established by Rabindranath Tagore, started functioning in 1921. The oldest and "solely native and purely national," research institute in Asia, the Indian Association for Cultivation of Science, supported by

Fig. 7.18. Coat of Arms of Daly College for the Children of the Nobility of Central India.

generous public contributions, was opened in Calcutta in 1876, "with a view to its advancement by original research and to its varied applications to the arts and comforts of life." About 38 years later, the Indian Science Congress Association was established in Calcutta (1914) with the aims of advancing and promoting the cause of science in India, holding an annual Congress at a suitable place in India, publishing proceedings, journals, transactions, etc., and popularising science.

All these institutions of learning not only spread modern Western education in the country but also generated a strong sense of nationalism and a burning desire to free the country from British Rule as students learnt about world history and various peoples' movements for democracy and the French, American, and Russian Revolutions, the American Civil War, as well as the Magna Carta.

Children of rich parents and nobility often travelled to England for higher education. Thus, M. K. Gandhi (1888), Mohammad Ali Jinnah (1892), J. L. Nehru (1905), and H. J. Bhabha (1928) were sent to England by their families for higher education. A few students won scholarships to study abroad. B. R. Ambedkar, the noted jurist, social reformer, and economist, for example, went to Columbia University, USA, in 1913, and from there to the London School of Economics in 1916 on a scholarship from Baroda Princely State. Mulk Raj Anand, the famous English writer, went to England in 1925 and initially worked in a restaurant to support his education. Quite a few aspiring students also went to Britain to study for the Indian (initially Imperial) Civil Service Examinations, once it was open to Indians. The Indian student community in India and abroad was to play a significant role in India's Freedom Movement, overtly and covertly.

Still, the literacy levels were low at around 12% with only about 0.2–0.3 million students in 20 universities and 500 colleges spread across the country, when India became independent in 1947 [20].

7.10 Slavery, Prisoners, Refugees, and Education

A related discussion involves slavery, which also led to the diffusion of knowledge and skills from one society to another. We have seen that in Ancient Rome, most of the teachers were Greek slaves. Communities defeated in wars or people captured during raids conducted for this purpose have been put into slavery since ancient times in almost all nations across the world. Debt slavery was also common in several places and has essentially continued as bonded labour into recent times after slavery was abolished. Slaves who were taken across long distances or civilisations naturally transfused new knowledge and learning to the society of the captors. It has been suggested that slaves were responsible for the prosperity of Athens and the leisure of the aristocrats, who had time to create the high culture now considered the beginning of Western civilisation and, paradoxically, may have shaped their ideas about freedom which is believed to be central to Western heritage.

The slave trade network was well established and in the 9th-century Baghdad; for example, one could get slaves from Western Europe, Eastern Europe, Central Asia, and Africa. The Mongols, in the 13th century, took slaves from their conquered lands to their capital Karakorum (and later Sarai) from where they were sold throughout Asia and Eurasia. There are suggestions that Mongols, while killing the entire populations of defeated lands, used to separate artisans to be taken to their capitol [21].

The Arab armies of Muhammad bin Qasim during their invasion of Sind at the beginning of the eighth century are reported to have enslaved tens of thousands of Indians. Most early Islamic invasions/rules in India resulted in a large number of slaves being taken/sent abroad who were used to maintain the gardens, cultivate, watch over livestock on plantation-style farms, maintain irrigation canals, work in brick factories, and involve in carpentry, pottery, and construction for Central Asia's wealthy families. It was formalised when Sultan Alauddin Khilji legalised the enslavement of those who defaulted in paying their taxes. This practice was continued during the Mughal rule.

Slaves from India were also especially sought after because of their skills in textiles, agricultural production, and architecture. There are reports to suggest that the phrase "slaves from India, horses from Parthia (modern Khorasan)" was about the sale of slaves and horses.

In his early 11th-century *Tarikh al-Yamini*, the Arab historian al-'Utbi recorded that in 1001, the armies of Mahmud of Ghazna conquered Peshawar and Waihand, "in the midst of the land of Hindustan" and captured some 100,000 youths. Some years later, following his 12th expedition into India during 1018–1019, Mahmud is reported to have returned to his capital with such a large number of slaves that their value was placed at only 2–10 dirhams each. This price was so low that, according to al-'Utbi, "merchants came from distant cities to purchase them, so that the countries of Central Asia, Iraq and Khurasan were filled with them, and the fair and the dark, the rich and the poor, mingled in one common slavery" [22].

When Taimur Lane invaded and sacked Delhi in 1398, he took a large number of artisans as slaves, even though he had ordered a general massacre of about 100,000 slaves captured earlier, along his route and in Delhi.

In Taimur's own words, "I ordered that all the artisans and clever mechanics who were masters of their respective crafts should be selected from among the prisoners and set aside, and accordingly, some thousands of craftsmen were bidden to await my command. All these I distributed among the princes and amirs who were present, or who were officially engaged in other parts of my dominions. I had determined to build a Jami' Masjid in Samarkand, the seat of my empire, which should be without a rival in any country; and for this reason, I ordered that all builders and stone masons should be set apart for my own special service" [23].

There are records to suggest that Timur Lane had taken (carried away) "Sculptors, stonemasons and stucco-workers from Azerbaijan, Isfahan and Delhi; Mosaic-workers from Shiraz; weavers, glassblowers and potters from Damascus." These slave artisans were to create the magnificent architecture of Samarkand, especially the Bibi Khanum Mosque (Fig. 7.19), and make it one of the most prosperous cities in the world through their labour and expertise. Records suggest that Central Asian slave markets had also swelled and then crashed with Indians following Shah Jahan's aborted attempt at the annexation of Balkh in 1646–1647 and the capture and enslavement of many of the retreating Indian soldiers [22].

There are reports of Portuguese and Dutch ships making slave raids in Bengal, along the Hooghly River and other delta regions. Australians indulged in what was known as "birding" to capture slaves from the islands in the Pacific to work on their extensive farms.

In more recent times, after the defeat of Germany during the Second World War, more than 1,600 German scientists, engineers, and technicians were taken to the USA under the Operation Paperclip, while the Soviet Union relocated more than 2,200 German specialists, under Operation Osoaviakhim, during one night on 22 October 1946. The US efforts continued under Operation Overcast. Other members of the Allied Forces also conducted several such operations. One of the most famous scientists taken to the USA under Operation Paperclip was Wernher von Braun

Fig. 7.19. Bibi Khanum Mosque, Samarkand (Credit: Wikipedia).

(1912–1977), who was working on the V2 rockets in Germany and who went on to develop the Saturn V rocket, which propelled the Apollo Spacecraft to Moon.

The persecution of Jews in most of Europe for centuries and the denial of rights to own land in several countries forced them to concentrate on education and industry in which they excelled. The gruesome persecution of Jews by the Nazis forced many to emigrate to the UK and USA, where they made valuable contributions to education, research, and industry. Nazis also killed more than 6 million Jews in one of the cruellest, most brutal, and most inhuman manners, aimed at the elimination of their entire populations, in the history of mankind. The persecution of Zoroastrians, after Islamic rulers overtook Persia, forced them to immigrate to India, where they made valuable contributions to trade and industry.

7.10.1 *Slave trade and knowledge*

Vikings had been conducting raids to capture slaves to be sold for centuries. Crimean Khanate from 1475 to 1783 thrived by raiding the neighbouring Slavic and Caucasian sedentary societies and selling the captives into the slave markets of Eurasia. The Ottoman Empire, in general, and its centre Turkey, in particular, thrived on slaves from both the White Slavic north and the Black African south for close to 500 years from the beginning of the fifteenth century. They constituted the janissary soldiers, who were mostly confiscated ("tribute") Christian children taken to Istanbul and raised as soldiers or even administrators. They ran the empire, manned its ships, generated much of its handicraft products, and served as domestic servants and in harems. We have seen how the Chinese slaves introduced the art of making paper to Arabs.

The most gruesome slave trade involved the capture of more than 12.5 million Africans who were taken to the Americas. It was part of the notorious Triangular Trade which operated between Europe, Africa, and the Americas from the sixteenth to nineteenth centuries and involved ships from Europe sailing to African ports loaded with goods (textiles, iron, and weapons), picking up slaves from Africa to sell to the Americas (with many slaves dying during the journey because of overcrowding and unhygienic conditions on the ships), and returning to Europe with sugar and cotton grown by these slaves. Daniel Dafoe (c. 1660–1731 CE), an English novelist, pamphleteer, journalist, and author of *Robinson Crusoe,* is believed to have famously remarked: "No African trade, no Negroes; no Negroes, no sugars, gingers, indicoes (*sic. indigo*) etc; no sugars, etc, no islands; no islands, no continent; no continent, no trade." It is not wrong to say that the wealth of these nations was created by slave labour and colonialism.

By now, it is well established that "slavery and the transatlantic slave trade shaped the science of the Atlantic world as deeply as it influenced its economy and demographics. The unfree migration of millions of enslaved Africans via the transatlantic slave trade also led to the movement of plants, animals, diseases, and natural knowledge throughout the Atlantic basin. Encounters between diverse groups of people and unfamiliar environments engendered new natural and medical knowledge. Enslaved Africans made significant contributions to Atlantic science as specimen collectors, guides, and informants" [24]. Coming from different nationalities, the slaves also often invented Creole and Pidgin languages [16], to be able to communicate with each other.

The slaves from Africa left a lasting influence on music, as they reconstructed their musical instruments like percussive, string, and wind instruments, from drums and banjos to the balafo or balafon (a kind of xylophone), the flute, the musical bow (a stringed instrument), and the panpipe (a tuned pipe). The slaves' spiritual and work songs went on to lay the musical foundations for the most popular forms of music in later American history — including the Blues and Jazz. This is a vast subject and is beyond the scope of this study.

Slaves in America, especially those in the southern states, were not permitted to learn to read and write. A heavy fine was imposed on anyone found to be teaching them. Yet, they kept their traditions, sufferings, origins, and culture somehow alive through "oral history," and several of their descendants were even able to locate the villages of their ancestors from such fragments [25]. Much later, the slogan, "Each One Teach One," was used to spread literacy among them, which resulted in keeping their memories organised and recorded.

Even though slavery was abolished in 1807 by Britain, the slaves in the British Colonies were not freed until 1838, and only after slave owners, rather than the slaves themselves, received compensation. It was to end in Dutch colonies in 1863 and in Portuguese colonies only in 1869.

After slavery was formally abolished, Indians were taken to the West Indies, Mauritius, South Africa, Fiji, etc., as indentured workers. The cost of the passage was supposed to be treated as "borrowed money," and they were to work off their "debt" by an agreement to "voluntarily work" for the "lenders" for a specified term. Extensive studies exist on their living and working conditions and their struggles, which were as bad as those for slaves before them, and their contribution to the countries to which they were taken. The famous, Nobel Prize-winning author V. S. Naipaul (1932–2018 CE), born in Trinidad, was a descendant of one such worker from India.

7.11 International Education

In the remaining part of this chapter, we will be more concerned with the spread of education in science and engineering and how it has benefitted from international collaboration.

Even though the British are credited with the introduction of modern education in India, as we saw above, science education was only at a very elementary level for quite some time, as the colonisers felt that "it was premature to give the natives a taste of European science." British persistence on oriental education for decades, despite the increasing demand for the introduction and promotion of modern science from a section of Indian society, remains "one of the most unintelligible facts in the history of English education in India" [19].

However, as the British Empire underwent stronger consolidation and extension, the need to train natives into more useful branches became not only stronger but also necessary to exploit India's natural resources, fully and systematically.

7.12 Setting Up Organisations to Help Explore and Exploit

The British started the Great Trigonometrical Survey in 1802 to survey the entire Indian subcontinent with scientific precision. Its notable achievements were mapping the Western Ghats and measuring the heights of Dapsang or Chogori Mountain peak (now called K2), Sagarmatha (now called Mount Everest), and Kanchenjunga peaks. Radhanath Sikdar (1815–1870 CE), a brilliant mathematician, who had earlier obtained his education at the Hindu College, is believed to have played an important role in determining the height of Mount Everest in 1852. Several Indians played a great role in the mapping of Tibet and the Brahmaputra River by going there disguised as lamas, as Europeans were not allowed to enter Tibet. Similarly, the navigability of the Indus from Karachi to Lahore and the connecting rivers was determined by the survey conducted by Captain Alexander Burnes in 1831, with the ruse of transporting five dray horses sent as a gift by the King of England to Maharaja Ranjit Singh.

The Archaeological Survey of India was established by Alexander Cunningham in 1861 to research, conserve, and preserve the cultural and historical monuments in the country. It went on to make a detailed survey of Buddhist monuments and excavate the Indus Valley Civilisation sites. Rakhal Das Banerjee (1885–1930 CE), a former student of the Presidency

College, Calcutta, was to play a prominent role in determining the antiquity of the Mohenjo-Daro.

George Abraham Grierson was to head the newly created Department of Linguistic Survey of India in 1898, and it went on to produce a comprehensive catalogue of 364 languages and dialects of India, in 11 volumes. Archaeological research and excavations worldwide have always remained truly international, with the active participation of local researchers.

Still, to effectively rule the country, it was essential to have even more intimate knowledge of the country's topography through an exhaustive physical survey of the land. A Survey School was opened in Guindy in the Madras Presidency in 1794; however, it trained only English boys and natives were only supposed to help them as the British feared that the maps could fall into the hands of other colonial powers like the Portuguese, the French, or the Dutch. Later, surveying was taught at some other colleges as well. Indians were trained as lower-grade technicians in the use of measuring and survey equipment needed by the army, navy, and other technical establishments. Similarly, Indians were trained (only) as overseers for the construction of roads, bridges, buildings, railways, canals, docks, etc., especially at the "Poona Engineering and Mechanical School" which was opened in 1854. The Government Science College, Jabalpur, the first science college in India, was established in 1836.

To be able to assist medical doctors in the army and civil services, elementary works of European medical science were taught at some schools. A medical school to train Indians to work "under" European doctors was opened in Calcutta in 1822. Two medical colleges were opened in Calcutta and Madras in 1835. Medical colleges elsewhere were opened subsequently. Madras Medical College became the first one to open its doors to women students in 1875, while Calcutta Medical College admitted its first woman student in 1884.

The first (civil) engineering college was opened at Roorkee in 1847. It was renamed Thomason College of Civil Engineering and became the University of Roorkee (the first engineering university of Independent India in 1949) and then the Indian Institute of Technology in 2001. As the need for trained engineers rose rapidly, the College of Civil Engineering, Calcutta, was opened in 1856; it was renamed Bengal Engineering College in 1857. The Survey School in Guindy (Madras) was to become the Civil Engineering School in 1858 and the College of Engineering in 1861, with the inclusion of mechanical engineering. The Poona Engineering College as an affiliate of Bombay University was started in 1866.

Subsequently, several important institutes, namely Victoria Jubilee Technical Institute, Bombay (1887); College of Engineering and Technology, started by the National Council of Education and now known as Jadavpur University (1906); and the Indian Institute of Science, started by Jamshed Tata (1911), were set up. The School of Mines, Dhanbad; the Harcourt Butler Technological Institute, Kanpur; and the School of Chemical Technology, Bombay, were also started in the period between 1921 and 1937. All these institutes initially employed a substantial number of European teachers. There are reports to suggest that even when native teachers were employed their salaries were considerably less than those of their European counterparts, as in many other government jobs. The case of the famous scientist and teacher, J. C. Bose, at Presidency College, Calcutta, is very well known, who, in protest, had not collected his salary for close to 3 years.

The Geological Survey of India was founded in 1851, for conducting geological surveys and studies of India. Its role in assessing the mineral wealth of India is well known. We reproduce here an 1870 photograph (Fig. 7.20) of its leading staff who were all Europeans. All its superintendents and directors continued to be Europeans till 1951.

Fig. 7.20. The Geological Survey of India in 1870 (Credit: Wikipedia).

7.13 Recording the Flora

India has had a well-established tradition of "true to life" paintings of flora and fauna, which reached its pinnacle during the rule of the Mughal Emperor Jehangir (1569–1627 CE), who was a great naturalist in his own right and maintained records of very accurate observations about the flora and fauna of the country. He also employed several excellent painters, including Ustad Mansur (c. 1590–c. 1630 CE), for making pictorial records (Fig. 7.21).

The tradition was continued during the colonial era, and there was a widespread interest in botanical collections in India, and the missionaries, the East India Company, and the Governments of Bengal, Madras, and Bombay were engaged in collections and making catalogues, along with excellent and detailed paintings of the flora [26]. In 1787, the East India Company set up a botanical garden at Sibpur, Calcutta.

The Botanical Survey of India started functioning in 1890 and brought out monumental catalogues (Fig. 7.22) on *the Flora of the Presidency of*

Fig. 7.21. The Tulip by Mansur (1621 CE).

Fig. 7.22. Semal (*Bombax Ceiba*, Cotton Tree) from the Collection of the Illustrations of William Roxburgh [26].

Bombay, *Flora of the Presidency of Madras*, *Bengal Plants*, *Flora of the Upper Gangetic plains and the Botany of Bihar and Orissa*, and finally *the Flora of British India* by J. D. Hooker [26]. Needless to add, all these works were done either by Europeans or under their supervision.

However, the Botanical Survey of India suffered due to a lack of funds and was revived by Janaki Ammal (1897–1984 CE), a botanist famous for creating more sweet cane sugar and studying magnolia in detail, when she was appointed as Officer on Special Duty to reorganise the Botanical Survey of India, after the independence.

The Agricultural Research Institute, popularly known as Pusa Institute, was established in 1905 at Pusa in Bihar, with a grant from an American philanthropist. It was established to revive the economically beneficial — for the British but disastrous and excessively exploitative for Indian farmers — crop of indigo. Its name was changed to Imperial Institute of

Agricultural Research in 1911 and to Imperial Agricultural Research Institute in 1919. Following a devastating earthquake in 1934, the institute was shifted to Delhi in 1936. After the independence, the institute was renamed as Indian Agricultural Research Institute. It helped usher in the Green Revolution in the country, in the 1970s. Till 1935, all its directors were Europeans.

7.13.1 *Exchanges in agriculture, forestry, and horses*

While we are at it, we recall that cassava (tapioca), cashew nuts, ground nut (peanut), pineapple, custard apple, chillies, papaya, sweet potato, potato, tobacco, and maize were introduced by the Portuguese in India in the Malabar Coast in the early 1700s, while mangoes along with apples, bananas, coffee, and wheat were introduced in the New World in what is popularly known as "Columbian exchange." Mangoes from India, when introduced in Brazil, spread rapidly. Mangoes of India are often juicy and of "sucking type" (one of its names in Sanskrit is "rasaal" which means juicy).

The Portuguese, who were exporting vast amounts of mangoes to Europe, thought of having a variety which could be "served on the table." Jesuits of Goa are believed to have introduced modern techniques of grafting and succeeded in developing many new cultivars of mangoes, which were given Portuguese names. This also resulted in the development of the greatly popular Alfonso mangoes (Fig. 7.23), named after the Viceroy

Fig. 7.23. Alphonso Mangoes: Created by Jesuits in Goa.

of Portuguese India, Alfonso de Albuquerque (1453–1515 CE), which are prized for their taste, fragrance, vibrant saffron colour, rich and creamy texture, and delicate, non-fibrous, juicy pulp that's smooth and buttery. Jesuits are also believed to have imparted the knowledge of modern grafting techniques to Mughal India, where it flourished.

Similarly, peas, black beans, beetroot, carrot, squash, lettuce, tomato, cauliflower, etc., and a great many varieties of fruits and flowers were introduced in India by the British. Many fruits, vegetables, cereals, and trees were earlier introduced by Arabs and Central Asians. All these introductions required teaching/learning and reorientation of the art and science of their cultivation to which Indian farmers rose quickly and magnificently.

The British also saw to clearing of large tracks of forests to plant tea, coffee, and rubber, in addition to exploiting them for timber. The story of the forced cultivation of indigo, opium, and other cash crops, instead of cereals, is well known and is known to have led to a series of devastating famines in various parts of India [27].

The knowledge and education of food production have been exchanged by mankind since time immemorial, and these have spread even after nations tried to stop it to protect their commercial interest, as in the case of coffee (Arabs) and silk (Chinese). It is known that Vasco de Gama could sail to India only because a sailor told him about the monsoon, which was a south-westerly wind that blew from East Africa to India in the summer (between May and September) and then abruptly reversed itself and became a north-easterly wind that blew from India to Africa in the winter (between October and April).

Before closing this part of the discussion, we recall that the humid climate of India has never been ideal for breeding horses (except in Rajasthan where Marwari horses were reared), while Central Asia and Arabia provided ideal conditions for them. Thus, India had been importing horses from ancient times. We have seen that the horses from Central Asia were mostly paid for with slaves from India, at least during the Islamic Rule in India. The horses from Arabia were mostly brought by ship to southern India (and thus called Saindhav, from the sea), while those from Central Asia (called Turki horses in India) were brought overland. The Vijayanagar Empire (1336–1646) owed its supremacy to the regular import of Arabian horses, yet it is said that to maximise the trade, Arabs kept a very strict control on horse doctors who were not allowed to travel to India, while the Empire even paid for horses which died during the journey.

7.14 Recording the Fauna

We start by recalling that Ustad Mansur (working during the reign of Akbar and Jahangir) and his colleagues made wonderfully detailed paintings of animals and birds. Mansur is also credited with the most accurate depiction of the flightless bird "dodo" from Mauritius from a living specimen (Fig. 7.24).

The Indian Museum was opened in 1875, and it continued to collect archaeological, zoological, and geological samples and preserve and catalogue them as part of the activities of the Asiatic Society. The Zoological Survey of India was formally started in 1916, with the famous zoologist Thomas Nelson Annandale as its first director to promote surveys,

Fig. 7.24. Dodo (centre) by Ustad Mansur (1625 CE).

Fig. 7.25. Spotted Dove (*Turtur suratensis*) [29].

exploration, and research, leading to the advancement of our knowledge of various aspects of the exceptionally rich animal life of India.

The Fauna of British India was a series of scientific books (Fig. 7.25) published by the British government in India starting in 1881. The vast literature has now been digitised and is made available by the Zoological Survey of India [28]. The British Rule in India was also responsible for its large-scale hunting of wild animals, especially tigers, leopards, and cheetahs, which faced extinction by the time India became independent. Unfortunately, copying their lifestyles and "adventure," the elites of independent India continued this process till the cheetah was completely extinct in the country, and the number of tigers was greatly reduced. Fortunately, at least the number of tigers could be revived by Project Tiger launched in 1973.

7.15 Understanding the Weather

One of the first steps at understanding the scientific basis of weather patterns, especially the summer monsoon of India, was taken by Edmond

Halley (1656–1741) who explained it based on differential heating of the continental and oceanic areas in 1686. We shall discuss it in detail, in a later chapter. This initial (classical) description has been brought to a high degree of sophistication now, with capabilities to accurately predict the progression of the monsoon and the formation and movement of cyclones. It has continued to build on observations, now supplemented with data collected by satellites and studies performed over three and a half centuries. The first meteorological observatories were established in India by the British East India Company in Calcutta (1785), Madras (1796), and Bombay (1826). The meteorological studies were also encouraged by the Asiatic Society, and Henry Paddington developed his theories of tropical storms and coined the word "cyclone." Following a devastating tropical cyclone that hit Calcutta in 1864, and famines in 1866 and 1871 due to the failure of the monsoons, the Indian Meteorology Department was established in 1875 at Calcutta. Sir John Eliot took over as the first Director General of Observatories in 1889. The Department was shifted first to Shimla (1905), then to Pune (1928), and finally to New Delhi in 1944.

7.16 Independent India: Hiatus and Acceleration after Early Steps

The progress of education, research, and development in India has recently been studied in detail [20], and several figures to follow are taken from there.

India accorded high priority to education, research, and development right from day one after independence. By 2021, it had more than 40 million students across its universities and other higher educational institutions. India also has about 0.35 million science and technology professionals across its educational and R&D institutions.

Starting with just four central universities (generally better funded and arguably better run) in the 1960s, this number has grown to more than 50 by now (Fig. 7.26). Yet, we see that for close to 40 years, no central universities were added to this list, depriving the meritorious students of the opportunity for quality education.

The setting up of the first five Indian Institutes of Technology along the lines of the Massachusetts Institute of Technology, in the early years after independence, was a major step in engineering education in the country:

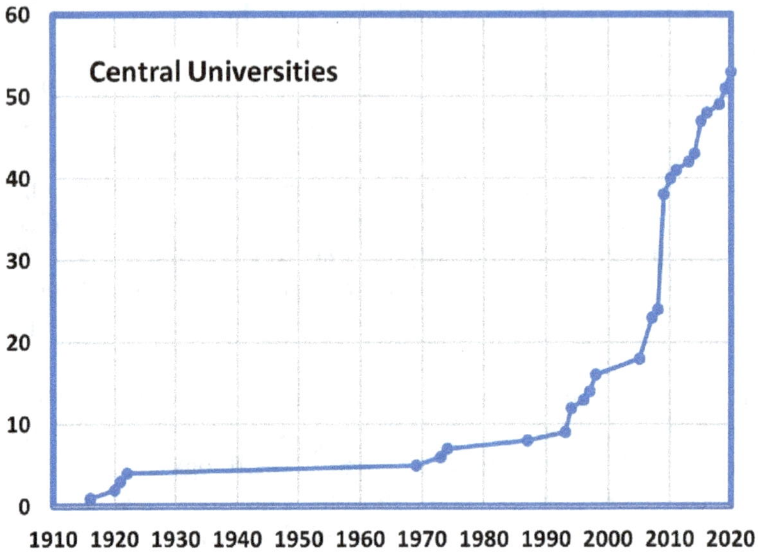

Fig. 7.26. Growth of Central Universities in India [20].

- The first IIT at Kharagpur in West Bengal established in 1951 drew faculty members from the US, the UK, Ireland, France, the USSR, Germany, Norway, Sweden, and Poland.
- The second IIT in Bombay (now Mumbai) was established in 1958 with assistance from UNESCO and funds contributed by the Soviet Union.
- The third IIT at Madras (now Chennai) was established in 1959 with technical and financial assistance from the former government of West Germany.
- The fourth IIT at Kanpur, Uttar Pradesh, was established in 1959 with the assistance of a consortium of nine US research universities as part of the Kanpur Indo-American Programme.
- The fifth IIT in New Delhi was established in 1961 with assistance from the UK.

These developed into the most prestigious engineering colleges in the world, known for their stringent standards of selection and education, rigorous academic programmes, and excellent career opportunities. Unfortunately, the total number of seats offered was limited to less than 3000 every year, by the time all the five IITs became operational. This led to a very tough competition for the very large number of meritorious students coming out of the rapidly expanding secondary education programme of the country and encouraged the mushrooming of very expensive coaching centres (now even operating online), which took away the joy of learning and put students under extreme pressure, some of them even failing to be able to bear with it.

These IITs have been blamed for encouraging brain drain. At the same time, it has also been suggested that the Indian industrial scene did not expand rapidly enough to accommodate these brilliant students.

No new IITs were opened for over 30 years, leaving a large number of meritorious students frustrated and forced to join the second-rung engineering colleges, many of which charge(d) very high capitation fees, keeping them out of reach of deserving but financially weak students, and yet do (did) not have "at par" levels of education and training. Only now, several new IITs have been opened (Fig. 7.27). However, we must have many more seats in them to provide a sufficiently large number of highly skilled engineers and scientists to meet the growing need for a knowledge economy and highly innovative production using high technology in a rapidly advancing country.

When India started establishing many large public sector undertakings for the industrial development of the country, it immediately faced a severe shortage of professional managers. George Robbins of the University of California, invited at the suggestion of the Planning Commission in 1959, advised setting up institutes of management for

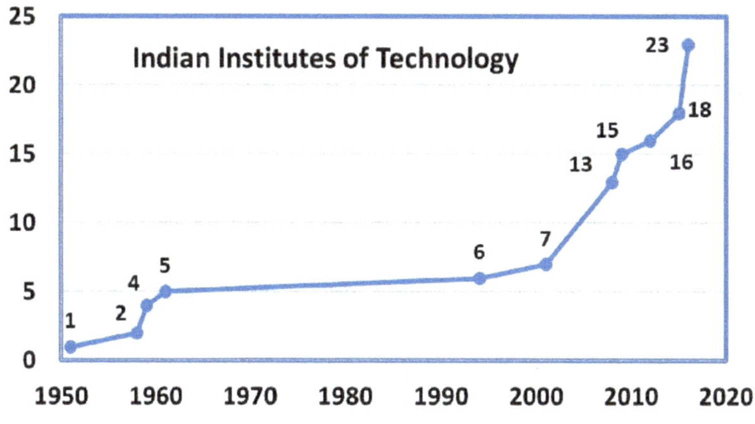

Fig. 7.27. Growth of Indian Institutes of Technology in India [20].

Fig. 7.28. Growth of Indian Institutes of Management in India [20].

providing training in essential skills in managing businesses and industries.

The first Indian Institute of Management (IIM) was established in Calcutta in 1961 in collaboration with the MIT Sloan School of Management, the Government of West Bengal, the Ford Foundation, and the Indian industry.

The second IIM at Ahmedabad was also established in 1961, with the active support of the Government of India, the Government of Gujarat, Harvard Business School, and prominent members of the Indian industry. Space scientist Vikram Sarabhai and businessman Kasturbhai Lalbhai played pivotal roles in setting up the institute.

Modern industries need to coordinate supplies and technologies from across the world, in intricate ways. This requires a modern management, alert to the global nature of any development today. Yet, India delayed the opening of the third IIM by 12 years and the fourth IIM by 11 more years, etc. (Fig, 7.28). The vast neglect of a large number of competent young men has possibly limited India's growth.

All this has resulted in limiting the number of researchers in the country (Fig. 7.29). A much larger number of them could have created a much more vibrant culture of innovation, discovery, and industry [20]. The same scarcity of seats in affordable-quality medical colleges had badly affected the number of medical practitioners in the country.

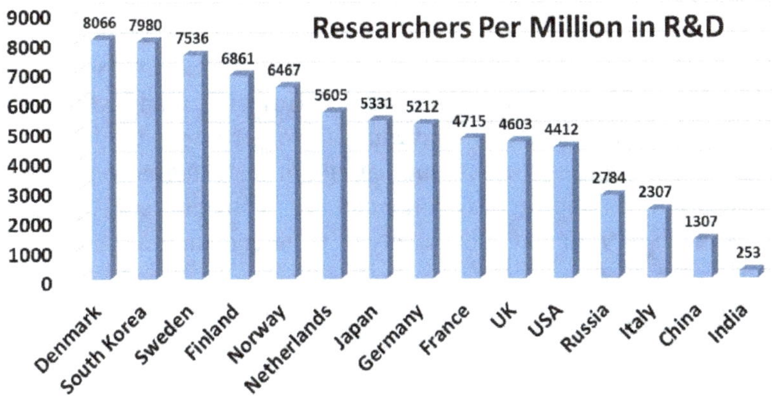

Fig. 7.29. Number of Researchers per Million Population in Some Selected Countries (*Source*: UNESCO) [20].

When a large number of National Institutes of Technology and private engineering colleges, which were opened in India, met the growing demands of engineering seats in the country (though not necessarily at the levels of education available to students at IITs), it led to a big drop in students interested in basic sciences, for which a need for a teaching-cum-research institute like Indian Institute of Science established way back in 1909 was felt. To meet this pressing demand and to provide an opportunity for research to brilliant students already at the undergraduate level, setting up of several Indian Institutes of Science Education and Research (IISER) was envisaged. Thus, the first two IISERs were set up in Kolkata and Pune in 2006. Next, IISERs at Mohali (2007), Bhopal and Thiruvananthapuram (2008), Tirupati (2015), and Berhampur (2016) were established. A sister institute, the National Institute of Science Education and Research, was opened in Bhubaneshwar in 2006, by the Department of Atomic Energy. It also offers a degree course in MSc in Medical and Radiological Physics Programme. These institutes have already started playing an important role in education and research in all aspects of basic sciences.

We hope that considering the size and the population of the country and the large number of meritorious and aspiring students, many more institutes of excellence of all hues are started so that no talent goes unnurtured and gets deprived of attaining its fullest intellectual and aspirational potential.

7.17 Indian Students Going Abroad

We have seen that during colonial times, Indian students, who could afford it, went to England for higher studies. However, many went for the study of law or medicine, as the avenues for other jobs were limited. In the past several decades, the extreme scarcity of seats in good engineering colleges and universities led to a large number of students seeking education abroad (Figs. 7.30 and 7.31), many of them choosing not to return to

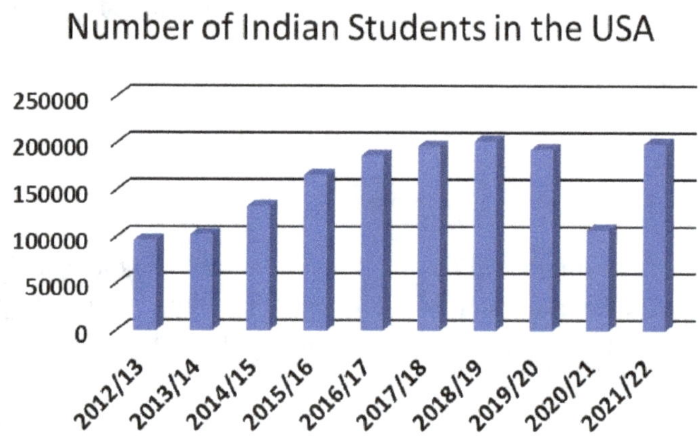

Fig. 7.30. The Number of Indian Students in the USA (*Source*: Opendoors).

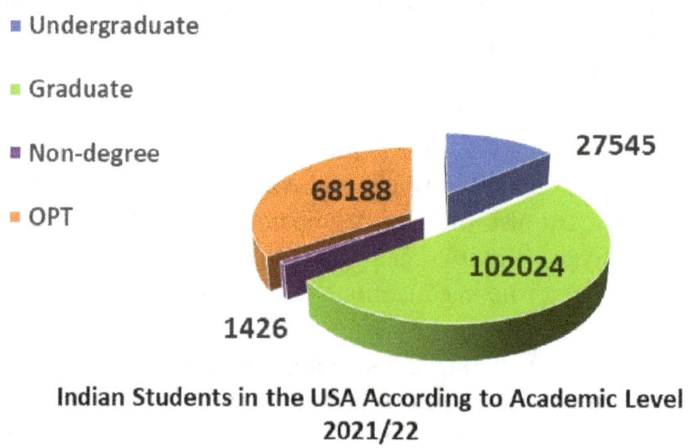

Fig. 7.31. The Number of Indian Students in the USA According to Academic Level (*Source*: www.opendoors.org).

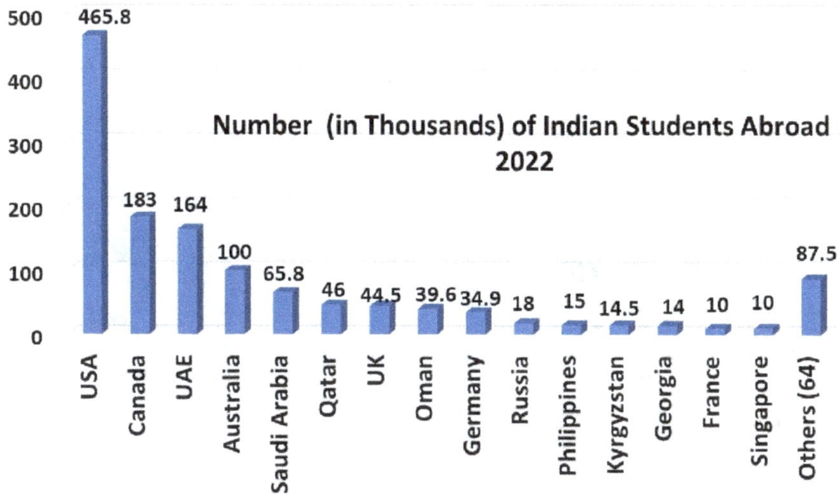

Fig. 7.32. Indian Students in Different Countries in 2022.

the country, adding to the brain drain. This has completely negated the advantage that their exposure to international education in a multicultural environment and modern technology could have brought to the country.

Another emerging trend among students going to the USA for higher education is that many are getting there already at the undergraduate level.

Before independence, most Indian students went to the UK for higher education. After the independence, the number of students going to the UK decreased and those going to the USA increased. In 2022, according to the Ministry of External Affairs, 1.3 million students (Fig. 7.32) were studying in 78 countries (excluding Ukraine, from where they had to be evacuated earlier because of the war).

As we had seen earlier, India has been attracting international students for centuries. This trend has continued and according to the All India Survey of Higher Education [30], up to 50,000 of them from over 170 countries travel to India every year for education (Fig. 7.33). This number is expected to increase as India improves its infrastructure for higher education.

Several prestigious universities and engineering colleges from India have opened their international campuses abroad. India is encouraging foreign universities of repute to open their campuses in India. This will give a whole new meaning to the advantages of international education

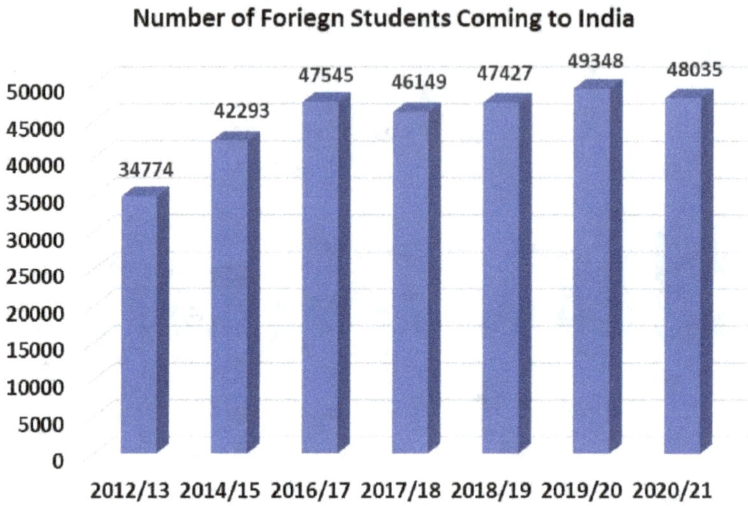

Fig. 7.33. Foreign Students Coming to India for Education.

and will benefit everyone by preparing our students to work in a truly global atmosphere by exposing them to a multicultural environment.

It is rather important to know the areas of study for which Indian students are going abroad (Fig. 7.34). We immediately see that most students are going for engineering, maths and computer science, and business management, the very same three areas for which we still do not have enough seats in premier institutions.

It has also been estimated that Indian students spent 25–35 billion USD on education abroad in 2019, and that this number may touch 75–85 billion USD in 2024 when the number of Indian students going abroad for education is expected to swell to 1.8 million or more. This increasing number of students going abroad for education has been variously attributed to increased awareness about the benefits of studying abroad, improvement in the finances of Indians due to the improving economy of the country, high standards of living abroad, and, as mentioned earlier, limitations of the education infrastructure of India. It should not surprise us that several universities from across the world actively woo Indian students by holding educational fairs in India.

On the other end, about 1 million students from 200 countries were studying in the USA (Fig. 7.35) during the academic year 2021–2022. Public universities in the USA generally charge higher tuition fees from

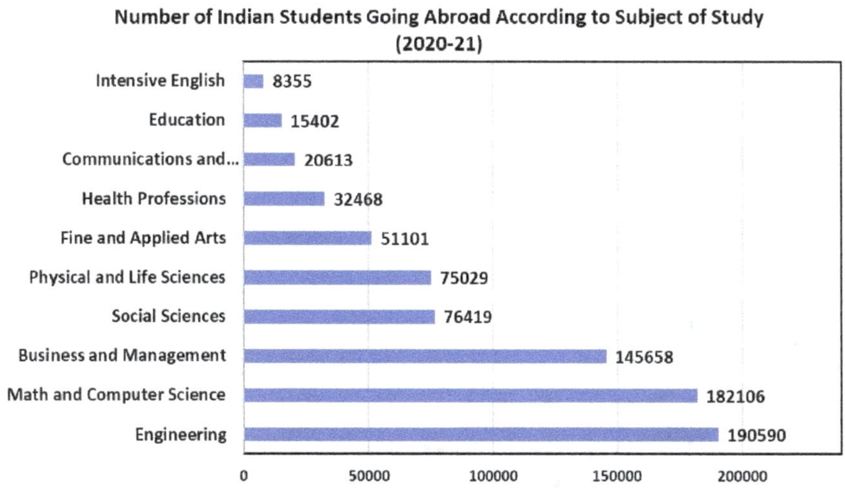

Fig. 7.34. Areas of Study of Indian Students Going Abroad (2020–2021) (*Source*: Oxford International Digital Institute).

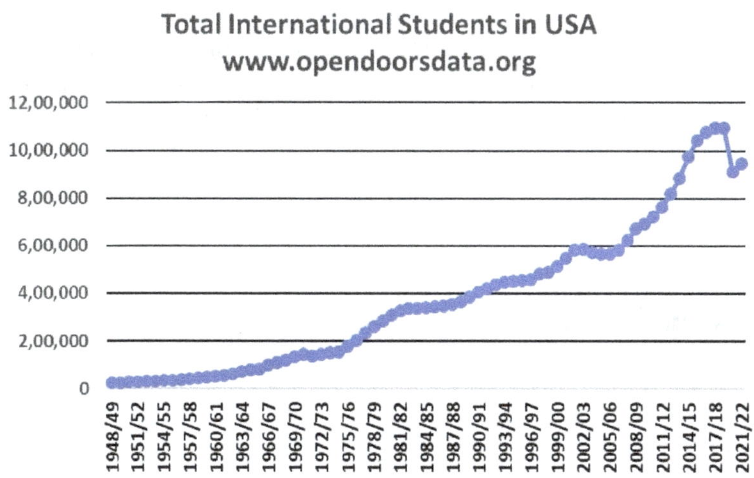

Fig. 7.35. International Students in the USA.

international students and their economic impact was about 34 billion USD.

There were about 700,000 international students, with students from China, India, and Nigeria accounting for about 45% of them, in the academic year 2021–2022 in the UK. They accounted for about 25% of the

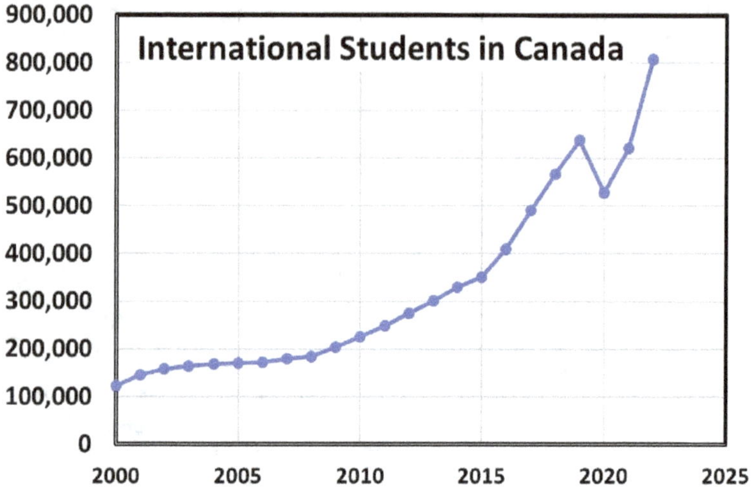

Fig. 7.36. International Students in Canada (*Source*: https://erudera.com).

total student strength in higher education in the UK, paid about 10 billion GBP as tuition fees, and had an economic impact of 42 billion GBP.

Canada is another one of the leading destinations for international students for higher education. It had about 800,000 international students in 2022, out of which close to 300,000 students were from India and 100,000 students were from China (Fig. 7.36). International students contribute about 22 billion CAD every year to the Canadian economy.

7.18 Conclusions

Spanish novelist Miguel de Cervantes (1547–1616), the celebrated author of *Don Quixote*, is credited with saying, "A proverb is a short sentence based on long experience." Proverbs in all the societies of the world celebrate the virtue of travelling for education. It is a pity that Indians, who were famous travellers and explorers, with their ships sailing to the islands of South Asia and Mesopotamia since prehistoric times, developed the "fear of losing their caste" at perhaps an early stage in their history. *Baudhayana Dharma-sutra* (II.1.2.2), composed more than 2,000 years ago, lists this "Samudrolanghana" or "Sagarollanghana" (crossing the seas) as the first of many reasons for the "loss of caste." This especially applied to Brahmins, the traditional teachers in India for millennia, as there was fear that travel abroad prevented a Brahmin from performing various rites and rituals in the prescribed manner at the prescribed time. The early travellers to England were severely criticised and advised/forced to undergo "purifying" rituals to be readmitted to their caste. This fear of losing caste is believed to have been used by the British to give the ultimate punishments to the rebels of the 1857 mutiny, by imprisoning them on the Andaman Island, which involved crossing the sea.

We wonder if this limited the timely exposure of our people to developments going on across the world, except for what they came to know from wave after wave of invaders. If true, this may also explain why we lagged in adopting technological developments sweeping the West, which ushered in the Industrial Revolution, as a precursor to colonialism which was to ravage India.

Even during the Mughal rule, with all its glitter, pomp, and splendour, India somehow had not responded creatively to the stimulus of resurgent Europe, which had reached out across great oceans and linked up with India, even before the Mughals arrived (1526 CE) on the scene. Many historians have argued that there was no ferment of ideas, and curiosity about the science and technology of Europe did not go beyond the dilettantism of a few Amirs. Except for a few minor gadgets, India showed little interest in European inventions; even in military technology, the adoption of improvements was slow and haphazard.

It is a great pity. The famed historian Abraham Eraly (1934–2015 CE) noted that during this very period, Europe had broken free and started to go adventuring across far horizons, to the ends of the Earth and the confines of the mind. He further argued that Columbus had discovered America when Babur was still a lad. By the time Babur arrived in Agra, Magellan's Expedition had circumnavigated the world, and the Portuguese had already been in India for a quarter of a century. Copernicus, Michelangelo, Leonardo Da Vinci, and Machiavelli were Babur's contemporaries, and Shakespeare, Kepler, and Galileo were the contemporaries of Akbar and Jehangir, Newton of Aurangzeb. We can only speculate how India could have benefitted from the exchange of knowledge and education with Europe during this period.

Getting back, we note that now it is estimated that there are 281 million migrants across the world, out of which about 18.68 million are Indians and 13.45 are non-resident Indians. We have already seen that 1.3 million Indian students were studying in 78 countries in 2022.

This exposes our people to massive international exposure and international education.

By now, curricula in sciences in most colleges and universities are at the same level as in most reputed universities across the world. This allows for a seamless movement of bright students to different countries for study and research. This has also made it easy for scientists and engineers to collaborate with their counterparts in different countries, which is very valuable. Moreover, an international education helps in broadening perspectives and acceptance of diversity and breakthrough ideas. Students with international experience acquire enhanced vocabulary, better etiquette, critical thinking skills, improved confidence, and the characteristics of a global citizen. We shall discuss some specific examples later in this monograph.

We know that before 1750 India and China were the most prominent trading nations and between them controlled 40% of the world's wealth. It has been suggested by some authors that the recent rise of China and India is gradually restoring the global share of wealth to a historic norm that existed before 1750 [31].

Recall that Indian textiles were exported from the time of Indus Valley Civilisation; it was used to wrap Egyptian mummies and to clothe Roman senators (who called it woven wind). The turbans of Central Asia and the Arab world were made of Indian textiles. Before the Europeans arrived by sea, most of this export was to Central Asia and overland, and often followed the so-called Silk Road, though sea trade was also common. The Silk Road (which existed for millennia, before it acquired that name), due to its busy traffic of monks, merchants, artists, astronomers, and scientists, is often described as creating "an empire of ideas."

Gurcharan Das has argued that [31], "After the Europeans 'discovered' India, trade with European countries surged, especially in the seventeenth and eighteenth centuries. Indian textiles, in particular, flourished and Indian textile producers fulfilled a quarter of the world's demand. This created jobs for tens of millions — from farmers who grew cotton and natural dyes, to spinners who turned cotton into yarn, to weavers who turned yarns into bolts of cloth, to printers who adorned the cloth, and so on to transporters, wholesalers and retailers — who sold it to customers. .... Only in the early nineteenth century did this picture change when the machines of the Industrial Revolution began to supplant handlooms."

Das even argues that "it is a mistake to believe that the British Raj destroyed the lives of India's textile workers. The truth is that handlooms died around the world and the vast number of workers associated with textiles became a victim of technological obsolescence. ….. India's mills would have destroyed handlooms, and in fact, they did just that after the second half of the nineteenth century. … Business is dynamic and old technologies constantly die as new ones are born. A country should not close its borders …. in the belief that it cannot compete with the world. When (India) opened its borders in 1991, its software and outsourcing industry rose to the occasion and responded brilliantly to the opportunity offered by globalisation. It was the stimulus of international demand for software and remote services from India that brought the country new respect, helped lift its economic growth rate, and raised millions into the middle class" [31].

International education, ideally, is poised to keep nations always at the forefront of development and take advantage of every technological advance, immediately. Ideally, it can also generate a more democratic, fair, and equal world and lead to vibrant international cooperation.

We must add that, the practice of the internationalisation of higher education has been called, "the process of commercialising research and postsecondary education, and international competition for the recruitment of foreign students from wealthy and privileged countries to generate revenue, secure national profile, and build an international reputation" [32].

We close this by adding that a paradigm shift has taken place in the internationalisation of education with the advent of the Internet, which has made the videos of lectures and even live lectures by the finest teachers at the finest universities available to all students across the world. A great many of these teachers even encourage or entertain email exchanges with students, which often leads to the start of a valuable collaboration, even before the teacher and the student meet in person!

Literature from across the world is full of praise for knowledge and those who impart knowledge. The teachers, use a process called education. Education is based on exchange, and the exchange of knowledge has brought us to where we are.

In almost all societies, education was often denied to a large body of the population, including women. This unfortunately continues even now and even when education is available to all, in principle, often only the rich can afford it. We have seen that it not only frustrates a large body of meritorious students but also deprives the world of their possible contributions. This needs to be corrected so that mankind can reach the zenith of its possibilities and capabilities.

Yes, education and knowledge have evolved over the ages, and yes,

"न चौरहार्यं न च राजहार्यं
न भ्रातृभाज्यं न च भारकारि।
व्यये कृते वर्धत एव नित्यं
विद्याधनं सर्वधनप्रधानम्॥"

"Thieves cannot steal it (the knowledge). Kings cannot snatch it away. It cannot be divided among brothers. It does not cause a burden. If spent, it indeed always keeps growing. The wealth of knowledge is the most superior wealth of all."

Nothing is more satisfying and rewarding than the transmission of knowledge and education. Recall the African proverb, which we shall invoke again, "It takes a village to educate a child." We are living in a global village now, and we must enjoy all the benefits that an international collaboration entails and promises. And we shall see later that "Human history becomes more and more a race between education and catastrophe," as remarked by H. G. Wells (1866–1946 CE). We need to remember that education is also the main equalising force — both between and within countries — by stimulating the diffusion of knowledge and skills.

No wonder that, one of the greatest thinkers of the last century, Rabindranath Tagore (1861–1941 CE) adopted the motto from the Shukla Yajur Veda, *Madhyandina Samhita* (Chapter 32, verse 8):

"यत्र विश्वं भवत्येकनीडम्"

i.e., "Where the entire world becomes a nest," when he established his university Visva Bharati at Shantiniketan, as mentioned earlier. It was his opinion and dream that "Visva Bharati represents India where she has her wealth of mind which is for all. Visva Bharati acknowledges India's obligation to offer to others the hospitality of her best culture and India's right to accept from others their best."

References

[1] https://www.worldhistory.org/article/2144/schooldays-sumerian-satire-the-scribal-life/, See also, Y. N. Harari, "*Sapiens: A Brief History of Humankind*" (New York: Harper, 2015).

[2] G. Gutherz, S. Gordin, L. Sáenzb, O. Levya, and J. Beranta, "Translating Akkadian to English with Neural Machine Translation", *PNAS Nexus* 2 (2023) 1–10, https://doi.org/10.1093/pnasnexus/pgad096.

[3] M. Danino, "*Lost River: On the Trail of the Sarasvati*" (India: Penguin, 2010).

[4] https://www.harappa.com/seal/1.html.

[5] https://peepalinspires.com/samavartana-smaavrtn-samskara-the-celebration-of-cognitive-activities/.

[6] R. Narasimha, "On the NIAS Logo" (Bengaluru: National Institute of Advanced Studies (Special Publication), 2021), NIAS/HUM/MHI/U/SP/08/2021.

[7] K. Plofker, "*Mathematics in India*" (New Jersey: Princeton University Press, 2001), p. 17.

[8] "An invocation of the seventeen great sagely adepts of glorious Nalanda", by His Holiness the Dalai Lama (December 2001), https://thubtenchodron.org/2001/12/nalanda-prayer/.

[9] V. Deshpande, "Ancient Indian Medicine and Its Spread to China", *Economic and Political Weekly* 36 (2001) 1078–1081.

[10] J. Kieschnick, "*The Impact of Buddhism on Chinese Material Culture*" (Princeton: Princeton University Press, 2003).

[11] Recall, e.g., the immensely popular song of Rabindranath Tagore, "Purano Shei Diner Katha", Inspired by Scottish bard Robert Burns' "Auld Lang Syne" (Old Long Since).

[12] H. Feldmann, "The Long Shadows of Spanish and French Colonial Education", *KYKLOS* 69 (2016) 32–64, https://doi.org/10.1111/kykl.12102, and references therein.

[13] C. Laye, "*L'Enfant noir*" (Paris: Plon, 1953), Translated by J. Kirkup and E. Jones, "*The Dark Child*" (New York: Noonday Press, 1954).

[14] N. C. Chaudhuri, "*The Autobiography of An Unknown Indian*" (McMillan, 1951), London.

[15] A. Sen, "*Home in the World: A Memoir*" (London: Penguins, 2021).

[16] P. Mohan, "*Wanderers, Kings, Merchants: The Story of India through Its Languages*" (Viking India, 2021), Gurgaon.

[17] L. Riberio, "A Review of the Educational System in Portuguese Goa", *Kruti* 2 (2016) 49–60, https://carmelcollegegoa.org/wp-content/uploads/naac/criteria/3/3.3.2/16-17/a_review_of_the_educational.pdf; V. Coutinho, "Goa's History of Education: A Case Study of Portuguese Colonialism", PhD thesis (Chicago: Loyola University, 1975). https://ecommons.luc.edu/luc_diss/1568.
[18] F. Anjum, "Education and Learning in Mughal India: A Critical Study of Colonial Perceptions", *Journal of the Research Society of Pakistan* 55 (2018) 83.
[19] See e.g., S. Sangwan, "Science Education in India under Colonial Constraints, 1792-1857", *Oxford Review of Education* 16 (1990) 81–95, and references therein.
[20] V. S. Ramamurthy and D. K. Srivastava, "How Long Is the Long Path for Developed India?" (Bengaluru: National Institute of Advanced Studies). Report No. NIAS/NSE/U/RR/11/2023.
[21] V. Yan, "*Jenghiz Khan*" (Moscow: Raduga Publishers (English translation), 1989).
[22] S. C. Levi, "Hindus beyond the Hindu Kush: Indians in the Central Asian Slave Trade", *Journal of the Royal Asiatic Society* 12 (2002) 277–288, and references therein.
[23] "Chapter 9 — Timur's Account of His Invasion of India and Sack of Delhi". In: A. V. Williams Jackson (ed.) "*History of India: The Mohammedan Period as Described by Its Own Historians*", The Grolier Society, London (1907) Vol. 5, https://www.ibiblio.org/britishraj/Jackson5/chapter09.html.
[24] K. S. Murphy, "The Slave Trade and Natural Science" (2016). DOI: 10.1093/OBO/9780199730414-0266.
[25] A. Haley, "*Roots: The Saga of an American Family*" (New York: Doubleday, 1976).
[26] J. D. Hooker, "*A Sketch of the Flora of India*" (London: Eyre and Spottiswoode, 1904), W. Roxburgh, "*Plants of the Coast of Coromandel*", in three large folio volumes (London: W. Bulmer and Co., 1795, 1798, and 1819), "*Hortus Bengalensis*" (Serampore, 1814), and "*Flora Indica*" (Serampore, 1832), available at https://www.biodiversitylibrary.org/. [The original set of Roxburgh's drawings made between 1780 and 1815, by Indian artists, and consisting of 2532 drawings with Central National Herbarium, Botanical Survey of India, Howrah.]
[27] V. S. Ramamurthy and D. K. Srivastava, "*Art and Science of Managing Public Risks*" (Singapore: World Scientific Publishing Company, 2022).
[28] https://faunaofindia.nic.in/; see also https://www.biodiversitylibrary.org/.

[29] W. T. Blanford (ed.) "*The Fauna of British India, Including Ceylon and Burma, Vol. 4: Birds*" (London: Taylor & Francis, 1888).

[30] https://www.aishe.gov.in/aishe/home.

[31] G. Das (ed.) "Introduction", S. C. Levi "*Caravans: Punjabi Khatri Merchants on the Silk Road*" (India: Penguin, 2016).

[32] A. K. Taskoh, "A Critical Policy Analysis of Internationalization in Postsecondary Education: An Ontario Case Study", PhD thesis (University of Western Ontario, 2014), https://ir.lib.uwo.ca/etd/1933.

[33] D. Arnold, "*Famine: Social Crisis and Historical Change*", New Perspectives on the Past (New York: Wiley-Blackwell, 1991).

Chapter 8

Where No One Has Gone Before: Space Exploration

> "The Earth is the cradle of humanity, but one cannot live in the cradle forever."
>
> — Konstantin Eduardovich Tsiolkovsky

8.1 Introduction

Humans have, forever, looked at the skies with fascination and wondered about stars, planets, and constellations. They even attributed forebodings to shooting stars and appearances of comets and solar and lunar eclipses, and imagined celestial beings in the arrangements of constellations. Over time, they accumulated enough observations to discern the systematics in them to take the first tentative steps and sow the seeds of astronomy.

In more recent times, the curiosity, interest, and excitement of generations of children and grown-ups from across the world in space exploration have been stimulated by excellent writers of science fiction like Isaac Asimov and Arthur C. Clarke, and science communicators like Carl Sagan. This interest has been kept alive by numerous films, radio plays (recall the radio drama, *The War of the Worlds*, based on the novel of H. G. Wells, first broadcast in 1938, and the excitement/panic it caused), and the television series *Star Trek* along with its sequels, which started in 1966, which chronicled the "Journeys of the Spaceship Enterprise" and opened with the journal entry,

> *"Space: the final frontier. These are the voyages of the Starship Enterprise. Its five-year ('continuing' in sequels) mission: to explore strange new worlds; to seek out new life and new civilisations; to boldly go where no man ('one' in sequels) has gone before!"*

delivered by William Shatner playing the role of Captain James T. Kirk in the original series.

There are suggestions that the phrase, "*where no one has gone before,*" was taken from *Introduction to Outer Space*, a White House booklet, published in 1958, to garner support for the space programme to be started in the face of the stunning success of the USSR with the Sputnik flight of 1957 and stated:

> *"The first of these factors is the compelling urge of man to explore and to discover, the thrust of curiosity that leads men to try to go where no one has gone before. Most of the surface of the earth has been explored and men now turn on the exploration of outer space as their next objective."*

During the time that has passed since the launch of the first artificial satellite in 1957, astronauts have travelled to the Moon, probes have explored the solar system and beyond, instruments in space have discovered thousands of planets around other stars, and several tens of thousands of satellites go around the Earth in low Earth orbits. These successes, — which started as a competition — mostly between the USSR and the USA and graduated into cooperation and collaboration, are built upon the dreams of science fiction, visions of space exploration, science, and mathematics worked out by people from across the world, and the technological efforts of thousands of engineers and technicians. The contributions of Konstantin Eduardovich Tsiolkovsky and Wernher Von Braun easily shine most brilliantly on the firmament of space exploration for providing the initial visions.

8.2 Konstantin Eduardovich Tsiolkovsky

Konstantin Tsiolkovsky (Fig. 8.1) was born to a Polish father and a Volga Tatar/Russian mother in 1857, who had relocated to Russia. He was 10

Fig. 8.1. Konstantin Eduardovich Tsiolkovsky (1857–1935 CE).

when he suffered a severe bout of scarlet fever which left him with a near-total hearing loss. It tragically led to the termination of any formal education for him. He became a reclusive, self-taught child, who spent much of his time reading books. Seeing his interest, he was sent to Moscow, where he spent 3 years at Moscow Library and was formally tutored, with the help of an ear trumpet, by a leading intellectual of that time. Later, he became a high school teacher because of his proficiency in mathematics and the sciences and lived and taught in Kaluga, near Moscow.

When he was still a teenager, he began to contemplate the possibility of space travel. Tsiolkovsky was the first person to determine the escape velocity from the Earth into orbit, which could be achieved by using a multistage rocket fuelled by liquid oxygen and liquid hydrogen.

His "Tsiolkovsky Formula," also known as the rocket equation, established the relationships between rocket speed, the speed of the gas at the exit, and the mass of the rocket and its propellant.

During his lifetime, he published over 500 works on space travel and related subjects. He was inspired by Jules Verne, the author of *From the Earth to the Moon* (1865) and *Around the Moon* (1870), and published several science fiction novels and wrote extensively about the future of space travel, including the possibility of colonising other planets. His most famous work, *The Exploration of Cosmic Space by Means of Reaction Devices*, was published in 1903, but this theoretical work was not widely influential outside Russia.

In 1918, he was made a member of the Soviet Academy and in 1920, he received a government pension, enabling him to devote all his time to research and writing. Konstantin Tsiolkovsky is generally considered the Father of Astronautics and Rocket Dynamics.

His main works include the theory of multistage rockets, which is still used today; the calculation of the escape velocity from Earth's gravity, which is necessary for any spacecraft to reach the orbit mentioned earlier; the proposal to use liquid oxygen and liquid hydrogen as rocket propellants, which are still the most efficient fuels used in rockets today; and the concept of a space elevator. He proposed the idea of space stations (Fig. 8.2) and touched upon concepts such as artificial gravity, airlocks, and life support systems that would become fundamental to later space missions.

Konstantin Tsiolkovsky, in a letter to the Polish science fiction author Jerzy Żuławski (1874–1915) in 1911, wrote, "The Earth is the cradle of humanity, but one cannot live in the cradle forever."

This along with his monumental theoretical works about space travel set generations of brilliant scientists and engineers on the course to space exploration. Tsiolkovsky's work was highly influential in the development of the Soviet space programme. His ideas were studied by many of the leading Soviet rocket engineers, including Sergei Korolev and Valentin Glushko, who often referred to him as "The Father of Soviet Rocketry."

Tsiolkovsky's legacy lives on, and he is considered one of the most influential figures in the history of space exploration, recognised for his pioneering contributions to the theoretical understanding of rocketry and his foresight into the possibilities of human space exploration.

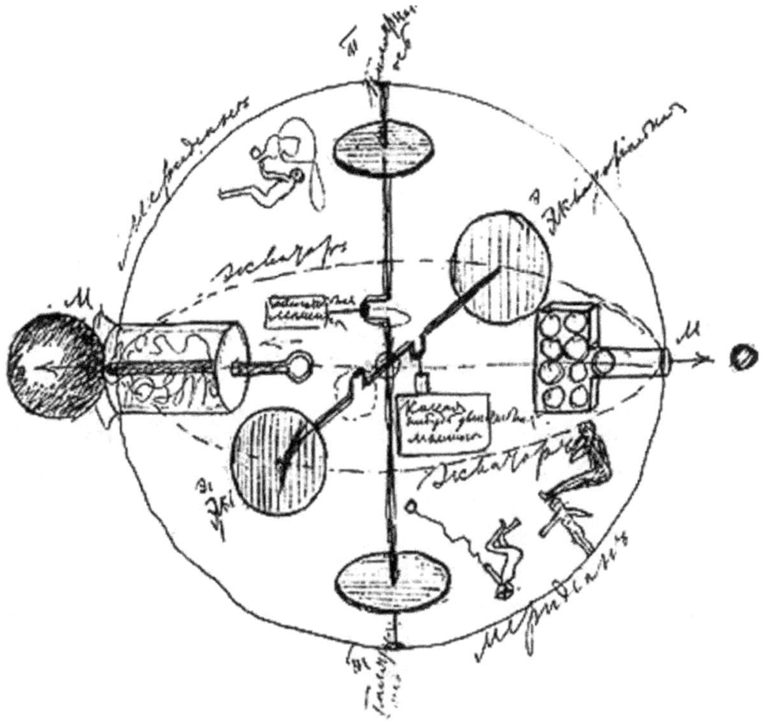

Fig. 8.2. Drawing of a Space Station by Konstantin Tsiolkovsky.

8.3 Wernher Von Braun

Wernher Von Braun (1912–1977 CE) was a German engineer, who played a prominent role in the development of all aspects of rockets and space travel, initially as a defence scientist in Nazi Germany and then in the USA. He was inspired by the work of Herbert Goddard of the USA and always maintained that "Until 1936, Goddard was ahead of us all."

Robert Hutchings Goddard (1882–1945 CE), the famous American Professor and inventor, is generally considered by many in the USA as "The Father of Rocketry" for his extensive studies on the science and engineering of rockets and for the development of rocket motor using liquid fuels (liquid oxygen and gasoline or ethanol) in 1925, which had a great influence on the design of the V-2 rockets of Von Braun in Nazi Germany, 15 years later.

Von Braun obtained a Ph.D. in 1934 from the University of Berlin, on his dissertation titled, "Design, Theoretical and Experimental Contributions to the Problem of the Liquid Fuel Rocket," on a grant from the German Army. Its contents were deemed so important to the future of Germany's military that it was hidden under a new title, "Regarding Combustion Experiments," and transferred to the Army Ordnance's custody and published only in 1960.

Several thousand V-2 (Vengeance Weapon 2) rockets — the world's first ballistic guided missiles, produced under his supervision — mostly employing prisoners from Nazi concentration camps, were used during World War II to bombard cities in the United Kingdom, Belgium, and France.

Von Braun and the members of his rocket development team surrendered to the US forces at the end of the war and were brought to the USA under Operation Paperclip. Soon after that, he first worked for the US Army and then as the Director of the National Aeronautics and Space Administration's Marshall Space Flight Centre. He oversaw the launch of the first US satellite, Explorer-1, using the Redstone rocket based on expanded designs of the V-2 rocket on 31 January 1958. In May and July of 1961, his Redstone rockets launched the first two American astronauts into space during Project Mercury. He led the development of large space launch vehicles, Saturn I, IB, and V. The Saturn I was the largest rocket ever built till 1961, the Saturn IB launched (1966) the Apollo spacecraft into Earth's orbit (1966), and the massive Saturn V transported the Apollo 11 module to the Moon in 1969.

Von Braun was an excellent communicator and wrote two popular books, *Conquest of the Moon* (1953) and *Space Frontier* (1967), and used his skills in public outreach to raise the interest of the American people in space exploration, where he also discussed his concept of a doughnut-shaped spinning space station (Fig. 8.3), with artificial gravity, housing several hundred to several thousand people.

8.4 The Space Programme of the USSR

The Union of Soviet Socialist Republics (USSR) started its work on rockets in 1921, but these were severely hampered and derailed due to the devastating Second World War. It picked up again from around 1955 and continued till 1991 when the Soviet Union was dissolved. Russia picked

Fig. 8.3. Wernher Von Braun with a Model of his Spinning Space Station.

up from that point onwards and continued to make excellent progress, as we shall shortly see. Like the space programme of the USA which benefited from the expertise of German scientists (along with many finished and semi-finished V-2 rockets and their components) taken away under Operation Paperclip after the defeat of Nazi Germany, the space programme of the USSR also benefited from the expertise of German scientists taken away under the Operation Osoaviakhim. Several V-2 rockets were also taken away or assembled by British scientists, who also launched and evaluated them under Operation Backfire.

The USSR had several firsts to its credit in space programmes. The Sputnik-1 (Fig. 8.4) was launched on 4 October 1957, using a modified version of its Inter Continental Ballistic Missile R7. The USSR launched the first animal (a dog named Laika) into Earth's orbit on 3 November 1957.

Next, it placed the first human, Yuri Gagarin, in space (Fig. 8.5), who spent 108 minutes in a tiny Vostok 1 space capsule, completing an orbit around the Earth on 12 April 1961. Valentina Tereshkova was the first

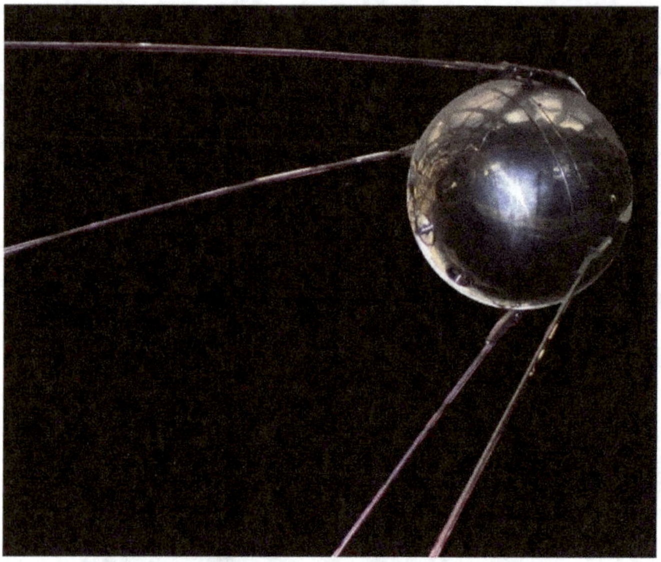

Fig. 8.4. A Model of Sputnik-1.

Fig. 8.5. Yuri Gagarin on the Way to the Historic Launch of Vostok on 12 April 1961 (Credit: NASA Images).

woman in space on 16 June 1963. This was soon followed by the first spacewalk by Alexei Leonov during a mission launched on 18 March 1965.

The launch of Sputnik-1 and Yuri Gagarin's travel to space rattled the USA. On 25 May 1961, the President of the USA, John F. Kennedy, proposed that the US "should commit itself to achieving the goal, before this decade is out, of landing a man on the Moon and returning him safely to the Earth" to prove its technological supremacy.

The USSR, however, continued to strengthen its computerised robotic missions, and its Luna-2 became the first human-made object to touch the Moon on 13 September 1959. On 7 October 1959, the Soviet spacecraft Luna-3 provided the first photograph of the far side of the Moon. Luna-9, launched on 31 January 1966, was the first spacecraft (Fig. 8.6) to achieve a lunar soft landing (a survivable landing) on 3 February 1966, 3 years before Apollo-11's landing, and to transmit photographic data from the Moon's surface to Earth.

The Lunokhod (literally Moonwalker) programme of the USSR achieved the first space rover deployment in 1966 and sent the first robotic probe, Luna-16, that automatically extracted a sample of lunar soil and brought it to Earth in 1970. The Luna-24, launched on 11 August 1976, was the last of the initial series of Luna programmes of the Soviet Union, which returned with lunar regolith (the layer of unconsolidated

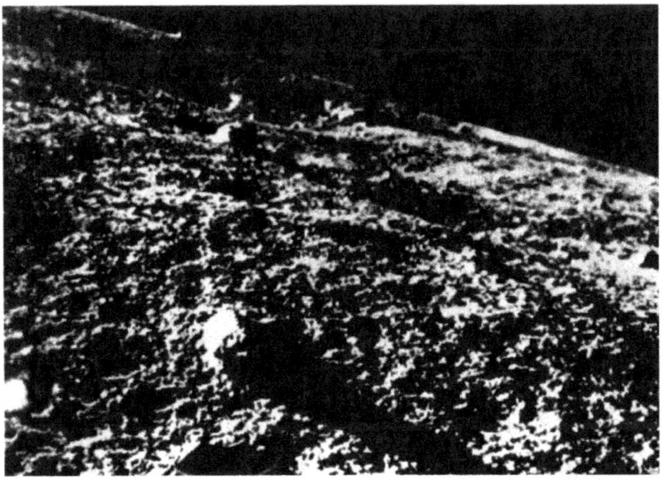

Fig. 8.6. The First-Ever Photograph Taken from the Surface of the Moon (Credit: NASA; Original Image from Luna 9, USSR).

solid material covering the Lunar bedrock). Luna-25 was a Russian lunar lander mission that failed in August 2023. The mission was Russia's first attempt to land a spacecraft on the Moon since 1976.

Under the Venera series of programmes, the USSR sent several probes to Venus from 1961 to 1984, out of which 13 entered the atmosphere and 10 landed on its surface. These also achieved several firsts, e.g., the first human-made devices to enter the atmosphere of another planet (Venera 3 on 1 March 1966), the first to make a soft landing on another planet (Venera 7 on 15 December 1970), the first to return images from another planet's surface (Venera 9 on 8 June 1975), the first to record sounds on another planet (Venera 13 on 30 October 1981), and the first to perform high-resolution radar mapping scans (Venera 15 on 2 June 1983).

The USSR was also the first to reach Mars with Mars 1, launched on 1 November 1962; the first to touch the Martian surface with Mars 2, launched on 19 May 1971; and the first to successfully land a spacecraft on Mars with Mars 3, launched on 28 May 1971.

The Soviet Union further launched the first space station, Salyut 1, on 19 April 1971. The first modular space station, Mir, was launched on 20 February 1986.

The space programme of the USSR was a collaboration programme between scientists and engineers from Russia, Ukraine, Belarus, Kazakhstan, and other regions, who combined their scientific, engineering, and technological resources to achieve significant milestones in space exploration.

8.5 The Space Programme of the USA

We have noted earlier that the US Army had captured and transported many V-2 rockets from Nazi Germany towards the end of the Second World War, along with the most important rocket scientist in Germany, Wernher Von Braun, and his team. The US Army conducted several test firings of these rockets, made many more, and started inviting American scientists to mount experiments for high-altitude research on these.

The International Geophysical Year (IGY) was to be held in 1957–1958 under the auspices of the International Council of Scientific Unions, and the planners of IGY, the Soviet Union and the USA planned to launch an artificial Earth satellite by 1957.

It has also been reported that the USA was keen on developing a technology to gather intelligence on military preparations within the Soviet borders, using an orbiting satellite with cameras and flying them over Soviet territory.

For whatever reason (arrogance?), there was widespread scepticism in the USA about the technical capabilities of the USSR to develop a sophisticated scientific satellite and a rocket powerful enough to put it into orbit. Thus, the successful launch of Sputnik-1 on 4 October 1957 — using the R-7 rocket, which had given the USSR its Inter Continental Ballistic Missile — also the first in the world, surprised and even shocked most of them. A larger second satellite carrying scientific instruments and the dog Laika, the first living creature in orbit, was launched on 3 November 1957, as we saw earlier. An even larger spacecraft, which was originally intended to be the first Soviet satellite, Sputnik-3, carrying a large array of instruments for geophysical research of the upper atmosphere and near space, was launched on 15 May 1958.

The USA succeeded in launching a satellite, Explorer, into space on 31 January 1958, using a more powerful rocket than V-2, called Jupiter or Juno C. One remarkable achievement of this satellite was the discovery of the James Van Allen radiation belts (Fig. 8.7), formed as high energy particles radiated by the Sun are trapped in the magnetic field of the Earth (outer belt) and cosmic rays interact with the atmosphere of the Earth (inner).

In response to the first Sputnik launches, the US government created the National Aeronautics and Space Administration (NASA) to manage

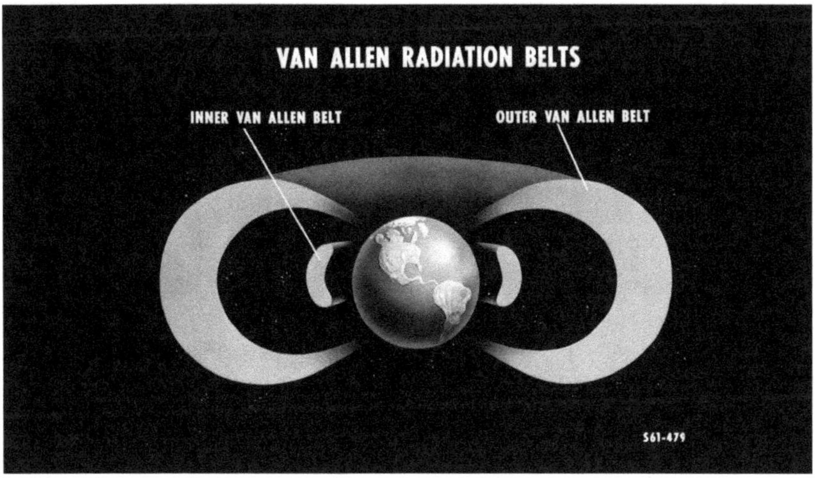

Fig. 8.7. Historic Schematic Image of Van Allen Radiation Belts (Credit: NASA).

the civilian segment of the space programme. NASA began its operation on 1 October 1958 and absorbed the National Advisory Committee for Aeronautics, which had been managing and conducting aeronautics research, experiments, flight tests, and simulations since 1915, the Jet Propulsion Laboratory, and Von Braun's rocket team, and became a formidable organisation covering almost all aspects of space exploration. At its peak (1966–1967), NASA employed more than 30,000 people and cornered 4.5% of the total expenditure of the US government. Even now, it accounts for about 0.5% of US expenditure and employs more than 18,000 people.

The success of the first crewed space flight by Yuri Gagarin, who travelled on Vostok 1 and completed one orbit of Earth on 12 April 1961, was followed by the launch of Alan Shepard, a NASA astronaut, on a suborbital trajectory, a flight that goes into space but does not go all the way around Earth, about 3 weeks later, and the full-scale space race between the two superpowers had started (see later).

Three weeks later, on 25 May 1961, President John F. Kennedy challenged the United States to an ambitious goal, declaring: "I believe that this nation should commit itself to achieve the goal, before the decade is out, of landing a man on the moon and returning him safely to Earth." He repeated and elaborated upon the commitment to space exploration in his speech at Rice University on 12 September 1962.

In what follows, we briefly recapitulate the major initial events in the space race, which culminated in collaboration in space:

- **2 August 1955:** The Soviet Union responded to the US announcement 4 days earlier of its intent to launch artificial satellites for the IGY by declaring that they would also launch the first artificial satellite "*shortly*."
- **4 October 1957:** The USSR successfully launches Sputnik-1, the first Earth-orbiting satellite in history.
- **3 November 1957:** The USSR successfully launches Sputnik-2, carrying a dog named Laika into space and becoming the first nation to successfully send a living organism into orbit.
- **31 January 1958:** The USA launches its first satellite Explorer 1. The programme launched 103 missions by 2023, out of which 99 were successful and contributed to a vast body of results in earth science, heliophysics, and astronomy.

- **1 October 1958:** The National Aeronautics and Space Administration (NASA) was created in the USA.
- **2 January 1959:** The USSR launches Luna 1, which accidentally escapes the orbit of the Moon and orbits the Sun instead.
- **12 September 1959:** The USSR launches Luna 2, the first spacecraft to reach the surface of the Moon.
- **4 October 1959:** The USSR launches Luna 3, which orbits the Moon and photographs the far side of the Moon.
- **12 April 1961:** Yuri Gagarin becomes the first man to reach space aboard Vostok 1, launched by the USSR.
- **5 May 1961:** Alan Shepard becomes the first American in space aboard the Mercury-Redstone 3 (or Freedom 7) crewed spacecraft. The Mercury Project carried out 20 uncrewed (one of them with a chimpanzee) and 6 crewed flights, till the end of the project in 1963.
- **16 June 1963:** Valentina Tereshkova becomes the first woman in space aboard Vostok 6, launched by the USSR.
- **18 March 1965:** The first spacewalk by Alexei Leonov who leaves his spacecraft, the Voskhod 2, launched by the USSR.
- **14 July 1965:** The USA spacecraft, Mariner 4, becomes the first to reach the planet Mars and return the first close-up images of the Martian surface. Other missions of the Mariner series studied Venus and Mercury.
- **1967:** Three Americans are killed during the launch of Apollo 1, and one Soviet cosmonaut is killed when the parachute on his Soyuz-1 capsule fails to open on his re-entry into Earth's atmosphere.
- **23 April 1967:** The USSR launches its spacecraft Soyuz-1 with the crew, following the first launch of an uncrewed mission in November 1966. The mission ended in tragedy as mentioned above. Soyuz-3, launched on 26 October 1968, becomes the first fully successful crewed mission. Soyuz-11, launched in 1971, is the only other flight of Soyuz to suffer an accident, in which all the crew members are killed. Soyuz spacecraft, along with the Soyuz rocket, has continuously evolved and adapted to new challenges and requirements over the years. The Soyuz programme remains one of the most iconic, enduring, and successful space programmes in history, having been in operation for over half a century. From 2011 — when the Space Shuttle was retired, to 2020 — when the SpaceX Crew Dragon was ready, Soyuz served as the only means to ferry crew and supplies/equipment to and

from the International Space Station (ISS). The Chinese spacecraft Shenzhou, servicing its space station, Tiangong, launched from 2021 to 2022, is believed to resemble it.

- **21 December 1968:** The USA spacecraft Apollo 8 becomes the first human-crewed spacecraft to reach the Moon, orbit it, and successfully return to Earth.
- **20 July 1969:** Neil Armstrong and Edwin "Buzz" Aldrin (Fig. 8.8) become the first men to walk on the Moon while their crewmate Michael Collins continued to orbit the Moon aboard **Apollo 11** launched by the USA. Armstrong makes the famous comment, "One small step for (a) man, one giant leap for mankind."

These were followed by **Apollo 12–17**, with Apollo 13 aborted mid-way through the flight. **Apollo 18** was dropped, and the module was later used in the **Apollo-Soyuz** Mission.

Fig. 8.8. Buzz Aldrin on the Moon in a Photograph Taken by Neil Armstrong, who is Seen in the Visor Reflection Along with Earth, the Lunar Module Eagle, and the Flag of the USA, During the Apollo 11 Mission.

Fig. 8.9. Rakesh Sharma: First Indian in Space Aboard Salyut-7.

- **19 April 1971:** The USSR launches **Salut-1**, which becomes the first crewed space station. The **Salyut programme** continued till 1986 and covered the launch of four scientific and two military space stations, while two launches failed. The first Indian astronaut **Rakesh Sharma** (Fig. 8.9) joined the **Salyut-7**, flying on board Soyuz T-11 on 3 April 1984.
- **14 May 1973:** The USA launches its first space station, **Skylab**.
- **August-September 1977:** The **USA** launches robotic space probes, Voyager 2 (August 20) and Voyager 1 (September 5), to study the outer planets of the Solar System. After flying by Jupiter (1979), Saturn (1980 and 1981), and Neptune (1986 and 1989), both spacecraft are now in interstellar space, collecting data about the environment beyond the Solar System.
- **12 April 1981:** The USA launches its first **Space Shuttle** Columbia. The programme continued till July 2011.

- **20 November 1998:** Russia launches the first segment of the **International Space Station**, the Zarya Control Module. Other launches followed, which built the Station in modules. It has been continuously occupied since November 2000 and became fully operational in May 2009.

8.6 The Space Shuttle Programme of the USA

The Space Shuttle programme (officially Space Transportation System) provided a remarkable reusable spacecraft system for efficient and cost-effective access to space. It consisted of an orbiter, which was launched with two reusable solid rocket boosters, and a disposable external fuel tank. The reusable rocket boosters provided 85% of the boost in the first 2 minutes and were jettisoned, parachuted down, recovered, refurbished, and reused. The external fuel tank used liquid hydrogen and liquid oxygen for fuel and operated for 10 seconds after the main engines were cut off. It was discarded and jettisoned after the orbiter was put in orbit. It re-entered the Earth's atmosphere, broke up, and got destroyed. The orbiter was a partially reusable orbital spacecraft system that could carry astronauts and payloads into a low Earth orbit. After the space operations, it used to re-enter the atmosphere and land as a glider, along with its crew and returning payload. It could carry up to eight astronauts and 23 metric tonnes of payload into a low Earth orbit.

Six Space Shuttles, Enterprise (prototype, used for approach and landing tests), Columbia (1981), Challenger (1983), Discovery (1984), Atlantis (1985) (Fig. 8.10), and Endeavour (1992), were commissioned. The Space Shuttle Challenger broke apart 73 seconds into its tenth flight, killing all seven crew members aboard on 28 January 1986. Columbia, which had successfully carried out 27 missions, disintegrated upon re-entry during its twenty-eighth mission, killing the seven-member crew, on 1 February 2003. Kalpana Chawla (1962–2003), an Indian-born American astronaut and aerospace engineer, who was the first woman of Indian origin (Fig. 8.11) to fly to space, was killed in this disaster. It was her second flight aboard Columbia.

The Space Shuttle programme was launched in 1981, and having flown 135 missions, lasted till 2011. It enabled NASA to launch and repair satellites, conduct scientific experiments in space, and build the ISS, by delivering the core components, modules, and numerous supplies. The Space Shuttle was instrumental in launching (24/25 April 1990),

Fig. 8.10. The Space Shuttle Atlantis in Flight 19, 2 December 1988. The booster Rockets are Firing, and the Orbiter Atop the External Fuel Tank is Seen in Red (Credit: NASA).

Fig. 8.11. Kalpana Chawla: First Woman of Indian Origin in Space (Credit: NASA).

servicing, and upgrading the Hubble Space Telescope, a revolutionary astronomical observatory built by the collaboration of NASA and the European Space Agency and used by thousands of astronomers from across the world. It has provided incredible images and scientific insights about the universe. It also deployed the Chandra X-ray Observatory.

Ultimately, a combination of issues, with the high cost of operations, ageing fleet, safety concerns, limit on the payload which could be carried, shifting priorities of space exploration, and availability of other options (e.g., Soyuz from Russia and other private operators) to transport astronauts and supplies, led to its discontinuation.

The total expenditure on this Mission has been estimated at 200 billion USD.

8.7 Cold War

The struggle for supremacy between the United States of America and its allies and the Union of Soviet Socialist Republics and its satellite states, which lasted for several decades after the Second World War, is called the Cold War. During this period, the soldiers of the USSR and the USA did not battle directly, but the continuous political manoeuvrings, deep distrust, espionage and counterespionage, overt and covert propaganda, arms race and arms buildup, political coalitions, military coalitions, selective economic aids, selective technology denials and embargoes, regime changes/toppling, political assassinations and attempts at political assassinations, support (and instigation) for or subversion of rebellions and agitations, etc., kept them in a high state of alert and tension.

The friction was inherent in their very association as they were united only in their opposition to the Axis Powers during the war but opposed to each other's political ideology.

The USSR created a buffer zone between its borders and Western Europe by setting pro-communist regimes in Poland, Hungary, Bulgaria, Czechoslovakia, Romania, Albania, and East Germany, while the United States tried to contain and prevent the spread of Soviet and communist influence in Western European nations, such as France, Italy, and Greece. The USA also implemented the Marshall Plan (1947) under which it provided \$12.7 billion, which would amount to about \$130 billion now, to the UK (26%), France (18%), West Germany (11%), Italy (9%), and the Netherlands (9%), and the rest to 13 other countries, as economic assistance to arrest the political instability which could have fuelled a communist takeover of their governments.

Soon thereafter, the USA joined the North Atlantic Treaty Organisation, which aimed to contain the USSR, and the USSR entered a collective defence treaty, the Warsaw Pact, with Albania, Bulgaria, Czechoslovakia, East Germany, Hungary, Poland, and Romania, from Central and Eastern Europe. The Cold War Era saw periods of severe ups and downs, e.g., the blockade of Berlin; the testing of the atomic bomb by the USSR; the capture of power by communists in China; the Korean War; the Cuban Missile Crisis; military intervention by the USSR in East Germany (1953), Hungary (1956), Czechoslovakia (1968), and Afghanistan (1979), to protect the communist regimes; and the brutal Vietnam War (1964–1975), involving the US forces. The two powers also intervened in several Asian, Arab, African, and Latin American nations, to safeguard their interests and enhance their influence.

The Cuban Missile Crisis brought the Soviet Union and the USA to the brink of a nuclear war, and this realisation made the two agree to the Nuclear Test-Ban Treaty of 1963, which banned aboveground nuclear weapons testing. A border conflict between China and the USSR in 1969 led to an easing of tensions between China and the USA (more as a measure to balance the USSR). There was further easing of the tensions of the Cold War in the 1970s, as the two superpowers initiated Strategic Arms Limitation Talks (SALT) and entered into the SALT I and SALT II agreements of 1972 and 1979, limiting the number of their antiballistic missiles and their strategic missiles capable of carrying nuclear weapons.

The Soviet-Afghan War, which had continued from 1977, to try to protect the pro-Soviet Rule in Afghanistan, ended in February 1989, as the Soviet forces withdrew. The Berlin Wall fell in November 1989, and East and West Germany were united by October 1990. The Soviet Union collapsed by the end of 1991, giving rise to 15 newly independent (democratic) countries, and the Cold War came to an end.

The space race played a significant part in the Cold War as the Americans and Soviets competed to prove their scientific, technical, intellectual, political, and economic superiority by having the first satellite to orbit Earth, the first human-crewed spacecraft, and the first person to walk on the Moon. The world's attention was captivated by this contest for dominance between the two superpowers, which filled it with despair and hope in turn.

8.8 The Advent of Space Collaboration

At a speech at the United Nations General Assembly on 20 September 1963, John F. Kennedy, President of the United States, declared, "In a field where the United States and the Soviet Union have a special capacity — space — there is room for new cooperation, for further joint efforts. I include among these possibilities," he added, "a joint expedition to the moon." He asked why should the United States and the Soviet Union conduct parallel efforts "that would include duplication of research, construction and expenditure?" and suggested a joint series of space missions, which if enacted, "will require a new approach to the Cold War." Soviet leader Nikita Khrushchev rejected the proposal, though there are suggestions that he did have second thoughts about it, afterwards. In the wake of Kennedy's assassination on 22 November 1963, both President Lyndon B. Johnson, who succeeded Kennedy, and Khrushchev abandoned the idea, and each nation continued with its space programme.

On 24 May 1972, President Richard Nixon, the first US President to visit Moscow, and Soviet Chairman Alexei Kosygin signed the Agreement Concerning Cooperation in the Exploration and Use of Outer Space for Peaceful Purposes [1].

This culminated in the 1975 linking of an Apollo spacecraft with a Soyuz command module.

The Apollo-Soyuz Mission was launched on 15 July 1975, in two separate flights, Apollo and Soyuz, which docked in space, and the two commanders Tom Stafford and Alexei Leonov exchanged the first international handshake. This act is seen to symbolically end the space race, paving the way for future joint missions, such as the Shuttle-Mir Programme and the ISS.

The Apollo-Soyuz Mission or the Apollo-Soyuz Test Programme (Fig. 8.12), as it was officially called, marked a significant milestone in space exploration and international cooperation, as it demonstrated that nations with different space systems and technologies could work together on complex space projects. It will be remembered for promoting peaceful collaboration between the United States and the Soviet Union during the Cold War era, leading to detente.

The Apollo-Salyut Test Mission was to be followed by the Shuttle-Soyuz Mission, under which the Space Shuttle(s) from the USA would have flown to the USSR space station Soyuz, but it did not materialise.

Fig. 8.12. A 1973 Artist's Conception of the Docking of Apollo and Soyuz (Credit: NASA).

On 25 January 1984, President Ronald Reagan used his State of the Union address to call for NASA to build a permanently manned space station. He further added: "A space station will permit quantum leaps in our research in science, communications, and in metals and lifesaving medicines which could be manufactured only in space. We want our friends to help us meet these challenges and share in their benefits. NASA will invite other countries to participate so we can strengthen peace, build prosperity, and expand freedom for all who share our goals," and set the stage for the **International Space Station**.

The collapse of the Soviet Union and the budgetary difficulties of the USA, which slashed their funds for the space programme, brought the USA and the Russian Federation, the successor of the USSR for these efforts, together to collaborate on the Shuttle-Mir programme, from 1994 to 1998, for which American President George H. W. Bush and Russian President Boris Yeltsin agreed to cooperate by signing the Agreement between the United States of America and the Russian Federation Concerning Cooperation in the Exploration and Use of Outer Space for Peaceful Purposes, in June 1992.

It has been suggested that the project gave a ruse to the American government to provide funds for the "cripplingly underfunded" Russian space programme, allowing the Russian government to keep Mir operating, the Russian space programme afloat, and ensuring that the Russian government "remained friendly towards the United States." It has also been suggested that the USA was expected to get an in-depth understanding of the technologies developed by the USSR/Russia through these collaborations.

As budgetary difficulties were also being faced by other nations in their space programmes, the USA started discussions with Europe, Japan, Canada, and Russia in the early 1990s. The plans for a new space station to be built under these intended collaborations, which became the **International Space Station**, were announced by American Vice-President Al Gore, Jr., and Russian Prime Minister Viktor Chernomyrdin in September 1993.

The space race started with the nuclear arms race between the USA and the USSR following the Second World War. Both were aided by the missile technology of Nazi Germany and scientists from their missile programme (V-2 rockets). Both worked under the belief that technological advantages required for such power were necessary for national security and political superiority.

The space race will be remembered for producing groundbreaking capabilities to launch artificial satellites, space probes of the Moon, Venus, and Mars, human space voyages in a low Earth orbit, and lunar missions, within a very short span of a few decades. It led to far-reaching achievements in science, space exploration, and technology.

8.9 The Shuttle-Mir Programme

We have seen how the agreement for collaboration between the American Space Shuttle and the Russian Space Station Mir (Fig. 8.13) was reached. It was a collaborative 11-mission space programme between Russia and the United States and involved American Space Shuttles visiting the Russian space station Mir, Russian cosmonauts flying on the Shuttle, and an American astronaut flying aboard a Soyuz spacecraft to engage in long-duration expeditions aboard Mir. The programme began in February 1994 and ended in June 1998. During that time, the Space Shuttle docked with Mir 10 times, and seven American astronauts spent nearly 1000 days living in orbit with Russian cosmonauts on board the space station Mir.

The Shuttle-Mir programme provided experience with long-duration spaceflights and put the compatibility of American and Russian spacecraft to test. The programme also allowed scientists from both countries to

Fig. 8.13. US Space Shuttle Atlantis with Russian Space Station Mir (1995).

conduct a wide range of experiments in space, especially on studies of the human body in space, the effects of microgravity on materials, and the development of new materials for use in space.

8.10 International Space Station

The ISS (Fig. 8.14) was launched in 1998. It is commonly referred to as the ISS and is a culmination of years of coordinated planning which brought international flight crews, multiple launch vehicles, globally distributed launch operations, training, engineering, development of facilities, communication networks, and the international scientific research community, together to build the largest modular space station in a low Earth orbit. The construction of the ISS began in 1998 and continued over several years, with various modules, components, and spacecraft being launched into space and connected.

ISS involves five space agencies: NASA from the United States, Roscosmos from Russia, Japan Aerospace Exploration Agency (JAXA), the European Space Agency (ESA), and the Canadian Space Agency (CSA). It serves as a microgravity and space environment research

Fig. 8.14. The International Space Station in 2018, as Photographed by Crew Members from a Soyuz Spacecraft (Credit: NASA/Roscosmos).

laboratory in which scientific research is conducted in astrobiology, astronomy, meteorology, physics, and other fields. The ISS is eminently suited for testing the spacecraft systems and equipment required for possible future long-duration missions to the Moon and Mars.

The first two modules of the ISS, Zarya (from Russia) and Unity (from the USA), were launched in 1998. Zarya served as the initial control module and provided propulsion and power during the early assembly stages, while Unity was the first connecting node that allowed future modules to be attached. After Zarya, several Russian modules were launched and added to the ISS, including Zvezda (2000), which served as the main Russian living quarters and control centre, and Pirs (2001), which was used as a docking compartment and an airlock for Russian spacewalks.

The United States provided the modules, Destiny (2001) — which housed research facilities for experiments in space, and Quest (2001) — as an airlock for the US spacewalks. The European Space Agency

Fig. 8.15. View of the Earth from the Cupola on the ISS (Credit: NASA).

launched the Columbus laboratory (2008), and the Japanese Aerospace Exploration Agency contributed the Kibo laboratory (2008–2009), which included multiple sections for various experiments, while the Canadian Space Agency provided Canadarm2 (2001), a highly advanced robotic arm, which plays a crucial role in the station's assembly and maintenance and which is used to capture and dock incoming spacecraft, assist astronauts during spacewalks, and move equipment and modules around the station. It also provided Dextre (or Special-Purpose Dexterous Manipulator), a sophisticated two-armed robot designed to perform maintenance tasks outside the ISS in 2008.

One of the most loved modules, Cupola (2010), a dome with several windows (Fig. 8.15), which serves as an observation and control point for robotics operations and provides a panoramic view of Earth and space, was also provided by Canada. The cupola also provides direct observation of developing storms (Fig. 8.16) as well as glaciers, agricultural fields, forests, cities, and coral reefs.

The Space Shuttle played an important role in the assembly of the ISS, by ferrying astronauts and components into space, which allowed the space station to be constructed piece by piece.

Fig. 8.16. Tropical Cyclone Bansi on 15 January 2015 (Credit: NASA).

8.11 Science Outreach of NASA

NASA, from its very inception, has always made a great, sincere, and very successful effort to get children, students, researchers, and the general masses interested in all aspects of sciences and space programmes, their progress, and their outcomes. It is not a wonder that NASA is perhaps the most coveted dream destination for students from across the world, and easily the most recognised laboratory, internationally.

Thus, during its Mars programme in 2009, NASA invited elementary school children to name its Mars Science Laboratory rover. The name chosen on their suggestion was Curiosity!

Again in 2019, NASA held a "Name the Rover" contest for the Mars 2020 Mission, which included the Mars Helicopter Scout. Thousands of students from around the world submitted their suggestions, showcasing imagination and a thirst for exploration. The name "Ingenuity" was chosen on the suggestion of a student in grade 11, who said that the mission embodied human ingenuity — the ability to invent, solve problems, and push boundaries and that "Ingenuity" was not just about the hardware, but the teamwork, perseverance, and daring spirit behind it!

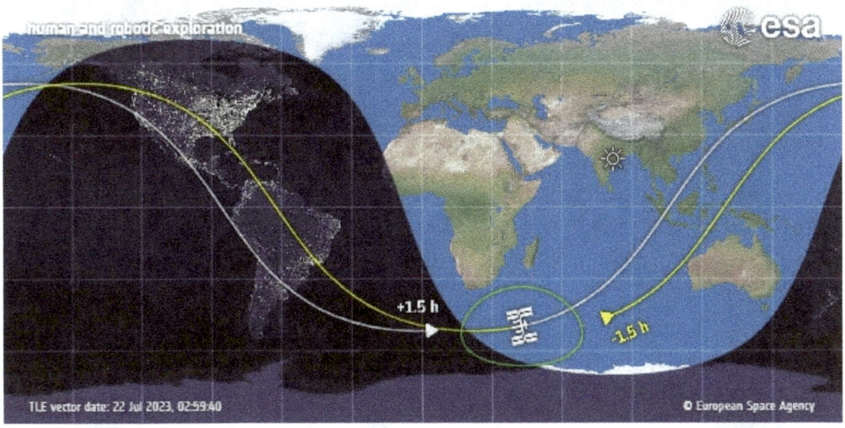

Fig. 8.17. Locating the ISS [2].

One interesting measure NASA/ESA has taken is to continuously update the information about the current location of the ISS (Fig. 8.17) so that anyone, especially the students, can see it in the sky with their naked eyes, as its trajectory passes over more than 90% of Earth's population, at an average height of 400 km, completing 16 orbits per day, at an average speed of 28,000 km/hour.

Similarly, International Space Station Ham Radio (also known as Amateur Radio on the International Space Station) has allowed groups of students in schools, camps, museums, and planetariums to hold a conversation with the people living in space, since the earliest space station expeditions. As the ISS passes overhead, students have about 9 minutes to ask crew members 10–20 questions (Fig. 8.18).

The electric effect of this opportunity and its lasting and wonderful imprint on young minds cannot be overestimated. ISS also encourages students to propose experiments to be conducted in space (Fig. 8.19).

More than 200 astronauts from 20 different countries have visited the ISS (Fig. 8.20). The Indian Space Research Organisation and NASA have agreed to develop a strategic framework for human spaceflight cooperation by the end of 2023, and NASA will provide advanced training to Indian astronauts to mount a joint effort to the ISS in 2024.

The Ukraine War, which started in February 2023, put the ISS collaboration under severe strain, but now, there is a general agreement that it will continue till 2030 or 2031.

Fig. 8.18. A Student Stands at a Microphone as he Asks NASA Astronaut Serena M. Auñón-Chancellor a Question During a Downlink Event (Credit: NASA).

Fig. 8.19. Japanese Astronaut Koichi Wakata in the Japanese Experiment Module at the International Space Station Carries out an Experiment Proposed by a Taiwanese Student (Credit: Japan Aerospace Exploration Agency).

Fig. 8.20. The International Crew of ISS (Credit: NASA, 2013).

8.12 Microgravity and Long-Duration Space Flights: Effects on Health

One of the primary purposes of the ISS is to study the effect of microgravity and confined stay on humans. Astronauts typically stay on the ISS for about 6 months, the rationale being that a journey to Mars may take up to 6 months. Microgravity can cause bone loss, muscle atrophy, and cardiovascular changes (due to fluid shift in the body, upwards and away from the feet and legs, which can change the volume of blood and blood pressure) in astronauts, which needs to be understood for developing countermeasures to protect astronauts on long-duration spaceflights, and even treat osteoporosis. Russian cosmonaut Mikhail Korniyenko and American astronaut Scott Kelly spent 340 days in orbit from March 2015 to March 2016. Scott Kelly's stay was also very interesting and important as scientists were able to use Mark, his identical twin brother, as a baseline to understand how the long spaceflight had changed him!

It has been established by now that the astronauts need to exercise for several hours a day using specialised equipment, offering resistance, to

Fig. 8.21. The Crew Enjoying the Peppers Grown in the Veggie Experiment [3].

help maintain muscle mass and bone density, and need to take a diet rich in protein and calcium.

Astronauts on long-duration space flights are likely to have difficulty sleeping due to the lack of gravity and the noise and vibration of the spacecraft, as well as due to the occurrence of 16 sunrises and sunsets within 24 hours, as the station orbits the Earth, which affects their circadian cycle. They may also experience psychological problems, such as anxiety, depression, and sleep disturbances, which need to be addressed by counselling and medication. Requirements like these have triggered rapid advances in telemedicine.

It has also been found that growing vegetables (Fig. 8.21, under the **Veggie** experiment to study plant growth in microgravity) and flowers (Fig. 8.22) creates a cheerful atmosphere in the Station. Veggie has successfully grown a variety of plants, including three types of lettuce, Chinese cabbage, Mizuna mustard, red Russian kale, and zinnia flowers, some of which were eaten and others sent to Earth for analysis. A major development has been water treatment under which 90% of the water is recycled.

Despite the protection provided by radiation shielding, the astronauts on long-duration flights are exposed to higher levels of radiation than people on Earth, with its attendant risks, which are being studied [4].

Fig. 8.22. Zinnia flower in the Veggie Facility on the ISS (Credit: NASA, 2016).

8.13 Scientific Experiments and Observations

Let us briefly discuss some of the other path-breaking experiments and innovative studies that have become possible with the facilities offered by the ISS:

- **The crystallisation of proteins:** Microgravity in the Space Station provides an excellent stage for the growth of crystals. Buoyancy and convection currents caused by the density difference between the crystal and the surrounding solution interfere with crystal growth in the environment of normal gravity. This leads to irregular crystal shapes, impurities, and difficulties in achieving large, high-quality crystals.

In the absence of gravity, crystals (Fig. 8.23) can grow more effectively because these disruptive forces are greatly reduced or eliminated, they grow more slowly, and the molecules have time to align on the surface of the crystal more perfectly, which leads to a more controlled crystal growth, resulting in larger, purer, and more uniform crystals.

This is very valuable for the study of proteins, which maintain the structure, function, and regulation of the body's tissues and organs and act as catalysts for chemical reactions, cell-to-cell signalling, and immune

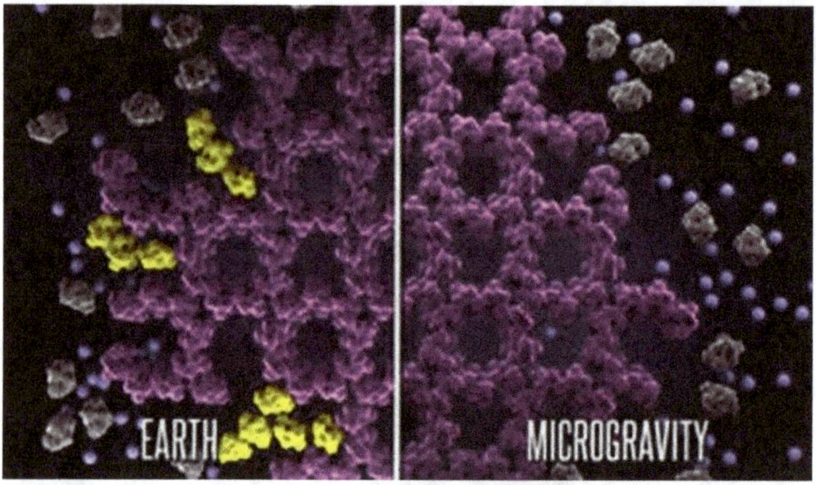

Fig. 8.23. Crystal Growth on Earth (left) and in Microgravity (right) (Credit: NASA).

responses. An understanding of protein structures, which requires high-intensity X-ray diffraction from their crystals, has now become possible, as good-quality large crystals are produced in microgravity.

In addition, microgravity also enables the growth of certain types of crystals that are challenging or impossible to grow on Earth due to gravity-related limitations. The absence of gravity also allows for the formation of complex crystal structures that may have unique properties and applications in various fields, such as materials science, pharmaceuticals, and electronics.

Thus, several hundred proteins and other macromolecules, such as DNA, RNA, and even whole viruses, have been crystallised in microgravity experiments for more than three decades, in the environment offered by space-based laboratories like the ISS [5].

Microgravity also alters processes of fluid mixing, a separation between gases and liquids, heat transfer, solidification, and combustion, thus providing avenues for advanced research in material sciences, by providing a stage for reactions that are not possible under conditions of gravity.

- **Bose-Einstein condensates (BECs):** These, often called the fifth state of matter, are formed when a dilute gas of bosons (particles with integer spin) is cooled to extremely low temperatures, typically just a fraction above absolute zero. In this state, the individual bosonic particles lose their identities and merge into a single collective quantum state, exhibiting wave-like properties on a macroscopic scale. These are being studied to develop and test technologies for future space-based applications as well as for quantum computing and to investigate the fundamental aspects of quantum mechanics.

On Earth, gravity limits the time that BECs can be studied because they quickly disperse. In microgravity, there is no gravity to pull the atoms apart, so BECs can last much longer. This allows scientists to study BECs in detail.

A very interesting observation was the study of flames and fire-extinguishing properties of various gases in the Flame Extinguishment Experiment **(FLEX)**, which found that flames in microgravity are spherical and at a lower temperature (about 600°C), while a typical candle (for example) burns at around 1,400°C, on the Earth, and is shaped like a teardrop.

When a candle is ignited, thousands of chemical reactions take place. In this process, molecules of hydrocarbon vaporise and get cracked apart by heat, and combine with oxygen to produce light, heat, carbon dioxide, and water molecules. Some of the hydrocarbon fragments form ring-shaped molecules and soot, which may be burnt or go away as smoke. The hot air rises and the flame follows it, drawing fresh cool air behind it. The flame acquires a teardrop shape, which flickers.

In microgravity, the hot air does not rise. Instead, the hot air expands in all directions, creating a spherical flame (Fig. 8.24). Moreover, in microgravity, the flame is not fed by the convection of hot air, but by the diffusion of oxygen, which also eliminates the turbulence of the flame. These studies are expected to help us design more efficient engines and fire suppression devices.

The ECOsystem Spaceborne Thermal Radiometer Experiment on Space Station (**ECOSTRESS**) provides the first-ever high space-time resolution thermal infrared measurements of the surface of the Earth (Fig. 8.25). These are likely to provide information on water stress in plants and how various regions may respond to future changes in climate.

Who can forget the adorable character of Data in the TV serial, *Star Trek: The Next Generation?* **Robonaut** is a humanoid robot (Fig. 8.26) designed with the versatility and dexterity to manipulate hardware, work

Fig. 8.24. Flame on Earth (left) and in Microgravity (right) on the ISS.

Fig. 8.25. Earth-Observing Payloads Attached to the Japanese Experiment Module (Credit: NASA).

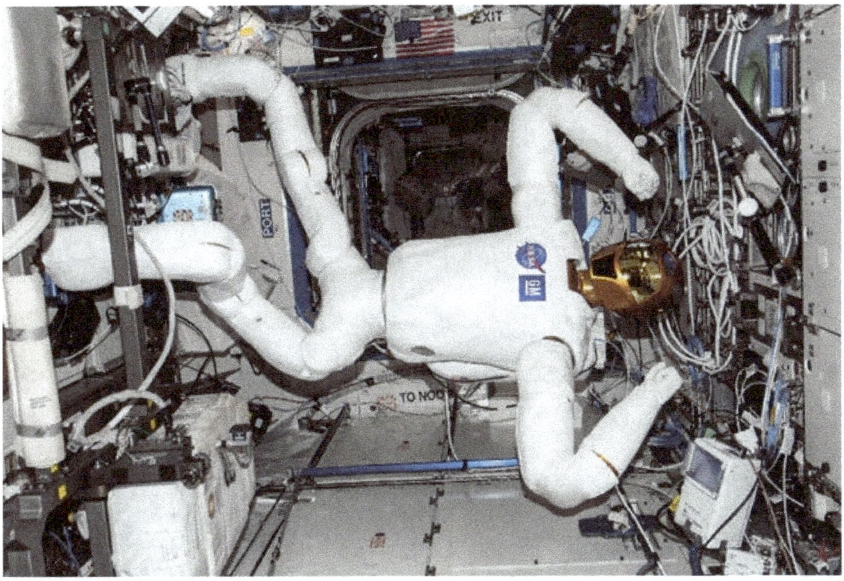

Fig. 8.26. Robonaut (Credit: NASA).

in high-risk environments, and respond safely to unexpected obstacles, which has proved invaluable on the ISS.

NASA has installed a Neutron star Interior Composition Explorer (**NICER**) to explore the structure, dynamics, and energetics of neutron stars through soft X-ray timing and spectroscopy.

The Station Explorer for X-ray Timing and Navigation Technology (**SEXTANT**) experiment has provided a most valuable result that millisecond pulsars could be used to accurately determine the location of an object moving at thousands of kilometre per hour in space.

The Alpha Magnetic Spectrometer (**AMS-02**) is a particle-physics detector (Fig. 8.27) that looks for dark matter, antimatter, and missing matter from a module attached to the outside of the ISS. It also performs precision measurements of cosmic rays.

8.14 The Space Programme of India

India's space programme has its origins in the early 1960s, shortly after gaining independence from British rule. Dr. Vikram Sarabhai (Fig. 8.28), often referred to as the "Father of the Indian Space Programme," played

Fig. 8.27. The Alpha Magnetic Spectrometer (2017) (Credit: NASA).

Fig. 8.28. Dr. Vikram Sarabhai (1919–1971 CE).

a crucial role in shaping the country's space ambitions. The Indian National Committee for Space Research (INCOSPAR) was formed by the Department of Atomic Energy. Its initial focus was on launching sounding rockets to study the upper atmosphere using rocket-borne instrumentation.

The Thumba Equatorial Rocket Launching Station (TERLS) was established near Thiruvananthapuram in South Kerala, in 1963. TERLS was ideally located for launching sounding rockets, as it is close to the equator, and the rocket could take optimum advantage of the Earth's substantial rotational speed of over 1650 km per hour relative to Earth's centre.

In 1969, INCOSPAR was reorganised and renamed the Indian Space Research Organisation (ISRO), with a mission to develop and apply space technology for the benefit of India. In 1975, India launched its first satellite, Aryabhata, using a Soviet rocket.

In 1980, ISRO launched the Satellite Launch Vehicle (SLV-3), which was the first Indian-made launch vehicle to successfully place a satellite in orbit. ISRO launched the Indian National Satellite System (INSAT) series of communication satellites in the early 1980s, which played a vital role in providing telecommunications and broadcasting, meteorology, and search and rescue operations services to India. In the 1990s, ISRO focussed on Indian Remote Sensing (IRS) to help with land use planning, natural resources management, and disaster monitoring. Polar Satellite Launch Vehicle (PSLV) was launched for the first time on 15 September 1993 and has become one of ISRO's reliable, versatile, workhorse launch vehicles, which enabled India to deploy a wide range of satellites into various orbits.

India launched its first lunar probe, Chandrayaan-1, on 22 October 2008 (Fig. 8.29), using its PSLV rocket, which orbited the Moon for 10 months and collected a wealth of data about the lunar surface. Its Moon Impact Probe hit the Moon, in a controlled manner, on 14 November 2008, and led to the pioneering discovery of ice and water molecules in lunar soil. It carried several instruments developed in collaboration with ESA, NASA, JPL, and laboratories in the USA, Japan, UK, Poland, Bulgaria, Norway, etc., and mapped the mineral content of the Moon's surface and sent several thousand pictures.

Chandrayaan-2 was the second lunar exploration mission by ISRO. It consisted of a lunar orbiter, a lander, and a rover. It was launched on

Fig. 8.29. Launch of Chandrayaan-1 (Credit: ISRO).

22 July 2019. The craft reached the Moon's orbit on 20 August 2019. The lander (which carried the rover) crashed. However, the orbiter from the Chandrayaan-2 mission continues to circle the Moon and carry out its functions, as it contains adequate fuel to remain in its orbit for almost 8 years, as estimated by ISRO in 2019.

India's third Moon mission, Chandrayaan-3 (Fig. 8.30), was launched on 14 July 2023. It consisted of an indigenous lander module (Vikram), a propulsion module, and a rover (Pragyan) carried inside the Lander Module. The Chandrayaan-3 craft landed (Fig. 8.31) successfully on the lunar south pole on 23 August, and India became the fourth nation to

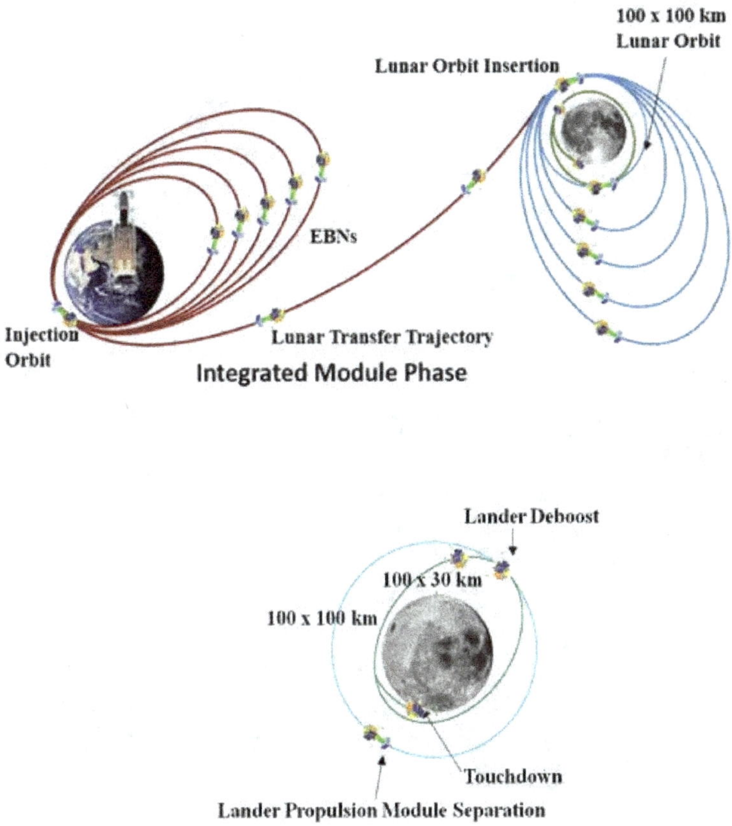

Fig. 8.30. Chandrayaan-3 Goes to the Moon (Credit: ISRO).

achieve a soft landing on the Moon following the USA, the former Soviet Union, and China, and the first to do so near the south pole, which is known for the abundance of hills, valleys, and boulders. On August 25, it released its Pragyan Rover (Fig. 8.32), which traversed over 100 m over 2 weeks, studying lunar rocks and soils with its five instruments. With this, the mission fulfilled all its stated objectives, namely: (i) to demonstrate a safe and soft landing on the lunar surface, (ii) to demonstrate a rover roving on the Moon, and, (iii) to conduct *in-situ* scientific experiments.

Chandrayaan-3 mission has led to the discovery of sulphur on the Moon, recorded what could be one of the first detections of "moonquake"

Fig. 8.31. Vikram Lander Near the South Pole of the Moon; Picture Taken by the Pragyan Rover (Credit: ISRO).

in over 50 years and discovered that while the surface temperature during the day was 50°C, just 10 cm below the ground, it dropped to –10°C. It has also made the first measurements of the density and temperature of the Moon's ionosphere and found "a relatively sparse" mix of ions and electrons in the 100-km-thick layer of electrically charged plasma that surrounds the Moon's surface near the south pole. The sparse plasma means that potential signals from the Moon would experience "minimal" delays in traversing the ionosphere and would not pose a problem for transmission [6]. The Chandrayaan-4 mission (to be launched around 2028) plans to bring back rock samples from the Moon, among other research objectives.

Fig. 8.32. Pragyan Rover on the Moon (Credit: ISRO).

Close on the heels of the Chandrayaan-3 mission, India launched its first spacecraft to study the Sun on 2 September 2023. The solar mission, named Aditya-L1, joined four other spacecraft circling a point in space known as Lagrange Point 1 (L1), after a journey of 1.5 million km on 6 January 2024. At the Lagrange point, the net gravitational pull of the Sun and the Earth precisely equals the centripetal force required for the craft to move with them and provides an uninterrupted view of the Sun. The Aditya-L1 mission is expected to help us unravel, among other things, three outstanding problems in solar physics:

1. how stars such as the Sun sustain their super-hot outer layer (the surface of the Sun has a temperature of about 5000°C–6000°C, while the corona [Fig. 8.33] has a temperature of several million degrees centigrade);

Fig. 8.33. The Corona of the Sun to be Studied by Aditya-L1.

2. how variations in the Sun's magnetic field affect Earth's atmosphere; and
3. how the Sun's magnetic fields create violent solar storms.

India's first Mars orbiter, Mangalyaan, was launched by ISRO on 5 November 2013. It successfully entered Mars orbit on 24 September 2014 and continued to send vast amounts of data from the study of the Martian atmosphere and surface till October 2022, when its batteries died.

Gaganyaan mission of India envisages a demonstration of human spaceflight capability by launching a crew of three members to an orbit of 400 km for a 3-day mission and bringing them back safely to Earth. The plans to send a spacecraft to study Venus, Shukrayaan, are proceeding smoothly.

India's space programme has benefitted from collaboration with the Soviet Union/Russia for the development of its satellite launch vehicle, and with several other countries for the development of the instruments on board its space missions. India collaborated with France on the development of satellite technology and space research, and France launched several satellites for India.

India's early space efforts did not involve significant collaboration with the United States due to geopolitical reasons. The USA imposed severe sanctions and technology denials on India because of *or with the excuse of* nuclear tests conducted by India in 1974 and 1998. However, the relationship has evolved with time. In recent years, the collaboration between ISRO and NASA has increased in areas, such as space science research, Earth observation, and satellite tracking.

India has been collaborating with ESA, Germany, Japan, Australia, etc., on various space projects, particularly in the field of Earth observation and satellite navigation, and has been launching satellites for several countries, for quite some time.

India has also helped several developing nations with their space programmes and space applications. It shares facilities and expertise through the Indian Institute of Remote Sensing and the UN-affiliated Centre for Space Science and Technology Education in Asia and the Pacific. India offers an international training programme on nanosatellite assembly and building, UNispace Nanosatellite Assembly & Training (UNNATI) by the Indian Space Research Organisation. It is building a large facility in Vietnam to provide a reliable operational space-based system for remote sensing in Southeast Asia. In 2017, India launched the South Asian Satellite in collaboration with Afghanistan, Bhutan, Bangladesh, Nepal, the Maldives, and Sri Lanka.

8.15 Conclusions

We have seen the enormous strides mankind has taken in space exploration. The future of space exploration, including the possibility of travelling to neighbouring planets — at least to Mars, a return to the Moon, using the Moon as a stop to proceed to Mars, setting up colonies on the Moon and Mars, harvesting of rare metals from asteroids — all seem to be within the reach of mankind. The search for extraterrestrial intelligent life continues with the expectation that the Universe is too vast and galaxies and stars far too many and far too old, not to harbour a planet where life may have evolved. It was this expectation which made NASA put a plaque (Fig. 8.34), designed by Carl Sagan, on each of the Pioneer 10 and Pioneer 11 spacecraft when these were launched to explore other planets in the early 1970s, which gives our location in the solar system, a sketch of a hydrogen atom, a symbolic picture of a man and a woman, and message of peace, envisaging a Universal Collaboration in the quest for knowledge.

The United Nations has a set of treaties to guide our behaviour in space [7] in place. The Artemis Accords, even though non-binding [8], aims to establish a framework of principles for cooperation among

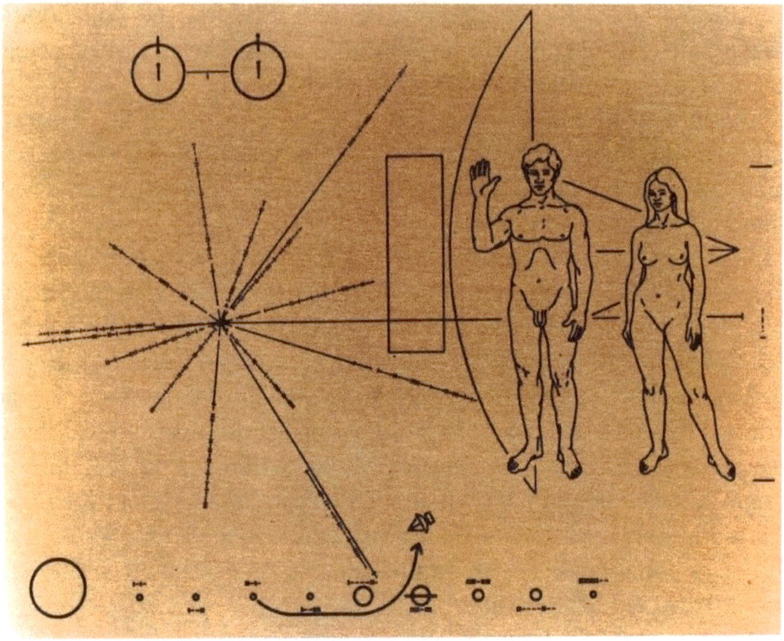

Fig. 8.34. Golden Plaque on Pioneer 10 (Credit: NASA).

countries and private entities in the exploration and use of outer space, with a specific focus on lunar exploration. NASA has stated that the international cooperation on Artemis, which aims to land on the Moon again, is intended to not only bolster space exploration but also enhance peaceful relationships between nations and that all activities will be conducted for peaceful purposes. The Accords envisage the following nine principles to be observed:

1. **Peaceful purposes:** The Accords reaffirm that space activities should be for peaceful purposes only.
2. **Transparency:** The Accords require countries to share scientific data and to cooperate on safety and sustainability.
3. **Interoperability:** The Accords require countries to develop and use compatible systems and technologies.
4. **Emergency assistance:** The Accords commit countries to providing assistance to astronauts in distress.

5. **Registration of space objects:** The Accords require countries to register all space objects that they launch.
6. **Protection of heritage:** The Accords commit countries to protecting the cultural and historical heritage of the Moon and other celestial bodies.
7. **Space resource utilisation:** The Accords recognise the right of countries to utilise space resources, but they also emphasise the need for sustainability.
8. **Deconfliction of activities:** The Accords commit countries to coordinating their activities in space to avoid collisions and other problems.
9. **Orbital debris and spacecraft disposal:** The Accords commit countries to taking steps to mitigate the risks posed by orbital debris and spacecraft disposal.

To further emphasise the message of peace and international collaboration, which science and space exploration convey, we would like to recall the "speech," which was to have been delivered by President Richard Nixon of the USA, in case the Apollo 11 Mission, which was being broadcast live on television, failed and the astronauts killed on the Moon (See Chapter 1). Fortunately, human efforts and technological competence did not let this happen, and the "speech" remained undelivered.

Yet, the sentiments contained there remain valid today and forever [9].

The Moon awaits our return and Mars awaits our arrival. And yes, man's search will not be denied!

Yet, there is no doubt that space and man's capacity to explore it have given rise to a realisation of the absurdities of conflict and the importance of international collaboration. Carl Sagan, looking at the picture of Earth taken by Voyager 2 from a distance of over 6 billion km, when it appears as a pale blue dot, exclaimed [10]:

> *"To my mind, there is perhaps no better demonstration of the folly of human conceits than this distant image of our tiny world. To me, it underscores our responsibility to deal more kindly and compassionately with one another and to preserve and cherish that pale blue dot, the only home we've ever known."*

Voyager 2 is more than 19 billion km away now (August 2023).

We would like to recall the scenes across India from 14 July 2023, when India launched Chandrayaan-3 from Sriharikota, to 23 August 2023, when it reached the Moon. The launch and the final moments of landing were watched by hundreds of millions of people, across India, who had closely followed the progress of the propulsion module to the Moon. It was a culmination of excitement and curiosity, which had started building even before July 14 and which reached a crescendo on the day of landing.

As the computerised images of the second-by-second delicate and intricate manoeuvres by the lander flashed across the television screen across the country, people from all walks of life, especially students, sat frozen, holding their breaths. The collective shout of joy across the nation, and among space enthusiasts across the world, was a sight to behold. People danced in the streets, played music, distributed sweets, started fireworks, and congratulated each other and the ISRO scientists and India for years of their painstaking work, and the process of learning from the failures of the Chandrayaan-2. It was a display of pure joy in scientific and technological achievement, and science outreach at its best.

If we can maintain this interest of the masses in space exploration, we are guaranteed a bright future for mankind.

We close this discussion by showing a mesmerising picture (Fig. 8.35) taken by NASA's Curiosity Mars Rover on 8 April 2023.

We do not doubt that glory in space awaits mankind.

Fig. 8.35. A Composite Photograph by Curiosity Rover on Mars (Credit: NASA).

References

[1] https://www.archives.gov/files/presidential-libraries/events/centennials/nixon/images/exhibit/agreement-of-cooperation.pdf.

[2] https://www.esa.int/Science_Exploration/Human_and_Robotic_Exploration/International_Space_Station/Where_is_the_International_Space_Station.

[3] https://astrobotany.com/plants-grown-in-space/.

[4] Z. Patel *et al.*, *Nature Partner Journals Microgravity* 6 (2020) 33, https://doi.org/10.1038/s41526-020-00124-6.

[5] A. McPherson and L. J. DeLucas, "Microgravity Protein Crystallization", *Nature Partner Journals Microgravity*, 1 (2015) 15010, https://doi.org/10.1038/npjmgrav.2015.10.

[6] T. V. Padma, "Four Things Chandrayaan-3 Has Taught us about the Lunar South Pole". *Nature, News Explainer* (8 September 2023), https://doi.org/10.1038/d41586-023-02852-7.

[7] "United Nations Treaties and Principals on Outerspace", ST/SPACE/11 (New York: United Nations, 2002).

[8] https://www.nasa.gov/specials/artemis-accords/img/Artemis-Accords-signed-13Oct2020.pdf.

B. R. Guruprasad, R. Nagappa, A. Mukherjee, U. Dasgupta, and V. Siddhartha, "Moon Missions: The Way Ahead" (Bengaluru: National Institute of Advanced Studies, 2023). Report No. NIAS/CSS/U/RR/10/2023; See also, S. K. Namdeo and N. Vera, "Contours of Space Diplomacy in the Global South", *AAAS Perspective* (16 May 2023), https://www.sciencediplomacy.org/perspective/2023/contours-space-diplomacy-in-global-south.

[9] https://www.archives.gov/files/presidential-libraries/events/centennials/nixon/images/exhibit/rn100-6-1-2.pdf.

[10] C. Sagan, "*Pale Blue Dot*" (USA: Random House, 1994).

Chapter 9

CERN: Hazaran Blbul — The Nightingale of a Thousand Songs

"He who enters CERN has lost his specific nationality and becomes a scientist of the world."

— Victor Weisskopf, during the 25th Anniversary Celebrations of CERN

"Places like CERN contribute to the kind of knowledge that not only enriches humanity but provides the wellspring of ideas that become the technologies of the future."

— Fabiola Gianotti, CERN's Director-General

9.1 Introduction

As the harsh winter comes to an end in the valley at the foot of Mount Ararat, the wild gardens bloom and thousands of birds sing their wonderful songs in unison. On this joyous occasion, the people of Armenia welcome the arrival of spring by recalling the legend of Hazaran Blbul. It is said that this Nightingale of a Thousand Songs, a tiny bird with bright plumage, always sang the loudest. It is also said that spring would never have come without its wonderful and sweet songs. People believe that its art revived withered gardens and mountains and valleys bloomed with verdant vegetation. The story of Hazaran Blbul is a story about the power of love, the importance of music, and the triumph of good over evil.

And so is the story of CERN.

Conseil Européen pour la Recherche Nucléaire (CERN) or the European Organization for Nuclear Research was born out of the despair, destruction, and desolation brought about by the Second World War. The war was witness to the loss of 75 to 80 million human lives, the largest numbers of them from the USSR, China, Poland, and Germany. The times were so desperate that 3 to 6 million Indians died due to starvation because of the colonial policies of Winston Churchill, who ignored pleas for help from even the British authorities who were administering the country and disallowed transport of food to the region of Bengal, even when food was available within India [1]. Moreover, 89,000 Indian soldiers were killed and 68,000 of them wounded in wars fought in Europe. Up to 6 million people of Jewish origin, 300,000 people with disabilities, and 250,000–500,000 Roma (or Romani/Gypsies) people were murdered during the Second World War.

The war also witnessed extreme brutality, cruelty, starvation, suffering, torture, death, and the murder of tens of thousands (possibly several times more) of prisoners of war. There was large-scale destruction of cities and industrial infrastructure, with iconic cities like London, Berlin, Warsaw, Stalingrad (now Volgograd), Tokyo, Hamburg, Dresden, Cologne, Rotterdam, Budapest, and Manila devastated. Hiroshima and Nagasaki were destroyed by the two nuclear attacks. The number of smaller cities destroyed ran into the hundreds. The war's impact on cities was profound, and the scars of destruction remained visible for years, even decades after the conflict ended.

The mayhem continued even after the war had subsided and estimates suggest that between 13.5 and 16.5 million German speakers fled, were evacuated, or later expelled from Central and Eastern Europe. There was a massive exodus of Jewish scientists, intellectuals, and industrialists from Europe, mostly to the USA, but also to other countries that were considered safe for them during this period. The turmoil caused by fascism and war completely depleted most of the resources of the continent of Europe, both material and intellectual.

The exodus of an alarming number of European physicists to countries with more advanced research equipment continued after the war ended.

In this atmosphere, scientists started discussing the possibility of a European physics laboratory. The idea was first voiced publicly in a message by the French physicist Louis de Broglie at the European Cultural Conference in Lausanne in December 1949. The idea received a lot of support as more and more scientists came to the realisation that further progress in physics required resources beyond those of individual European nations. The added advantage was that the statesmen of Europe were eager to promote suitable projects that symbolised the "new spirit of European unity."

A series of conferences in 1950 and 1951, organised with the help of UNESCO, prepared the ground for the establishment of an international laboratory for nuclear physics. Eleven governments signed an agreement setting up a provisional "Conseil Européen pour la Recherche Nucléaire" in Geneva in February 1952. It gave rise to the acronym CERN, which has been retained ever since. Soon thereafter, an offer from Switzerland, which had remained neutral during the war and which was centrally located, for a site near Geneva for the laboratory was accepted.

As 1955 dawned, this decision was ratified by twelve countries that became member states: Belgium, Denmark, the Federal Republic of Germany, France, Greece, Italy, the Netherlands, Norway, Sweden, Switzerland, the United Kingdom, and Yugoslavia. Austria joined the convention in 1959. Spain was a member from 1961 to December 31, 1968, when it withdrew due to financial reasons but rejoined on January 1, 1983. Yugoslavia withdrew in 1962 due to financial reasons.

As of now, CERN has 23 member states: Austria, Belgium, Bulgaria, Czech Republic, Denmark, Finland, France, Germany, Greece, Hungary, Israel, Italy, the Netherlands, Norway, Poland, Portugal, Romania, Serbia, Slovak Republic, Spain, Sweden, Switzerland, and the United Kingdom.

Croatia, India, Latvia, Lithuania, Pakistan, Türkiye, and Ukraine are Associate Member States, which is the pre-stage to membership. Japan and the United States of America hold Observer status for the Large Hadron Collider (LHC), while the United States of America also holds Observer status for the High Luminosity Large Hadron Collider (HL-LHC). The European Union and UNESCO currently have Observer status at CERN.

Consequent to the Ukraine war, the Observer status of the Russian Federation was suspended on 8 March 2022. The Observer status of the Joint Institute of Nuclear Research (JINR), Dubna, was also suspended on 25 March 2022. CERN as well as other labs have come up with different innovative mechanisms to publish papers that have Russian co-authors, either by omitting the names of their institutes and funding agencies or by listing their identities on websites like ORCID. This is rather frustrating for all and till such intermediate steps were adopted, the publication of important results was stalled. This issue continues to simmer and affect activities at CERN.

9.2 Accelerators and Experimental Results

A document was ceremoniously signed by the President of the Swiss Federation, the President of the Council of CERN, the President of the Council of the Republic and Canton of Geneva, and the first Director-General of CERN. The document was then put in a steel cylinder, which was encased in the foundation stone of CERN, which was laid by Felix Bloch (Fig. 9.1), its first Director-General [2], on June 10, 1955, on 40 hectares of land at Meryn in Geneva. The document had the following text in English and French:

Fig. 9.1. Felix Bloch, Director-General of CERN, Laying the Foundation Stone.

*"EUROPEAN ORGANIZATION
FOR NUCLEAR RESEARCH
ON THIS TENTH DAY OF JUNE, ONE THOUSAND NINE HUNDRED AND FIFTY-FIVE, ON GROUND GENEROUSLY GIVEN BY THE REPUBLIC AND CANTON OF GENEVA WAS LAID THE FOUNDATION STONE OF THE BUILDINGS OF THE HEADQUARTERS AND THE LABORATORIES OF THE EUROPEAN ORGANIZATION FOR NUCLEAR RESEARCH, THE FIRST EUROPEAN INSTITUTION DEVOTED TO CO-OPERATIVE RESEARCH FOR THE ADVANCEMENT OF PURE SCIENCE."*

The people associated with CERN had already started working on plans to build the biggest accelerator in the world (25 GeV, proton synchrotron) as well as a smaller, less powerful machine (a 600-MeV synchrocyclotron) to gain experience in making accelerators and start experiments. The synchrocyclotron began operating on August 1, 1957 and was soon being used 24 hours a day. The proton synchrotron (PS) reached an energy of 24 GeV on 24 November 1959 and became the most powerful accelerator in the world.

Several bubble chambers were built and were in use, especially for the study of neutrinos produced at the PS. In 1965, CERN and the French Atomic Energy Authority agreed to build a very large heavy-liquid bubble chamber, which came to be known as Gargamelle. In 1967, CERN, France, and the Federal Republic of Germany agreed to construct a very large hydrogen bubble chamber, known as "Big European Bubble Chamber." Gargamelle is famous for being the chamber in which neutral currents were discovered in 1973, which was an indication of the existence of Z bosons.

The 600-MeV synchrocyclotron was largely rebuilt in the 1970s to give higher proton beam intensities. From 1967 till 1990, it remained the site of one of the finest facilities for the study of short-lived nuclei, using the Isotope Separator Online (ISOLDE) facility. The ISOLDE was transferred to the Proton Synchrotron Booster in 1990, and the synchrocyclotron was closed.

Since the 1970s, the PS has been principally used to inject particles into new machines. Several modifications, over the years, have led to a 1000-fold increase in the intensity of its proton beam. It has been accelerating the nuclei of helium, oxygen, sulphur, argon, xenon, and lead, in addition to electrons, positrons, protons, and antiprotons.

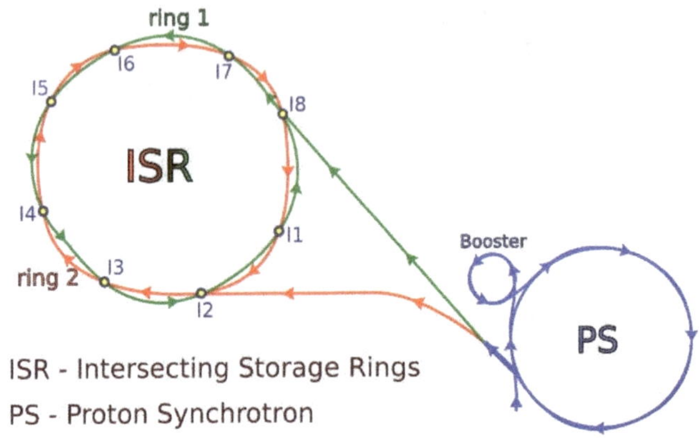

Fig. 9.2. The Proton Synchrotron and the Interacting Storage Rings at CERN.

The first hadron collider in the world, the Intersecting Storage Ring (ISR), was proposed in 1964, began operating on 27 January 1971, and ran till 1984. The ISR performed the first-ever proton–proton and proton–antiproton collisions, and it became the laboratory where stochastic cooling invented by Simon van der Meer was first developed. In between, the French government made the area adjoining the existing laboratory available to CERN for the installation of additional accelerators. Thus, "*CERN became the first international organization which crossed a frontier physically, as well as in spirit* [3]."

During the last phase of the ISR (Fig. 9.2), the energy of individual proton beams was raised to 31.4 GeV, leading to a centre of mass energy of more than 62 GeV for the collision, which would have required a beam of 2000-GeV protons hitting a stationary target (of protons).

A decision to build a 400-GeV Super Proton Synchrotron (SPS) was taken in 1971 and it started operation in June 1976. It is now the second-largest machine in CERN's accelerator complex. It measures about 7 km in circumference and operates up to an energy of 450 GeV (for protons). It provides different kinds of particles: heavy nuclei, electrons, positrons, protons, and antiprotons. SPS takes particles from the Proton Synchrotron, accelerates them, and feeds them to the Large Hadron Collider for further acceleration and use. It also provides beams to the NA61/SHINE, NA62, and the Common Muon and Proton Apparatus for Structure and Spectroscopy (COMPASS) experiments.

In 1983, the SPS, running as a proton–antiproton collider, led to discovery of W and Z bosons, which along with the discovery and implementation of the stochastic cooling won the Nobel Prize for Carlo Rubbia and Simon van der Meer in 1984. It also helped scientists study the formation of quark–gluon plasma in the relativistic collision of heavy nuclei, the matter in the first instants of the universe, about 15 years later.

The construction of a Large Electron–Positron (LEP) Collider, one of the largest particle accelerators ever built and having a 27-km circumference, to utilise the clean interaction process of point-like electrons and positrons, was approved in 1981. The challenging precision work on the construction formally started in February 1985, and the work on the detector systems (L3, ALEPH, OPAL, and DELPHI) started across the world. The LEP was officially inaugurated on 13 November 1989. The LEP initially reached a collision energy of 91 GeV, which was raised to 209 GeV by the end of 2000. The experiments at the LEP led to a precise and detailed study of the electroweak interaction and a precise value for the mass of W and Z bosons. The experiments at the LEP also provided proof that there are three — and *only* three — generations of particles of matter and gave a hint about the mass of the Higgs boson. The LEP was shut down on 2 November 2000 to make way for the construction of the Large Hadron Collider in the same tunnel.

The construction of the LHC, the world's largest and highest-energy particle collider, was completed in 2008 with the collaboration of over 10,000 scientists and hundreds of universities and laboratories, as well as more than 100 countries, at a cost of about 7 billion USD (Fig. 9.3).

It has four experimental stations: ALICE, ATLAS, CMS, and LHCb (Fig. 9.4). While ALICE (Fig. 9.5) specialises in the study of collisions of heavy nuclei and looks for quark–gluon plasma, the LHCb experiment specialises in investigating the slight differences between matter and antimatter by studying the "beauty/bottom quark" or "b quark." ATLAS (Fig. 9.6) and CMS (Fig. 9.7) study the Higgs boson, extra dimensions, and particles that can make up dark matter, with different magnet designs and techniques for the detection of particles.

The first proton–proton collisions, at a collision energy of 7000 GeV, were observed in 2010. The long-awaited (50 years) discovery of the Higgs boson (Fig. 9.8), which explains how elementary particles acquire mass, was announced by the ATLAS and CMS experiments on July 4, 2012 to a highly excited and expectant world. François Englert and

Fig. 9.3. The LHC Tunnel (Credit: CERN).

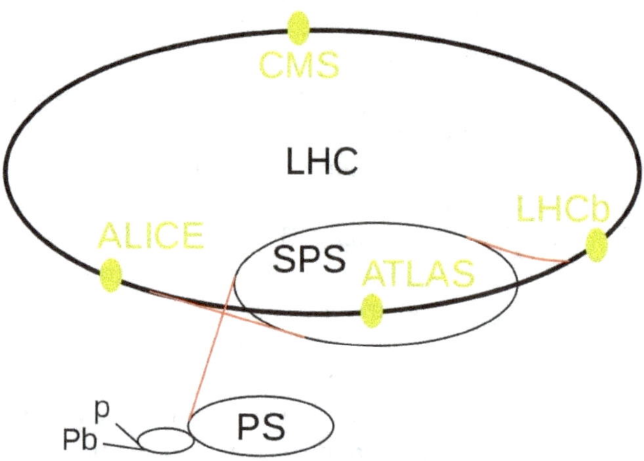

Fig. 9.4. The Schematic of the LHC at CERN along with the Experimental Stations.

Peter W. Higgs were jointly awarded the Nobel Prize for proposing the so-called Higgs Mechanism in 2013.

The LHC and its detectors have already undergone upgrades, which have increased the energy and intensity of the beams and improved the precision and efficiency of the detector systems.

Fig. 9.5. ALICE Experiment (Credit: CERN).

Fig. 9.6. ATLAS Detector at CERN (Credit: CERN).

Fig. 9.7. The CMS Detector: One of the Largest International Scientific Collaborations in History, Involving about 5500 Particle Physicists, Engineers, Technicians, Students, and Support Staff from 241 Institutes in 54 Countries. (Credit: CERN)

As indicated earlier, the LHC also continues to study the collision of heavy ions, such as lead nuclei, at extremely high energies. This has helped recreate conditions like the early universe (in a "mini bang") and the formation of a state of matter called the quark–gluon plasma, which existed just moments after the Big Bang. The LHC studies the decay modes of the Higgs boson with extreme precision, looking for its excited states, if any, and continues to search for new physics beyond the standard model, such as supersymmetry and extra dimensions, as well as hints of dark matter.

The LHC is likely to complete its mission in 2040. CERN has initiated studies to build a Future Circular Collider (FCC) to be installed in a tunnel measuring some 91 km in circumference at a depth of between 100 and 400 m on French and Swiss territory, passing under Lake Geneva (Fig. 9.9). It aims to push the energy and intensity frontiers of particle colliders to collision energies of 100 TeV in search of new physics [4]. An international collaboration of more than 150 universities, research institutes, and industrial partners from all over the world is developing

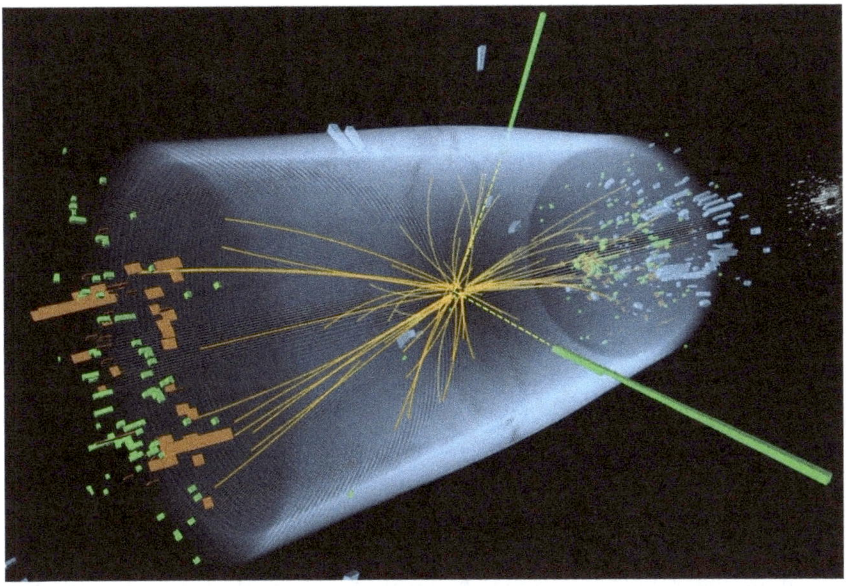

Fig. 9.8. An Event Recorded with the CMS Detector in 2012 at a Proton–Proton Centre of Mass Energy of 8 TeV. The Event Shows Characteristics Expected from the Decay of the Standard Model Higgs Boson to a Pair of Photons (Dashed Yellow Lines and Green Towers) (Credit: CERN).

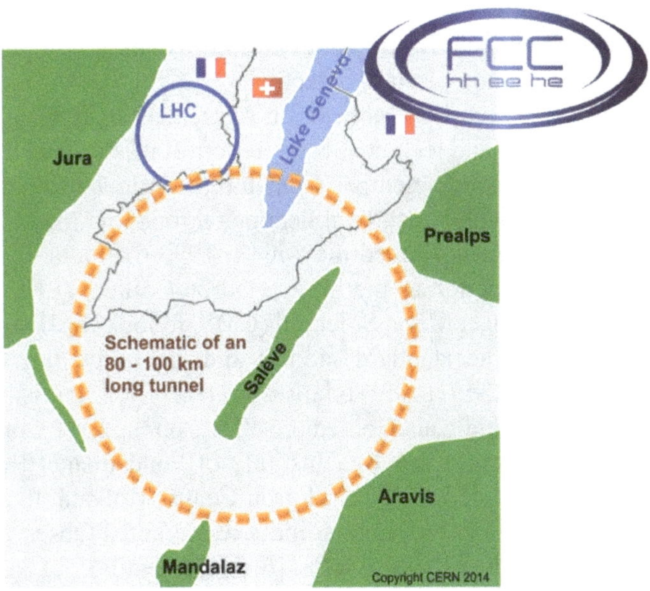

Fig. 9.9. Schematic of 100-km Tunnel for the FCC (Credit: CERN).

possibilities for circular colliders, new detector facilities, the associated infrastructure, cost estimates, global implementation scenarios, and the appropriate international governance structures.

The FCC examines scenarios for three different types of particle collisions: proton-proton and heavy-ion — heavy-ion collisions, electron–positron collisions, and the possibilities of proton–electron collisions and proton–heavy-ion collisions.

We add that there has been some spirited opposition to the building of the FCC [5], with people calling it a "waste" of money, resources, and effort — an accusation that has attracted mixed responses.

9.3 India at CERN

India's relationship with CERN started in the 1960s when high-energy physicists from the Tata Institute of Fundamental Research started actively participating in experiments at CERN [6, 7]. They were later joined in the 1990s by scientists from the Raja Ramanna Centre for Advanced Technology, Indore, who along with researchers from other institutes and universities built components for the Large Electron–Positron (LEP) collider and the L3 (to measure W particles with unprecedented accuracy, to determine important properties of the Z boson, and to set limits on the possible mass of the Higgs boson), and the WA93 and WA98 (used to study relativistic heavy-ion collisions and look for the formation of quark–gluon plasma) detectors.

India and CERN signed a Cooperation Agreement in 1991, which set priorities for scientific and technical cooperation. It was followed by several agreements/protocols. Scientists from India continued to participate in important physics analyses and publications throughout these years.

All these developments paved the way in 1996 for India to agree to take part in the construction of the Large Hadron Collider (LHC) and to contribute to the Compact Muon Solenoid (CMS, to look for Higgs boson and physics beyond the standard model) and "A Large Ion Collider Experiment" (ALICE, to study relativistic heavy-ion collisions and quark–gluon plasma). India also joined the efforts on the LHC Computing Grid with Tier-2 centres at the Tata Institute of Fundamental Research, Mumbai, and the Variable Energy Cyclotron Centre, Kolkata. In recognition of these substantial contributions, India was granted Observer status to the CERN Council in 2002 (see Refs. [6, 7] for details).

These successes also led to cooperation on Novel Accelerator Technologies through India's participation in CERN's Linac4, Superconducting Proton Linac (SPL), and Compact Linear Collider Test Facility (CTF3) projects. CERN is also participating in accelerator programmes in India. India also participates in the COMPASS, Isotope mass Separator OnLine facility (ISOLDE), and neutron Time of Flight (nTOF) experiments.

India became an Associate Member State of CERN on 16 January 2017.

To celebrate its association with CERN, which started in the 1960s and which is continuing ever more strongly, India gifted a 2-m-tall statue of the deity Lord Shiva engaged in Ananda Tandava, as Nataraja — the King of Dancers (Fig. 9.10). It was unveiled on 18 June 2004. It is believed that Lord Shiva danced the universe into existence. He motivates (preserves) it and will eventually extinguish it. Fritjof Capra [8] and Carl Sagan have repeatedly drawn a parallel between the cosmic dance of Nataraja and the modern study of the "cosmic dance" of subatomic particles, which are continuously created and annihilated.

9.4 Societal Benefits, Science Outreach

The construction and operation of the LHC have driven advancements in cutting-edge technologies for superconducting magnets, cryogenics, computing, and data storage. These innovations have applications in various fields beyond particle physics, including medical imaging, aerospace, and materials science. Medical imaging benefits from new types of fast, bright, and dense scintillating crystals for PET scanners, for which CERN initially made pioneering contributions [9].

The Accelerator-Driven Subcritical System (ADSS) is an interesting concept that has been developed at CERN with several laboratories across the world working on it [10]. It uses a high-intensity beam of high-energy protons from an accelerator to produce neutrons from the spallation of heavy nuclei to induce nuclear fission and even incinerate nuclear waste in an inherently subcritical and safe nuclear reactor.

The technologies developed at CERN, such as particle detectors and accelerators, have been adapted for medical purposes. Hadron therapy aims to treat tumours with beams of protons and light ions to reduce the radiation exposure of healthy tissues. Several therapy centres using

Fig. 9.10. Nataraja Statue at CERN (Credit: R. Sahoo).

hadrons were built in close cooperation with CERN for accelerator design, construction, testing, and further technology developments. CERN has been providing support for the development of miniature linear accelerators for proton therapy.

Moreover, the software for simulating particle interactions in detectors used routinely at CERN is valuable for space applications and for calculating the precise radiation dose for cancer treatment planning systems.

The pixel detector technologies used in medical diagnostics, industrial processes, X-ray-based material analysis, and space missions on the International Space Station were developed at CERN.

There is a very interesting case involving the resolution of a mystery associated with the famous painting "Madonna and Child," attributed to Raphael (1483–1520 CE). It had changed hands several times, during which it became the property of different popes and was looted by Napoleon during his Italian campaign. In the 1930s, it ended up in a private collection in Prague, thus disappearing from public view. A particle detector system developed at CERN was used to take several very high-resolution images of the painting at different wavelengths of X-ray, examining which experts confirmed that the painting contained the unique brushstrokes employed by Raphael and was indeed his creation [11].

By far the most visible technology is the World Wide Web invented by Tim Berner Lee at CERN in 1989, which has had the most far-reaching impact on our daily life. It was driven by the need for better communication and sharing of vast quantities of data among scientists around the world. CERN pioneered yet another breakthrough, the touchscreen, which is now used extensively in mobile phones, iPads, computers, and almost all walks of life.

It is well known that working on the projects at CERN has raised the level of competence of the almost all industries — civil, mechanical, cryogenics, micro-electronics, electrical power, very large scale integration chips, imaging, machine learning, computing, and data transfer, etc.

CERN is one of the most successful models for international cooperation and a living testimony of it. It is estimated that CERN hosts about 70% of the world's particle physicists, who use one of the biggest scientific endeavours in the history of humankind — the Large Hadron Collider and the detectors there — to study the most basic constituents of nature and their interactions. About 16,000 scientists of more than 110 different nationalities work together at CERN, towards a common goal, regardless of ethnic, cultural, political, or religious differences.

Like NASA in the USA, CERN has had a vibrant science outreach programme since its inception, thereby improving science education from secondary school to the postgraduate level and imparting a broader understanding of science to the public. It has been publishing and distributing *CERN Courier*, a lively journal with news about CERN, accelerator laboratories around the world, conference reports, and important scientific developments, since 1959, free of cost to anyone, anywhere in the world.

Every summer, about 300 undergraduate students from CERN member and non-member states participate in an 8–13-week internship programme. This is in addition to the several hundred young physics students, engineers, and technicians who receive higher degrees from universities for work carried out at CERN. It is understood that 600 students obtain a doctorate every year on work done using results obtained at CERN.

As a part of its science outreach, CERN organises guided tours for about 120,000 visitors per year — among whom 70,000 are school students. It also organises travelling exhibitions, which have been seen by more than one million visitors in the member states.

Richard Feynman is believed to have said the following: "*If you can't explain your research to your grandmother, then you don't understand it yourself.*" With this exalted and difficult job in mind, CERN has operated a highly popular permanent exhibition Microcosm since September 1997. CERN has also opened another permanent exhibition, called Science Gateway, to show the importance of fundamental research and its applications to society, infuse everyone who visits it with curiosity and a passion for science, and inspire young people to take up careers in science. At the Science Gateway, scientists and the public can interact daily. It is designed to provide an immersive hands-on experience of the joys of science to people of all ages (participants should be older than five years of age). The brochures at this exhibition are another lovable feature, with wonderful cartoons (Fig. 9.11) explaining almost all the work being done at CERN (see, e.g., Ref. [12]). Because of all these, CERN has by now entered our everyday lexicon.

9.5 The Working of CERN

There is no doubt that CERN has been a successful collaborative international venture that has attracted brilliant young minds into science. It is expected that it will help all regions and contribute to building a peaceful global society.

What we have not sufficiently emphasised is that each accelerator planned by CERN was completed often before time, mostly within budget, and most exceeded their designed energies and intensities. This phenomenon has been studied and contrasted with the International Thermonuclear Experimental Reactor (ITER), which is considerably behind schedule and has greatly exceeded its budget.

Over the years, CERN's administration has evolved to become somewhat flexible and accommodate scientists and politicians as much as possible. The CERN Council has two members each from the twenty-three member states, one representing his or her government's administration and the other representing national scientific interests. The Council is not supposed to micromanage the projects, leaving it to the CERN management.

Fig. 9.11. A Brochure for the ALICE Experiment at CERN [12].

Countries contribute to the central infrastructure of the laboratory, and the financial contributions of the member states are used to fund the operation of CERN, including the construction and operation of the LHC, the salaries of the staff, and the purchase of equipment. The Council considers several factors when deciding the contribution of an associate member state, including the country's net national income, its scientific and technological capabilities, and its contribution to CERN's research programme [13].

The associate member states are not required to contribute as much as the member states. However, they are expected to contribute a significant amount to support CERN's research programme.

However, while the money for the infrastructure of the laboratories comes from the pooled fund, the money for the experiments does not. Thus, for example, while the LHC was built by CERN using the money contributed by the member states, the four giant detectors, ALICE, ATLAS, CMS, and LHCb, were funded, designed, and built by an independent collaboration of nations. This ensures that if one country falls behind, it does not mean the entire project suffers.

Inspired by its success, several international collaborations closely follow the CERN model: the European Southern Observatory (ESO), the Human Genome Project (HGP), the International Space Station (ISS), the Laser Interferometer Gravitational-Wave Observatory (LIGO), the International Thermonuclear Experimental Reactor (ITER), the Global Ocean Sampling Expedition (GOS), the Square Kilometre Array (SKA), the International Fusion Materials Irradiation Facility (IFMIF), the Global Initiative on Sharing All Influenza Data (GISAID), the European Molecular Biology Laboratory (EMBL), the European Space Agency (ESA), the Joint European Torus (JET), the International Institute for Applied Systems Analysis (IIASA), and the Synchrotron-Light for Experimental Science and Applications in the Middle East (SESAME).

The SESAME project, for its part, represents an audacious effort to use science for diplomacy in the heart of the Middle East. Unfortunately, peace in the Middle East is under severe strain as we write this. However, we hope that peace will prevail before long. After all, it is a unique joint venture that brings together scientists from Bahrain, Cyprus, Egypt, Iran, Israel, Jordan, Pakistan, the Palestinian Authority, and Turkey. The facility provides an intense source of light, including infrared and X-ray wavelengths, allowing researchers from the region to investigate the properties of advanced materials, biological processes, and cultural artefacts. SESAME Council members share a vision to unite scientists from many Arab and Muslim-majority nations, along with Israel, to work together towards common scientific goals.

It is important to realise that in some countries, funding for international collaborations like CERN comes from their Department of Foreign Affairs, while in others it comes from the Department of Science. In the latter cases, it leads to friction with domestic science projects which must compete with international collaborations for funding. As more and more science projects shift to large international collaborations, a more equitable solution to this irritant has to be found.

We must add that of all the international collaborations, ITER is most often compared with CERN. The fusion project is years behind schedule and projected to be billions of dollars over budget. Several commentators have attributed it to the "politics" of equal division of responsibilities, where different nations picked different components to build based on national interests, rather than efficiency. The central management of ITER is also blamed for not providing flexibility to adapt to design changes and delays.

Chris Llewellyn Smith, former Director-General of CERN, has suggested that "ITER is being built in a way that is closer to the way the LHC experiments were built, rather than the way the collider itself was built." Moreover, the fusion reactor is essentially a single giant experiment, and all the components must be in place and ready before it fires up. Thus, "A major problem with one of the members will impact the project as a whole [14]."

Russia had been contributing to CERN extensively. About 8% of the workforce at CERN was of Russian nationality. The Russian invasion of Ukraine in 2022 led to objections to the participation of Russians and Russia at CERN. CERN "intends to terminate" its cooperation agreements with Belarus and Russia when they expire, respectively, in June and December 2024 and take additional steps as warranted. In October 2024, CERN, confirmed that it will suspend access to its state-of-the-art facilities for roughly 500 scientists affiliated with Russian institutions from 30 November 2024. Belarus's five-year agreement was not renewed when it expired last 27 June 2024.

However, CERN has confirmed that these decisions do not impact employees of Joint Institute of Nuclear Research, Dubna, as it is considered an international institute.

Finally, we must add that almost all countries that have joined the CERN efforts have had to face criticism directed against this international collaboration from domestic research with which it competes for funding, as mentioned earlier.

9.6 Conclusions

A ceremony to mark the twenty-fifth anniversary of CERN was held there on 23 June 1979. Victor Weisskopf, one of the most respected theoretical nuclear physicists of his time, who had served as Director-General of CERN from 1961 to 1966, spoke about "The Significance of CERN."

He said, "*For me, the development of CERN in the last three decades is not only an impressive story of success but also a fulfilment of a dream. Our dream was to see a great and active laboratory of fundamental physics in Europe that transcends national boundaries and is a symbol of a bright future, when humanity will be united and when national pride does not refer to any specific country but to the whole of our great planet Earth.*"

He recounted the successes of CERN till that day, and while commenting on the study of nuclear reactions taking place in the stars, said, "The nuclear realm becomes active (apart from the laboratories like the CERN) only in the great cataclysms such as the star explosions or the big explosion in which the universe was born (the 'big bang'). To study the relevant phenomena, we must create conditions that do not exist on Earth."

And then, he almost presciently added, "I cannot repress a certain pride as a scientist that we were able to realise such cosmic environments in our laboratories." The LHC, which truly created the conditions of the Big Bang, was still more than three decades away.

It is well known that most fundamental discoveries and ideas about the structure of matter originated in Europe in the first third of the twentieth century. After the Second World War, this type of research took place mostly in the USA as Europe was suffering from the ravages of the war. Moreover, the size of establishments that could cater to these requirements was too large for any single country in Europe. We know that the enthusiasm and motivation to explore particle physics at the high-energy frontier transcend boundaries between nations and regions of the planet Earth. The same enthusiasm and motivation are shared by physicists of widely different cultures (Fig. 9.12) and origins, and international collaboration is the most natural strategy as "we all have the same dream and the same fundamental goal." CERN made this possible. It has also been remarked that elegance in construction, a perfectionist's care for precision and reliability, and pioneering inventions and development of new principles of instrumentation are the hallmarks of the laboratories, accelerators, and detectors at CERN. "All this was done by a group of engineer-physicists that has no parallel in any other institution and is unique in the world [3]."

Hazaran Blbul, the Nightingale of a Thousand Songs from an Armenian fairy tale, is believed to live beyond "the seven seas and the seven mountains." It is said that to reach there, the hero must pass through

Fig. 9.12. Particle Hunters (from Cartoon Album "Hunting Particles," CERN).

untold difficulties. It is also said that he must be "reborn" and "must have a kind and generous heart, a virtuous, honest, and noble soul, a strong will, and an unbreakable spirit, and transform to become his perfect, highest version." And only then can the hero find and bring back the Hazaran Blbul whose songs make the destroyed gardens green again. Hazaran Blbul sings in a thousand languages. It has a thousand ways and a thousand forms, and its songs of wonderful, inexplicable, untold, blinding beauty awaken the minds of the listeners.

We have seen that CERN is Hazaran Blbul and much more.

References

[1] M. Mukerjee, "*Churchill's Secret War: The British Empire and the Ravaging of India during World War II*" (New Delhi: Tranquebar Press, 2011).

[2] CERN, "Laying of Foundation Stone: Signature of the Agreement with the Swiss Federal Council" (10, 11 June 1955), http://cds.cern.ch/record/59219/files/1955%20CERN%20foundation%20stone%20booklet.pdf.

[3] "A Brief History of CERN", *CERN Courier* (25th Anniversary Issue) 19 (1979) 228.

[4] A. Abada *et al.*, "FCC Physics Opportunities: Future Circular Collider Conceptual Design Report Volume 1", *European Physical Journal C* 79 (2019) 474, "FCC-hh: The Hadron Collider", *European Physical Journal Special Topics* 228 (2019) 755–1107, and references therein.

[5] S. Hossenfelder, "The World Doesn't Need a New Gigantic Particle Collider", *Scientific American* (19 June 2020).

[6] D. K. Srivastava and V. S. Ramamurthy, "Chapter 51: India's Road to HEP". In: P. Charitos, T. Arabatzis, H. Cliff, G. Dissertori, J. Forneris, and L. Li-Ying (eds.) "*Big Science in the 21st Century: Economic and Societal Impacts*" (IOP Publishing, Bristol, UK, 2023), pp. 51-1–51-18.

[7] D. K. Srivastava, V. S. Ramamurthy, A. Shrivastava, and A. K. Mohanty, "Homi Bhabha and His Legacies with Specific Reference to Nuclear and High-Energy Physics Research in Independent India", *Current Science* 123 (2022) 429–440.

[8] F. Capra, "The Tao of Physics" (Boulder, Colorado: Shambhala Publications, 1975), p. 242.

[9] https://home.cern/sites/default/files/2018-07/CERN-Brochure-2016-005-Eng.pdf.

[10] D. K. Srivastava and V. S. Ramamurthy, "*Climate Change and Energy Options for a Sustainable Future*" (Singapore: World Scientific, 2021).

[11] A. Le Gall, "CERN Technology Helps Rediscover Lost Painting by Raphael", *CERN Courier* (21 September 2020), https://home.cern/news/news/knowledge-sharing/cern-technology-helps-rediscover-lost-painting-raphael#:~:text=For%20a%20long%20time%2C%20it,in%20Prague%20during%20the%201930's.

[12] https://cds.cern.ch/record/1456256/files/ALICE%20Brochure%20Eng.pdf.

[13] CERN Report No. CERN/FC/6628, CERN/3689 (1 December 2022). https://cds.cern.ch/record/2847387/files/English.pdf.

[14] M. Lucibella, "CERN: 60 Years of Collaboration", *American Physical Society News* 23(8) (2014), https://www.aps.org/publications/apsnews/201409/cern.cfm.

Chapter 10

ITER: The Way

"If, indeed, the sub-atomic energy of the stars is being freely used to maintain their furnaces, it seems to bring a little nearer to fulfilment our dream of controlling this latent power for the well-being of the human race — or for its suicide."

— Arthur Eddington (1882–1944 CE)

10.1 Preamble

We have known for close to a century that the sun and all the stars are enormous balls of burning plasma (Fig. 10.1). This burning plasma (mainly) involves a set of fusion reactions known as the proton–proton chain (Fig. 10.2). The self-heating property of hydrogen ions (protons) sustains the temperature of the ions above the level required for fusion to occur.

Our sun has been radiating heat and light for billions of years and is expected to do so for billions more before it becomes a red giant and swallows Mercury, Venus, and perhaps even the Earth. The sun uses about 600 million tonnes of hydrogen per second to generate 595.72 million tonnes of helium. The difference in the mass amounting to 4.48 million tonnes is converted into energy, and the sun radiates 3.8×10^{26} W of power. The fusion of protons is facilitated by the temperature of 15 million degrees centigrade at the core. The temperature is only about 6000 degrees centigrade at the surface. The enormous mass of the sun (about 2×10^{30} kg)

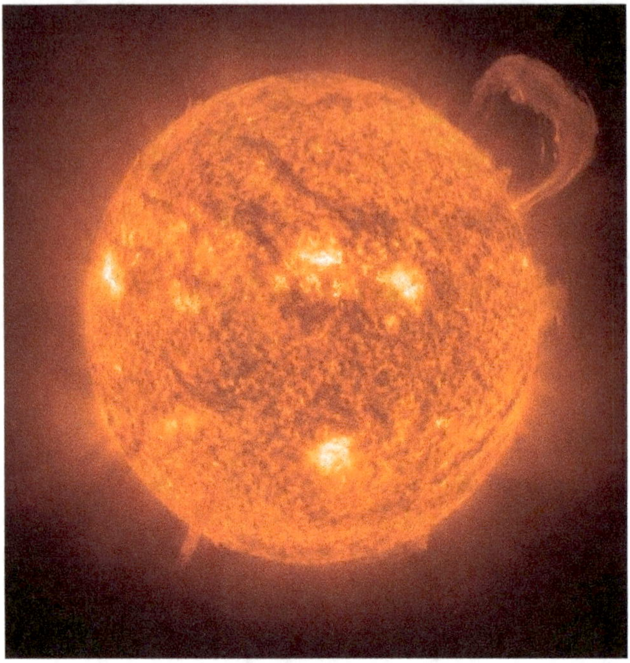

Fig. 10.1. The Sun (Credit: NASA).

and the resulting gravity (28 times that of the Earth) confine the plasma in a fireball with a radius of 696,340 km [1].

Once the source of the energy of the sun and stars was identified, Arthur Eddington (1882–1944), speaking at the annual meeting of the British Association in 1920 at Cardiff, expressed the following [2]:

> "If, indeed, the sub-atomic energy of the stars is being freely used to maintain their furnaces, it seems to bring a little nearer to fulfilment our dream of controlling this latent power for the well-being of the human race — or for its suicide."

The realisation of "uncontrolled fusion" was achieved within 32 years, when first the United States of America and later several other countries tested the hydrogen bomb, which brought humans closer to what Eddington called "suicide," with only the spectre of Mutually Assured Destruction (MAD), resulting from the race to acquire nuclear weapons, saving us from that terrible fate.

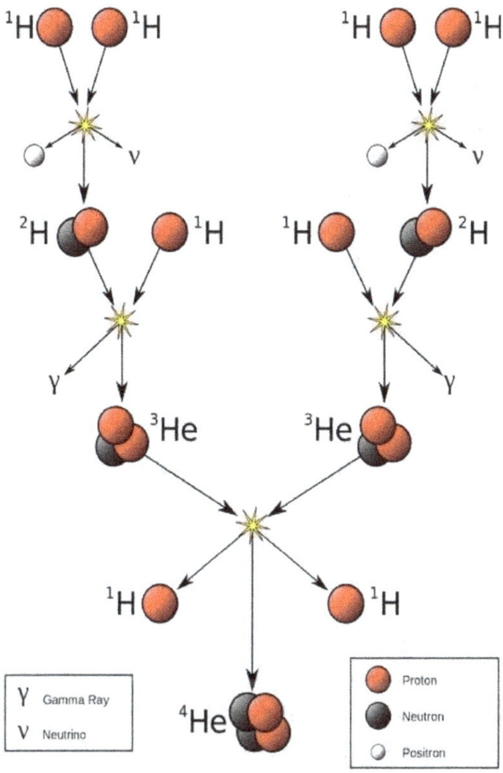

Fig. 10.2. The pp Chain in the Sun (Credit: Wikipedia).

The very high temperature at the core of the sun provides enough (kinetic) energy to the (positively charged) protons to overcome their mutual Coulomb repulsion and undergo collision. However, the pp chain, which accounts for most of the sun's energy production, proceeds via weak interaction (the first step is generally $p + p \rightarrow d\ (^2H) + e^+ +$ electron-neutrino) and has a very low probability. The reaction of choice in terrestrial efforts is $^2H + {}^3H \rightarrow {}^4He\ (+3.5\ MeV) + n\ (+14.1\ MeV)$, which has a much larger cross section at the collision energies that can overcome the mutual Coulomb repulsion of the two nuclei [1]. It is expected that the ^4He nuclei would deposit their energy in the plasma (due to collision with ions), while neutrons would escape and heat the plasma vessel.

This requires the plasma to be heated to about 150 million degrees centigrade. This is expected to be achieved with a combination of the following:

- Neutral beam injection, which involves injecting high-energy particles into the plasma. The particles are accelerated in a linear accelerator, neutralised, and then injected into the plasma. The neutral particles collide with the plasma particles and transfer their energy to them.
- Ion cyclotron resonance heating, where radio waves are used to heat the ions in the plasma. The radio waves are tuned to the ion cyclotron frequency at which the ions rotate in the magnetic field. The radio waves absorbed by the ions transfer their energy to the ions and heat them.
- Electron cyclotron resonance heating, where radio waves are used to heat the electrons in the plasma, as in ion cyclotron resonance heating.
- Ohmic heating, where an electric current is passed through the plasma. The passage of the current creates a magnetic field, which in turn creates resistance in the plasma, which heats it.

The advantages of nuclear fusion as an energy solution for our planet are well known due to its safety, fuel availability, and environmental protection. Planet Earth has an abundant supply of fuel (0.02% of the seawater is heavy water), needed to produce deuterium (^2H). A steady supply of tritium (^3H) is possible from heavy-water reactors or the use of neutrons produced during the fusion from their reaction with ^6Li (Fig. 10.3).

Most importantly, a fusion reactor would produce virtually no emission of carbon dioxide or atmospheric pollutants, there would be no chance of a meltdown, and its radioactive waste products would mostly be very short-lived compared to those produced by conventional nuclear reactors (fission reactors).

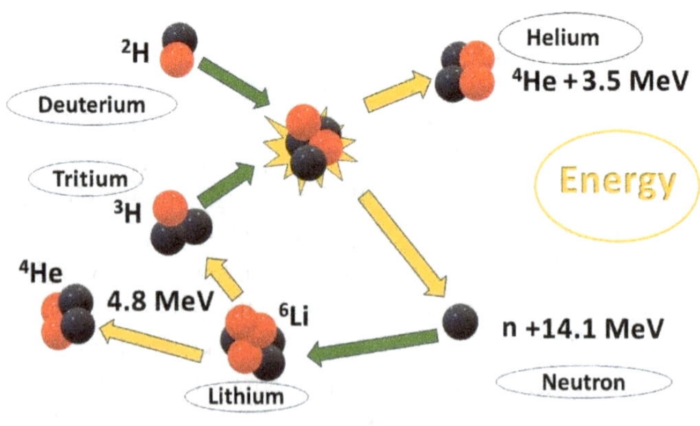

Fig. 10.3. Deuterium–Tritium Fusion and Tritium Breeding.

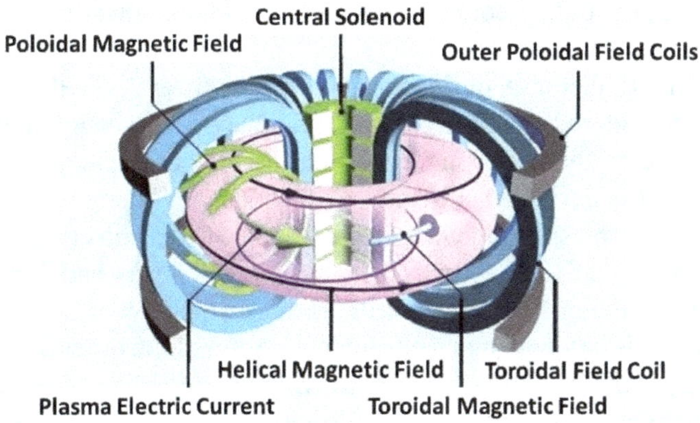

Fig. 10.4. Basic Components of a Tokamak: The Toroidal Field Coils (in Blue), the Central Solenoid (in Green), and Poloidal Field Coils (in Grey). The Total Magnetic Field (in Black) Around the Torus Confines the Path of Travel of the Charged Plasma Particles (Credit: EUROfusion and DOE).

The plasma can be confined using a magnetic field, and the fusion reactor is inherently safe. It does not create high-level radioactive waste or greenhouse gases, and the problems of global warming, acid rain, and toxic gas emissions do not arise with nuclear fusion.

The most common type of magnetic confinement fusion device is the tokamak (Fig. 10.4). Invented by two Soviet physicists, Igor Tamm and Andrei Sakharov, in the 1950s, it is a doughnut-shaped chamber with a strong magnetic field that encircles the plasma.

The other pathway to fusion is inertial confinement, which uses lasers or particle beams to compress a small pellet of fusion fuel to extremely high densities so that the high pressure and temperature created by the compression cause the fuel to fuse. The most common type of inertial confinement fusion device is the laser fusion device, which uses a series of lasers to compress the fuel pellet. Other types of inertial confinement fusion devices include particle beam fusion devices and shock wave fusion devices [1].

Both magnetic confinement and inertial confinement have their respective challenges. Magnetic confinement fusion is difficult because it is challenging to maintain the plasma in stable confinement for a long period. Inertial confinement fusion is difficult because it is challenging to create the high pressure and temperature needed to fuse the fuel.

However, despite the challenges, tokamaks are believed to be the most promising way to achieve fusion power. Scientists are confident that they will be able to overcome the obstacles and make fusion power a reality in the future.

The quality factor Q, defined as the output power divided by the input power, is considered a measure of the efficiency of the fusion device. The other parameter characterising a tokamak is the confinement time (τ) of the plasma. It is defined as the ratio of the plasma volume divided by the flow rate of the plasma and controls how long the plasma can be confined in the tokamak before it escapes. The earliest tokamaks had confinement parameters of the order of a few milliseconds. It is now often of the order of seconds. It is believed that a confinement time of the order of 100 s may be needed for the device to be economically viable. It can be increased by increasing the volume of the plasma and decreasing the flow rate (by increasing the magnetic field strength by employing superconducting coils to generate the magnetic fields).

Tokamaks have steadily, though quite slowly due to the challenges involved, improved in quality since the initial efforts in the 1950s. The first tokamak, T-1, built in 1958 in the former Soviet Union had a Q factor close to 2. The concept was further developed across the world, and the Joint European Torus (JET) tokamak was built in the United Kingdom in the 1970s, with a Q factor of 10. The International Thermonuclear Experimental Reactor (ITER, see later) tokamak, with a Q factor of 10, was proposed in the 1990s.

Several countries, e.g., China, the European Union, India, Japan, South Korea, and Russia, have started planning and working on a DEMO (demonstration) tokamak, a prototype fusion power plant, with a Q value of about 1000, which would be economically feasible. Even though its construction would be quite challenging, the confidence of the scientists stems from the fact that the magnetic fields used to confine the plasma have become stronger and the materials used to build tokamaks have become more resistant to heat and radiation.

10.2 Setting up an International Collaboration for ITER

The relentless pursuit of stockpiling more and more advanced nuclear weapons and missiles, which were sufficient to incinerate and obliterate life on Earth several times over, had brought about a realisation that a nuclear war leads to Mutually Assured Destruction (MAD) and can never be won. Calling the Mutually Assured Destruction a suicide pact, Ronald Reagan announced the Strategic Defence Initiative, popularly known as Star Wars, in March 1983, which was prohibitively complex and expensive to build and guard, both for the United States of America and the Union of Soviet Socialist Republics. The time was ripe to step back from the brinkmanship in which the Soviet Union and Western nations led by the United States of America had engaged for decades.

On the other side of the so-called Iron Curtain, Mikhail Gorbachev, a graduate student of law at the Moscow State University, became a close friend of Evgeny Velikhov, who was studying physics. The latter progressed to pilot the Soviet fusion research programme in 1973. Velikhov had long been convinced that progress in this massively challenging field of research *could only be achieved through international collaboration.* His conviction was shared by his Western colleagues, and soon by Gorbachev himself (Fig. 10.5).

Scientists from the two blocs had always maintained contact, and scientists researching fusion were no different; in fact, scientists from both sides of the East–West divide had been communicating and exchanging notes on research since the late 1950s. As early as 1973, the United States, the Soviet Union, Europe, and Japan had formally engaged in the

Fig. 10.5. Evgeny Velikhov and Mikhail Gorbachev: Friends from Their Student Days (Credit: ITER).

INTOR (INternational TOkamak Reactor) project, whose concept and organisation were similar to what ITER was to be. However, due to the lack of strong political determination, the INTOR never reached the design phase.

Mikhail Gorbachev had by 1985 become the Secretary General of the Communist Party of the Soviet Union. He supported collaboration with the West, rather than pursuing the ruinous arms race of the Cold War.

Gorbachev first revived interest in a collaborative fusion project in an October 1985 meeting with French President François Mitterrand. Gorbachev and Reagan met in Geneva in November 1985 (Fig. 10.6), and their discussion revolved around easing tensions between the two blocs. Against this background, a large international scientific and industrial collaboration that would open the way to a new energy source "for the benefit of all mankind" emerged as a powerful symbol of the post-Cold War world, that the two leaders wanted to shape [3].

At their second meeting in Reykjavik, Iceland, in October 1986, the two leaders reiterated their commitment and officially launched an international fusion initiative. The project was initially called International Thermonuclear Experimental Reactor and abbreviated as ITER.

Fig. 10.6. Ronald Reagan and Mikhail Gorbachev in Geneva, 1985 (Credit: ITER).

After some time, the name ITER was identified as the Latin word for "The Way!"

The compulsions of different nations/groups can now be summarised in the following. The EU is responsible for the largest portion of construction costs (45.6%); the remainder is shared equally by China, India, Japan, Korea, Russia, and the US (9.1% each).

- The United States and the European Union are both eager to maintain their leadership in fusion energy research. They see ITER as a key to this goal, as it is the first major fusion reactor that is expected to produce more energy than it consumes.
- Russia (the inheritor of the USSR mantle) and China are both investing heavily in fusion energy research. They see ITER as an opportunity to collaborate with other leading countries in this field, which may help accelerate their fusion programmes.
- India, Japan, and Korea are all looking for ways to reduce their reliance on fossil fuels. ITER offers them a chance to develop a new source of clean energy that could help them meet their climate goals.
- By working together on a common goal, the member states can build trust and understanding, and they can reduce the risk of conflict.
- We add that in addition to these seven groups/nations, the United Kingdom participates through the European Union's Fusion for Energy (F4E), Switzerland participates through Euratom and F4E, and the project has cooperation agreements with Australia, Canada, Kazakhstan, and Thailand.

10.3 ITER Progress

The ITER agreement was formally signed on 21 November 2006 (Fig. 10.7). The ITER tokamak is a major step towards the development of fusion power. It is the largest and most powerful tokamak ever built, and of course, it is expected to produce more fusion power than it consumes. The results from ITER will be essential for the development of DEMO, the prototype fusion power plant.

Construction of the ITER complex at Saint-Paul-lès-Durance in France started in 2013 and assembly of the tokamak began in 2020. The initial budget was estimated to be about 6 billion euros. However, delays, the rising price of raw materials, and changes to the initial design have

Fig. 10.7. The ITER Agreement is Signed, Elysee Palace Paris, 21 November 2006 (Credit: ITER).

seen the official budget estimate rise to between 18 billion and 20 billion euros. COVID-19 and damage to supplied parts during the transit have also added to the delay and rising costs. It was originally envisaged that ITER would test its first plasma in 2020 and achieve full fusion by 2023. However, the schedule is now to test the first plasma in 2025 and full fusion in 2035, which may be further pushed back, adding fuel to the cynical comment that "fusion power is always 30 years away," made by people who are ignorant of the complexities involved.

The design of ITER along with the integrated plant systems is given in Fig. 10.8.

The main contributions of various groups/nations (Fig. 10.9) are as follows:

- F4E (EU) is responsible for contributing to the design and manufacture of components such as the vacuum vessel, the diverter, and the magnets.
- China is responsible for components such as the correction coil, magnet supports, the first wall, and the shield blanket.

Fig. 10.8. The Design of ITER along with the Integrated Plant Systems (Credit: ORNL).

Fig. 10.9. The ITER Nations (Credit: ITER).

- India is responsible for the cryostat, in-vessel shielding, cooling, and cooling water systems. The Institute of Plasma Research, Department of Atomic Energy, coordinates these activities.
- Japan is responsible for collaborating with the ITER Organization and ITER members to help design and produce components for the tokamak, including the blanket remote handling system, the central solenoid coils, the plasma diagnostics systems, and the neutral beam injection heating systems.
- Korea is expected to provide the four sectors of the vacuum vessel, the blanket shield block, thermal shields, and the tritium storage and delivery system.
- The Russian Federation's contribution lies in the manufacture and supply of high-tech equipment and basic reactor systems, e.g., the supply of 22 km of conductors based on 90 tonnes of superconducting Nb3Sn strands for winding coils of a toroidal field, 11 km of conductors based on 40 tonnes of superconducting NbTi strands for winding coils of a poloidal field of the ITER magnetic system, and the manufacture of 179 of the most energy-intensive (up to 5 MW/sq. m.) panels of the first wall. The panels are 2 m wide, 1.4 m high, and weigh about 1000 kg, and are to be covered with beryllium plates soldered to CuCrZr

bronze and connected to a steel base. The Russian Federation will also conduct thermal tests of ITER components that are facing the plasma.

- The USA is responsible for the design and manufacture of components for the ITER project, and for contributing to the tokamak cooling system, the diagnostics systems, the electron and ion cyclotron heating transmission lines, the toroidal and central solenoid magnet systems, and the pellet injection systems.

The construction of the ITER tokamak has been called an assembly of "a giant three-dimensional puzzle" because the parts are manufactured around the world and then shipped to France for assembly according to the required precision and design. This has required very open and critical collaboration among scientists, engineers, technicians, and industry personnel from all the countries involved.

The ITER project is a global collaboration, and its success will depend on continued support from each one of its member countries. If ITER is successful, it will be a major victory for humanity, and it will pave the way for a clean and sustainable energy future. The success of the ITER project will be a major milestone in the development of fusion energy as a viable and sustainable source of electricity.

The ITER members represent three continents, over 40 languages, half of the world's population, and 85% of the global gross domestic product. Moreover, there are thousands of people working towards the success of ITER in the offices of the ITER Organization, the domestic agencies, laboratories, and industries.

The promise held by ITER, the complexity of its design, the international resolve to see that it works, the delays, and the cost overruns have attracted considerable attention [4] and criticism [5, 6], sometimes even bordering on scepticism. Some commentators have also argued that the funds utilised for ITER could have been better utilised to establish solar power and wind energy plants to combat climate change. However, we cannot be confident that the funds, if not provided to ITER, would have been automatically made available for climate change mitigation programmes.

There is also speculation about unforeseen technological challenges — for example, due to neutrons, possibly inadequate shielding against radiation, incompatibility or defects in components supplied by different member states, or escape of scorching plasma and (radioactive) tritium — though the scientists involved with the project are confident about being able to handle them.

10.4 Conclusions

The construction of ITER is undoubtedly a daunting and humbling task. And yet, this international collaboration is surely one of the highest points in the history of mankind and will pave the way for the demonstration project on which several countries have started work because of the confidence shown by the scientists. We can surely expect to have safe, carbon-free, and plentiful energy in the future if international collaborations continue, which would, once and for all, relieve us from the spectre of climate change and global warming.

References

[1] D. K. Srivastava and V. S. Ramamurthy, "*Climate Change and Energy Options for a Sustainable Future*" (Singapore: World Scientific, 2021).

[2] A. S. Eddington, "The Internal Constitution of Stars", *Nature* 106 (1920) 14–20.

[3] "Gorbachev: ITER is Part of His Legacy", *ITER Newsline* (12 September 2022).

[4] R. Khatchadourien, "A Star in a Bottle", *New Yorker* (3 March 2014).

[5] C. Seife, "World's Largest Fusion Project Is in Big Trouble, New Documents Reveal", *Scientific American* (15 June 2023).

[6] D. Kramer, "Further Delays at ITER Are Certain, but Their duration Isn't Clear", *Physics Today* 75 (2022) 20–22.

Chapter 11

International Centre for Theoretical Physics

"Science is a common cultural heritage to be shared by all humanity."

— Prof. Abdus Salam

11.1 Background

We have seen that the early 1960s were a time of great political upheaval and change around the world. The Cold War between the United States of America and the (former) Soviet Union was at its height, which led to a tense rivalry between them for global influence. Decolonisation was underway in Africa and Asia, as many countries gained their independence from European colonial powers. These events were accompanied by the rise of new social movements such as the civil rights movement and (anti-establishment) counterculture, which led to a period of great political and social ferment.

In the United States, John F. Kennedy was elected president in 1960 and ushered in a new era of hope and optimism. After his tragic assassination in 1963, Lyndon B. Johnson continued his legacy by enacting sweeping civil rights legislation and launching the Great Society programme to address poverty and inequality. It specifically aimed to reduce poverty levels, racial injustice, and crime, improve the environment, provide aid for education, attack disease, provide Medicare, and initiate urban renewal, beautification, conservation, and development of depressed regions.

On the other side, Nikita Khrushchev, a more reform-minded leader than his predecessor Joseph Stalin, was the leader of the Soviet Union. He initiated a period of de-Stalinisation and economic liberalisation. The Cuban Missile Crisis, which he faced in 1962, brought the world to the brink of nuclear war.

Europe was witnessing a period of economic recovery in the early 1960s. The European Economic Community, also known as the Common Market, was established in 1958 and began to promote economic integration among its member states. However, Europe was also divided by the Cold War, with the Iron Curtain separating communist Eastern Europe from democratic Western Europe.

In 1960 alone, seventeen African countries gained their independence from European colonial powers. This led to a period of political and economic instability, as the new African nations struggled to establish themselves and build democratic institutions. India and several other countries in Asia and the Middle East had become independent about a decade earlier.

In the 1960s, China was in the midst of the Great Leap Forward, a radical economic experiment that led to a devastating famine. In Vietnam, the United States was becoming increasingly involved in the civil war between the communist North and the anti-communist South.

The political landscape of Latin America in the 1960s was characterised by a tumultuous mix of social upheaval, revolutionary movements, and military coups. The region was caught in the crossfire of the Cold War, where the United States and the Soviet Union vied for influence and ideological dominance.

Moreover, in the early 1960s, many developing countries were facing a hard currency crisis due to several factors, including problems of export/import, debt repayment, and trade deficits. This made it difficult for them to import essential goods, such as food and medicine. It also made it difficult for them to invest in their economies and create jobs. In some cases, the hard currency crisis led to political instability and even coups d'état. Attempts were made to resolve the hard currency crisis through devaluation, import restrictions, foreign aid, and trade agreements.

Overall, the political scene across the world in the early 1960s was one of great change and upheaval. The Cold War, decolonisation, and the rise of new social movements were all leading to a new world order.

Amid all this, two scientists, Prof. Abdus Salam (1926 CE–1996 CE) and Prof. Paolo Budinich (or Budini) (1916 CE–2013 CE), conceived of

Fig. 11.1. Prof. Abdus Salam (1926–1996 CE).

the International Centre for Theoretical Physics (now rechristened as the Abdus Salam International Centre for Theoretical Physics) following the example of other scientists like Robert Oppenheimer (Institute of Advanced Studies, Princeton) and Niels Bohr (Institute of Theoretical Physics, now known as Niels Bohr Institute, Copenhagen), as well as the Nordic Institute of Theoretical Physics, Copenhagen (later Stockholm), and dreamt of building a bridge between the East and the West based on scientific collaboration. Abdus Salam (Fig. 11.1) was more concerned about North–South disparities and envisioned an institution capable of acting as a communication link between the two. Paolo Budinich was more concerned about initiating dialogue between scientists from the Eastern and the Western Blocs. Soon after its establishment in 1964 (initially with major funds from the Italian Government), physicists from the East and the South recognised the Centre as a pathway towards Western Europe and the United States.

At present, the ICTP receives funds from the Italian government, the International Atomic Energy Agency (IAEA), the United Nations Educational, Scientific and Cultural Organization (UNESCO), and many donors [1]. It has sometimes been suggested that international collaboration should take place in a neutral space to "avoid the psychological host-guest feelings invariably associated with national centres" [2], and the ICTP being run under the aegis of the United Nations fully satisfied this criterion.

11.2 Mission

The mission of the ICTP as envisioned in its agreement with the IAEA and UNESCO has remained "To foster, through training and research, the progress of all branches of theoretical physics... [paying] special attention to the needs of developing countries" and " to train young physicists for research..., to help in fostering the growth of advanced studies..., to conduct original research..., and provide an international forum for personal contacts... — again with emphasis on responding to the needs of science in developing countries [3]."

Abdus Salam viewed science as a common cultural heritage to be shared by all humanity and envisioned the ICTP as a place where young scientists from developing countries would be treated with dignity and respect and be given the chance to participate in the great adventure of expanding human knowledge.

He thought that it should not be *a centre for development*, with the patronising connotation that such a term carried, but instead a *centre of excellence in theoretical physics*, with substantial participation by scientists from the third world. He also felt that its contribution to the less developed countries would derive from the high quality of the research carried out there; its aim ought to be keeping third-world physicists "scientifically alive."

Prof. Abdus Salam also envisioned that the centre would demonstrate that scientists from the South could do physics of the same quality and relevance as their colleagues from the North. He felt that if the ICTP was to be revolutionary, it would be for its capacity to help the third world develop its "own" scientific resources. In his view, *only scientists, and theoretical physicists in particular, were essential for training good technicians, without having to depend on Western know-how.* It had already been realised that the complexity of the technology required high-quality teachers in basic sciences. Moreover, he felt that any differential treatment of these scientists, such as classifying them as scientists for development, and not just scientists, would be discriminatory [4].

11.3 Outcomes

The ICTP has rapidly developed into an institution for training and research of the highest academic standards for young third-world physicists and an emblematic centre of international collaboration, where scientists from third-world countries find a temporary home and to which high-level physicists from everywhere flock.

To achieve this, the ICTP runs several long-term programmes which offer scientists from developing countries the opportunity to remain up to date with rapid developments in science taking place throughout the world. It holds conferences, workshops, advanced workshops, and school, and effectively supports a Research Associateship Programme. For example, in 2022, the ICTP had 7127 visitors and held 60 conferences, schools, and workshops, with participants from 157 countries, 29% of whom were women. It also hosted 174 long-term researchers (staff, consultants, and long-term visiting scientists) from 41 countries. It is interesting to note that 57% of the visitors to the ICTP in 2022 were from developing countries, 33% from developed countries, and 10% from the least developed countries [1].

The ICTP's research topics have expanded to include high energy, cosmology and astroparticle physics; condensed matter and statistical physics; mathematics; earth system physics; quantitative life sciences; and science, technology, and innovation.

By now, the ICTP is widely regarded as an extraordinary environment for advancing knowledge in the physical and mathematical sciences, with a permanent faculty of distinguished scientists who conduct rigorous world-class, curiosity-driven research in frontier and interdisciplinary science ranging from string theory, cosmology, and black holes to quantum computing, climate science, and quantitative life sciences. These scientists teach and mentor hundreds of students and young researchers every year, equipping them to go on to study, teach, and conduct research in the world's finest universities and to contribute to the development of science in their home countries.

The ICTP has opened four regional centres of excellence in Brazil, Mexico, China, and Rwanda. For the last 10 years, it has been conducting an outreach programme, Physics Without Frontiers, for university students. It has, during this period, helped about 10,000 students through more than 300 activities in at least 50 countries, many of whom have gone on to obtain master's and PhD degrees, following their participation. It also conducts several diploma, master's, and joint doctorate programmes with various universities.

Realising that the research institutions in third-world countries face chronic shortages of funds and access to foreign currencies, such that they are unable to procure scientific equipment and books or subscribe to journals, the ICTP took the initiative to start a book donation programme to provide books and journals to physics and mathematics libraries, and later also equipment for basic research and teaching.

As a very large part of the funding has traditionally come from the Italian government, the ICTP has also paid special attention to the relationship it has with Italy. This relationship is reflected in the Training and Research in Italian Laboratories (TRIL) programme, which hosts many Italian visitors.

11.4 Impact

Dramatic changes have taken place in the world since the ICTP started. The fall of the Berlin Wall and the collapse of the (Former) Soviet Union ended a period of East–West confrontation. However, North–South disparities have in some cases become more acute. In addition, there has been a dramatic intensification of local conflicts. Conflicts whether in the Balkans, Middle East, North Africa, or anywhere else eventually become conflicts between cultures. The ICTP, while not claiming to solve the world's problems, acts under the assumption that science, as an activity and body of knowledge, is an integral part of human culture, representing some of its most universal and exportable components, as mentioned earlier.

Visitors to the ICTP experience its multi-ethnic population: Physicists from Algeria, Argentina, China, Cuba, India, Israel, Jordan, Mexico, Pakistan, Vietnam, etc., all participate collegially in the same scientific activities, and through their mutual apolitical discussions, rediscover the multidimensional complexity of reality.

The ICTP endeavours to help developing countries acquire, through science, the necessary instruments for their emancipation. It helps them break the isolation of their scientists; counteract the "brain drain" (though some commentators have called its perceived role in counteracting brain drain as naive [2]); and address problems caused by the lack of scientific books, journals, modern research equipment and a stable operational funding base. Furthermore, the Centre helps the least developed countries build and maintain their scientific communities — a necessary prerequisite for their social and economic development. It assists all developing countries in their efforts to upgrade their skills pool through the high educational standards that theoretical physics provides to inquiring minds of all ages [2].

One of the most successful programmes of the ICTP has been its Associateship Membership Programme. It directly addresses the problems faced by researchers of the third-world countries by allowing them to remain up to date with the rapid pace of scientific developments across the world, thus helping developing countries build and maintain their scientific communities. These programmes have a proven history of helping to counteract scientific isolation and brain drain.

The Associate Membership Scheme is a logical extension of the Centre's courses, schools, workshops, and seminars. During their stay at the ICTP to attend such training activities, hard-working, bright young scientists become involved in high-level scientific activities. However, once they return to their home institutions, they often lack colleagues with whom they can discuss, share, and transmit their newly acquired knowledge. They are also unable to find the necessary equipment, computing facilities, and up-to-date literature. This leads to their losing, little by little, the scientific patrimony that they built with hard work. "Under such circumstances, the temptation to move to a better environment may become too strong" [3]. Associate members are scientists from, and working in, developing countries who have the right to pay 3 scientific visits to the Centre in 6 years for periods of 6 weeks to 3 months. Some of these associates with exceptional scientific records may be appointed as senior associates for 5 years. A fixed sum is reserved for them out of which they can draw in order to subsidise their visits to the Centre.

The scheme also supports the junior associateship of younger scientists, which provides small grants for purchasing scientific books or subscriptions to periodicals.

The ICTP supports a Federation Arrangements Scheme which was established "to enable scientific institutions in developing countries, such as a university, research institute or governmental laboratory, to send their young scientists to the ICTP to attend important courses or to participate in other scientific activities relevant to their studies. The agreements are for a limited time (initially one year), with related expenses cost-shared between the ICTP and the partner institution. The Federation's contrasting characteristics make it an interesting complement to the Associateship programme: It is addressed to an institution, not an individual; it often serves as the basis for long-term fruitful cooperation between that institution and the ICTP; and it stimulates local authorities to invest human and financial resources to keep alive this link with the Centre [4]."

11.5 Legacy

The ICTP has continued to win praise for its contribution to the development of science across the world. In its first year, it won wholesome praise from Robert Oppenheimer (quoted in [5]):

> *"It seems to me that the Centre has been successful in these eight or nine months of operation in three important ways. It has cultivated and produced admirable theoretical physics, making it one of the great foci for the development of a fundamental understanding of the nature of matter. The Centre has obviously encouraged, stimulated, and helped talented visitors from developing countries who, after rather long periods of silence, have begun to write and publish during their visit to the Centre in Trieste. This is true of physicists whom I know from Latin America, from the Middle East, from Eastern Europe, and from Asia. It is doubtless true of others. The Centre has become a focus for the most fruitful and serious collaboration between experts from the United States and those from the Soviet Union on the fundamental problems of the instability of plasmas, and of means for controlling it. Without the Centre in Trieste, it seems to me doubtful that this collaboration would have been initiated or continued. In all the work at the Centre of which I know, very high standards prevail. In less than a year it has become one of the leading institutions in an important, difficult and fundamental field."*

The success of the ICTP spawned at least two other scientific enterprises in Trieste. The first one is the International School for Advanced Studies. It is an autonomous department of the University of Trieste, founded in 1978 by Professor Paolo Budinich, former Deputy Director of the Centre. It is a graduate school and awards the degrees of M.Sc. and Ph.D. in several specialities of physics and mathematics. A number of its students are from developing countries.

The second is the new International Centre for Genetic Engineering and Biotechnology, created under the aegis of the United Nations Industrial Development Organization. It runs over 45 state-of-the-art laboratories in Trieste, Italy; New Delhi, India; and Cape Town, South Africa and forms an interactive network with almost 70 member states, with operations aligned to those of the United Nations system. It plays a key role in biotechnology promoting research excellence, training, and technology transfer to industry, to contribute in concrete terms to sustainable global development.

The SESAME (Synchrotron-light for Experimental Science and Applications in the Middle East) facility in Jordan was also inspired by the same vision and was set up with the participation of CERN and UNESCO.

11.6 Conclusions

We have already discussed the rationale behind the setting up of the ICTP and its role in training a large number of researchers from third-world countries. We conclude this chapter by recalling an episode involving "Michael The Scot," which Prof. Salam often mentioned in his talks and writings [3, 6]:

> *"Seven hundred and fifty years ago, an impoverished Scotsman left his native glens to travel south to Toledo in Spain. His name was Michael, and his goal was to live and work at the Arab universities of Toledo and Cordova... His interests lay in the sciences of astrology and alchemy, then fashionable in Scotland. But once in Toledo, Michael formed the ambitious project of introducing Aristotle to Latin Europe, translating not from the original Greek, which he did not know, but from the Arabic translation taught in Spain [...] Toledo's school, representing as it did the finest synthesis of Arabic, Greek, Latin and Hebrew scholarship, was one of the most memorable of international essays into an international collaboration."*

Salam felt that in the present situation, the roles had been reversed [3] and "it was now the new Michaels from the south, who had to come to the north to learn science." They needed to return to their country, which the associateship scheme facilitated, so that "a deserving young man may spend his period of vacation in an invigorating environment .... and charge his batteries with new ideas, still spending the remaining nine months of his academic year back at home working in his own country [6]."

The ICTP has continued to foster international collaboration, which has especially benefitted third-world countries in general and basic sciences in particular.

References

[1] "ICTP: A Year in Review 2022", https://www.ictp.it/sites/default/files/attachments/ICTP_AnnualReport_2022_web-compressed.pdf.

[2] A. De Greiff, "Abdus Salam: A Migrant Scientist in Post-Imperial Times", *Economic and Political Weekly* 41 (2006) 228–234.

[3] J. G. Roederer, "The Abdus Salam International Centre for Theoretical Physics (ICTP) Today: An Overview of ICTP Aims and Activities" (ICTP, 1999).

[4] A. De Greiff, "Supporting Theoretical Physics for the Third World Development. The Ford Foundation and the International Centre for Theoretical Physics in Trieste (1966–1973)". In: G. Gemelli (ed.) *American Foundations and Large-Scale Research: Construction and Transfer of Knowledge* (Bologna: CLUEB, 2001), pp. 25–50. https://unesdoc.unesco.org/ark:/48223/pf0000145311.

[5] A. Hamende, "ICTP in Trieste: Creative Research for Global Development", *The IAEA at 30*, https://www.iaea.org/sites/default/files/publications/magazines/bulletin/bull29-3/29306086064.pdf.

[6] A. Salam, "The United Nations and the International World of Physics", *Bulletin of Atomic Scientists* (February 1968) 14–15. See also Idem, "Science and Technology in the Emerging Nations"; Idem, "The Advancement of Science for the Developing Countries". In: A. Tiselius and S. Nilsson (eds.) "*The Place of Values in a World of Facts*" (Sydney: Wiley Interscience Division and Almqvist and Wiksell, 1969), pp. 269–280.

Chapter 12

Facility for Antiproton and Ion Research

> "In giant planets, stars, and during stellar explosions and collisions, matter is subject to extreme conditions such as very high temperatures, pressures, and densities. FAIR will enable scientists to create such conditions in the laboratory."
>
> — Research at FAIR

12.1 Introduction

The Facility for Antiproton and Ion Research (FAIR) is under construction at GSI, Darmstadt, Germany. It is designed to provide intense ion and antiproton beams for experiments that will shed light on supernova explosions, neutron star mergers, formation of elements, the structure of hadrons, and intense plasmas, provide opportunities for extremely precise determination of fundamental constants, and offer an opportunity to simulate the effect of long exposures to high-energy particles during space travel.

By now, it is understood that (only) the lightest elements, like hydrogen and helium, were synthesised immediately after the Big Bang; as the quark–gluon plasma, the primordial deconfined strongly interacting matter which filled the nascent universe hadronised. It is also understood that elements up to iron were created during nuclear reactions in the cores of stars, formed by clouds of hydrogen gas becoming increasingly dense and hot and triggering nuclear reactions under gravity. However, it is believed that all the heavy elements including and beyond gold and platinum were produced via complex production paths during extreme astrophysical events, e.g., neutron star mergers.

It will be possible to study these processes, which underlie and sustain the synthesis of heavy elements in the universe in detail. These studies will be facilitated by nuclear reactions and production of rare radioactive ion species using intense heavy-ion beams produced at FAIR. FAIR will provide beams, from protons up to uranium ions, with a wide range of intensities and energies, in addition to secondary beams of antiprotons and rare isotopes.

The relativistic heavy-ion capabilities of FAIR will complement those of the CERN's Large Hadron Collider or Brookhaven National Laboratory's Relativistic Heavy Ion Collider, which produce quark-gluon plasma at low baryonic density but high temperatures. FAIR, on the other hand, will produce the quark-gluon plasma at high baryonic density and low temperature, which is likely to be present in the core of neutron stars (Fig. 12.1). In some sense, it is more like the Super Proton Synchrotron at CERN but at much higher intensities and will permit physicists to produce and study the dense hadronic matter and its transition to quark matter. The NICA (Nuclotron-based Ion Collider fAcility) accelerator complex under

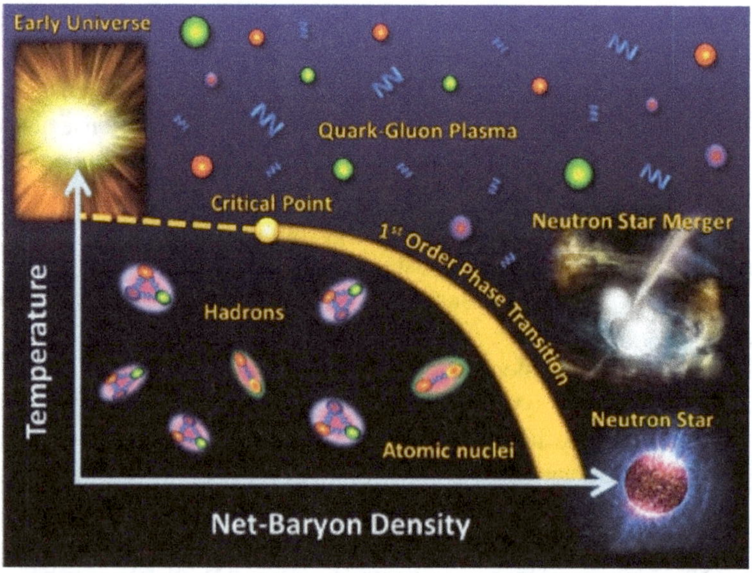

Fig. 12.1. The Putative Phase Diagram of Nuclear Matter from High Temperature and Low Baryonic Density to Low Temperature and High Baryonic Density (Credit: GSI/FAIR).

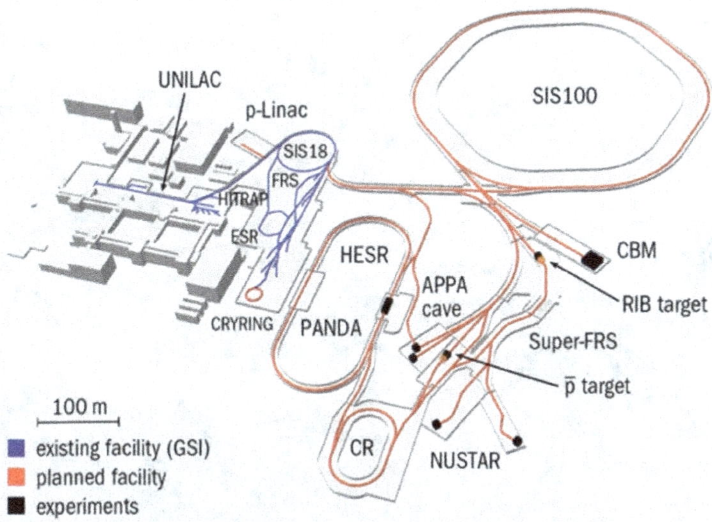

Fig. 12.2. The Existing GSI Accelerators UNILAC, SIS18, and ESR (Blue) and the FAIR Facilities (Red). FAIR Comprises the SIS100 Synchrotron; the Antiproton Separator and the Super Fragment Separator (Super-FRS); the Collector Ring (CR); the High-Energy Storage Ring (HESR); and the Experimental Stations for APPA, CBM, NUSTAR, and PANDA. The Proton Linac and the CRYRING Also Belong to the FAIR Instrumentation Network (Credit: GSI/FAIR).

construction at the Joint Institute for Nuclear Research (Dubna, Russia) is also expected to provide heavy ions at similar collision energies in collider mode.

The extremely high intensities of the beam at FAIR (Fig. 12.2) will permit scientists to produce and study reactions involving rare exotic hadronic states or rare, very short-lived radioactive nuclei. It will also enable us to investigate processes under the extreme temperatures and pressures that prevail in large planets, stars, and stellar explosions. As indicated earlier, FAIR will permit tests of quantum electrodynamics in the regime of very strong electromagnetic fields, especially by studying atoms with very high proton numbers but stripped of all but a very few (even just one) electrons!

Figure 12.3 gives a rendering of the completed GSI-FAIR facility. More details can be found in Ref. [1].

Fig. 12.3. A rendering of the completed GSI-FAIR facility (Credit: GSI/FAIR).

In brief, FAIR's scientific programme will cover hadron physics, nuclear structure and astrophysics, atomic physics, plasma physics, materials research, and radiation biophysics and its applications in cancer therapy and space research. These will be covered by the four experimental *pillars* [2]: Atomic and Plasma Physics and Applied sciences (APPA); Condensed Baryonic Matter (CBM); NUclear STructure, Astrophysics, and Reactions (NUSTAR); and the antiProton ANnihilation in DArmstadt (PANDA).

12.2 The Four Experimental Collaborations at FAIR

We have already introduced the four experimental programmes that will cover FAIR's scientific programmes. About 3000 scientists including more than 500 Ph.D. students from around the world are expected to carry out experimental studies at FAIR. In the following, we discuss these in a little more detail.

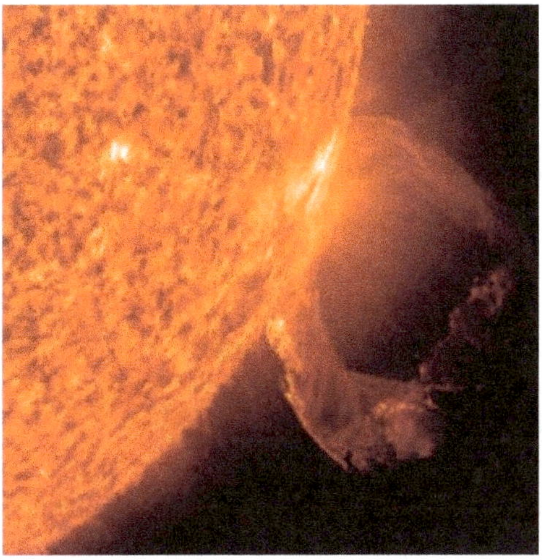

Fig. 12.4. The Arena of APPA.

12.2.1 *Atomic and Plasma Physics and Applied Sciences (APPA)*

APPA is an umbrella collaboration [3], with about 700 researchers in atomic physics, plasma physics, and applied sciences, with specific programmes in biophysics, medical physics, and materials science (Fig. 12.4). The following are the details of APPA:

i. It will provide opportunities for high-precision tests of bound-state QED in the non-perturbative regime. Additionally, it will determine fundamental constants such as the fine-structure constant to unprecedented accuracies, which will involve extremely precise measurements of the bound-state g-factors in medium- to high-Z hydrogen-like ions confined in a trap.
ii. Plasma physicists will be able to create and probe dense plasmas to test models of planetary and stellar structures.
iii. High-energy components of galactic cosmic radiation will be simulated to assess the risk of space missions for astronauts and electronic equipment by dedicated irradiation experiments.
iv. Finally, material science and geoscience researchers will be able to see how materials respond to the simultaneous application of irradiation and pressure, which is of interest in understanding processes in

the Earth's mantle and the synthesis of new materials under highly non-equilibrium conditions.

12.2.2 *The Compressed Baryonic Matter (CBM) Experiment*

The CBM experiment [4] has more than 500 participants. It is organised in a manner similar to the LHC experiments at CERN, which will use high-energy nucleus–nucleus collisions in a fixed-target mode to investigate the highly compressed nuclear matter (Fig. 12.5). The following are the details of the CBM experiment:

i. The CBM experiment is designed to recreate the matter densities found in supernova explosions, the cores of neutron stars, and neutron star mergers. The energies of the FAIR beams are perfectly suited to study the QCD phase diagram of strongly interacting matter at large net-baryon densities and low temperatures. This contrasts and complements the conditions of the very high temperatures and low net-baryon densities reached at the Relativistic Heavy Ion Collider in Brookhaven and the Large Hadron Collider at CERN, which are like the conditions that prevailed microseconds after the Big Bang.
ii. In doing this, the CBM experiment is expected to explore the QCD phase diagram which is predicted to exhibit a rich structure with a

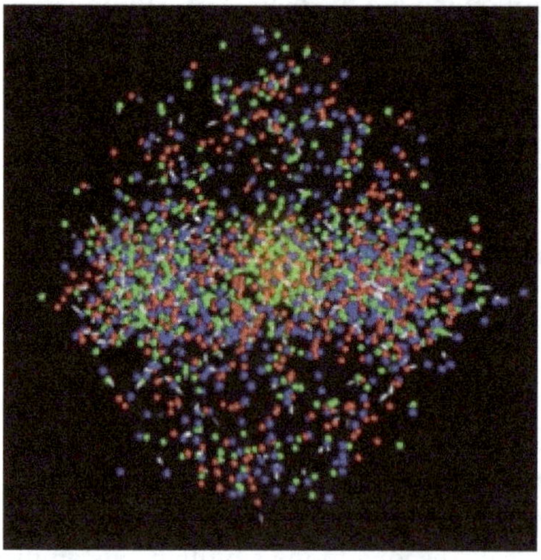

Fig. 12.5. The Arena of CBM.

critical point, a first-order phase transition between hadronic and partonic matter, or new phases such as quarkyonic matter.

iii. The CBM experiment is designed to run at interaction rates of up to 3–4 orders of magnitude higher than the rates reached in other high-energy heavy-ion experiments, which will be valuable to study rare processes with great precision.

12.2.3 *NUclear STructure, Astrophysics, and Reactions (NUSTAR)*

The NUSTAR collaboration [5] at FAIR has more than 800 participants from 180 institutes located in 38 countries. NUSTAR includes several sub-collaborations across different experimental setups, specifically suited to study various aspects of secondary radioactive ions, such as mass and lifetime measurements (Fig. 12.6). The following are the details of NUSTAR:

i. It is expected to play a major role in improving our knowledge of the synthesis and abundance of chemical elements, by exploring the structure and reaction properties of *very rare* radioactive ions produced for the *first* time by FAIR.

Fig. 12.6. The Arena of NUSTAR.

ii. It will offer an opportunity to understand the formation of very heavy elements through reactions involving rare nuclei at the limit of stability. FAIR will be able to artificially produce nuclei that occur as radioactive intermediate products in the formation of stable isotopes, directly measuring the different processes involved in the laboratory.
iii. The Super-FRS will use the highly intense beams at high energies to separate beams of the heaviest and most neutron-rich nuclei. The sophisticated network of FAIR's complex storage rings will then allow mass and lifetime measurements.

12.2.4 *The PANDA Experiment*

The antiProton ANnihilation in DArmstadt (PANDA) experiment [6] is a cooperation of more than 400 scientists from 19 countries, like the LHC experiments at CERN but smaller. Its goal is to understand hadrons using the power of an antiproton beam on fixed hydrogen or other nuclear targets (Fig. 12.7). The following are the details of the PANDA experiment:

i. Antiproton–proton annihilations have enormous advantages compared to proton–proton collisions. These involve conditions of small

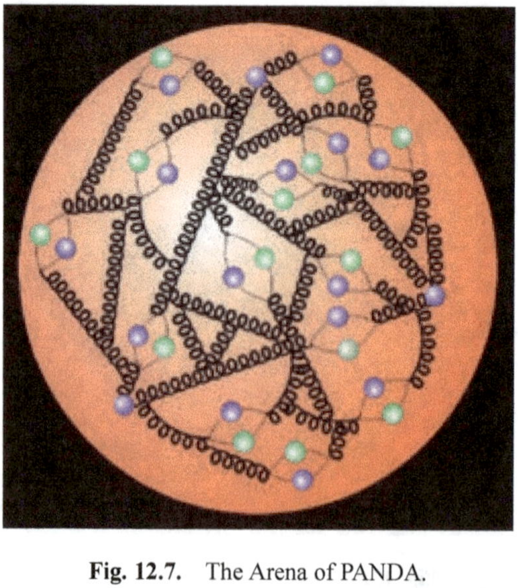

Fig. 12.7. The Arena of PANDA.

momentum transfer at maximum released energy with well-defined initial states and high-precision mass scanning.

ii. The vast difference in mass between the proton and its quark constituents is a result of the binding among quarks in the confinement regime. Exotic hadrons such as tetra- and pentaquarks, hybrids, and glueballs are expected to reveal uncharted properties of this binding.

iii. PANDA will help us understand the nature of hadrons and their spin degrees of freedom.

12.3 Evolution of the Proposal for FAIR and Delays

FAIR was developed through intense discussions, evaluations, and simulation studies by the international scientific community and the GSI laboratory (the Helmholtz Centre for Heavy Ion Research) around the turn of the century. GSI was founded in 1969. It was a natural foundation on which to develop the next generation of accelerators and experiments for these fields as it has a long and illustrious tradition in nuclear and atomic physics and heavy-ion research. Superheavy element research has been a strong pillar of the research programme at GSI Darmstadt since its foundation. Six new elements with atomic numbers 107–112 — bohrium, hassium, meitnerium, darmstadtium, roentgenium, and copernicium — were discovered along with many new isotopes in cold fusion reactions using targets of the doubly magic lead nucleus and its neighbour bismuth at GSI.

The initial start date of FAIR was 7 October 2010, when nine partner countries (Finland, France, Germany, India, Poland, Romania, Russia, Slovenia, and Sweden) signed an intergovernmental agreement for its construction and operation. Since then, the UK has joined FAIR as an associate member (2013), while the Czech Republic is an aspirant partner. FAIR is operated as a company, in which different countries are shareholders, decided by their contribution. Thus, for example, India holds 3.5% of the shares. Indian contribution is both in cash and kind. Indian researchers are involved in building detectors for the major experimental programmes of FAIR, i.e., NUSTAR, CBM, and PANDA. Another area of major participation is building equipment to be used at the heart of the FAIR accelerator, e.g., power converters, vacuum chambers, and beam stoppers.

Towards late 2014, several construction requirements forced major modifications of the civil construction design. This resulted in an inordinate delay and cost increase in the overall project.

Construction of the FAIR started in 2017. It was originally supposed to cost 1.3 billion Euros (2005 prices). It was revised to 2.9 billion Euros (2005 prices) in 2019. A further increase of 850 million Euros (2019 prices) has now been recommended to complete the facility. As mentioned earlier, the project has been delayed due to construction issues, changes in the details of beamlines, and cost escalation, requiring the raising of additional funds. The Ukraine War led to the suspension of Russian collaboration and the corresponding contributions.

An international review committee has recommended prioritisation of certain steps for a speedy start to the experiments. The completion of the accelerator complex is expected by 2025 along with the commencement of the first experiments at NUSTAR and APPA. The timely completion of the SIS100 complex is also recommended so that the CBM experiment can start. The committee advised delaying the PANDA experiment as it has no competition [7].

12.4 The Impact of the Ukraine War

This discussion would be incomplete unless we discuss the unfortunate developments due to the Ukraine War. We reproduce here statements released by the management of FAIR, its European partners (except Russia), and the Alliance of Science Organisations in Germany [8]. These statements reveal the limits of goodwill generated by international collaborations. The FAIR management has also established a task force to handle issues arising from this situation. One immediate outcome of these decisions was a need to raise additional funds for equipment, which were to be supplied by Russia to avoid further delay in the commissioning of FAIR.

12.4.1 *Statement on the War of Aggression by Russia on Ukraine*

"4 March 2022.

Researchers from all over the world have been working together for decades at the accelerators and experimental facilities at GSI and FAIR.

They work together on peaceful, non-military scientific objectives, independent of political, religious and ideological aspects.

We condemn the war of aggression of Russia and the breach of international law by the Russian government. That is why we fully stand behind the sanctions imposed by the German government and its international partners. We are aware that they will have a strong impact on our own activities, but we believe that these measures are necessary in the current situation.

In accordance with the Alliance of German Science Organisations, GSI/FAIR will immediately suspend all cooperation with Russian state institutions and business enterprises. Ongoing bilateral cooperation projects with researchers from Russian institutions will be suspended with immediate effect, furthermore, we will not conclude any new bilateral cooperation projects. For multilateral projects involving Russia, which include the FAIR project, GSI/FAIR will coordinate with the other partners regarding further implementation of the international agreements. Adjustments of the measures will be made depending on the further development of the situation.

We are very saddened and concerned by the tragic events in the Ukraine. Also at GSI/FAIR, employees are affected by the war in Ukraine, whether directly, because families or friends live in the contested areas, or through professional or personal ties to Ukraine or Russia. Our thoughts go to all the people who are affected directly or indirectly, with our deepest sympathy and support in these difficult times."

12.4.2 *Statement of the European Shareholders of FAIR GmbH, Except Russia*

"March 17, 2022.

The European Shareholders state that the invasion of the sovereign state of Ukraine is an unacceptable violation of the international law by the Russian Federation. This is incompatible with the peaceful intention of the FAIR Project and its spirit of international cooperation. The European Shareholders condemn the ongoing war against Ukraine and as a reaction suspend the existing, long-standing cooperation immediately, even though this cooperation is fundamentally in the interests of all sides. All current and planned activities with Russia are being frozen and subjected to critical review. There will be no new activities until further notice.

We express our personal appreciation towards our Russian colleagues with whom we cooperated successfully and on a trustful basis for the past years. We hope that at some point in the future, we will be able to meet again to continue our fruitful and peaceful cooperation in FAIR."

12.4.3 *Statement of the Alliance of Science Organisations in Germany: Solidarity with Partners in Ukraine — Consequences for Science and the Humanities*

"February 25, 2022.

The Alliance of German Science Organisations strongly supports the Federal Government's consistent action against Russia's belligerent attack on Ukraine, which is in violation of international law.

The Alliance regards the Russian invasion as an attack on the fundamental values of freedom, democracy and self-determination, which in turn provide the basis for academic freedom and opportunities for academic cooperation.

[… …]

The Alliance is aware of the consequences of these measures and at the same time deeply regrets them for science. Many research projects have been planned to take place over a period of several years and are massively affected by the current war situation. We live in a multidimensional world, and only by means of close international academic cooperation can the crises facing humanity, such as climate change, species extinction or infectious diseases, be overcome.

[….]"

12.5 Conclusions

FAIR will be a unique and very advanced international institution for research in atomic and nuclear physics for decades to come and will provide an opportunity for researchers to understand the fundamental structure of matter, explore its exotic forms, and understand how the universe evolved from its primordial state. FAIR is a monumental undertaking and a testament to human curiosity and collaboration. It is not just a facility; it is a launchpad for discovery, a place where the future of physics will be written as the experiments start. A very rich harvest of precise results and deep insights in physical sciences is awaited!

References

[1] https://www.gsi.de/en/researchaccelerators/fair.

[2] O. Boine-Frankenheim, C. Trautmann, P. Senger, N. Kalantar-Nayestanaki, K. Peters, and K.-D. Gross, "FAIR Forges Its Future", *CERN Courier* (10 July 2017), https://cerncourier.com/a/fair-forges-its-future/.

[3] T. Stöhlker *et al.*, "APPA at FAIR: From Fundamental to Applied Research", *Nuclear Instruments and Methods in Physics Research Section B: Beam Interactions with Materials and Atoms* 365 (2015) 680. https://fair-center.eu/user/experiments/appa.

[4] B. Friman, C. Höhne, J. Knoll, S. Leupold, J. Randrup, R. Rapp, and P. Senger, "*The CBM Physics Book: Compressed Baryonic Matter in Laboratory Experiments*", Lecture Notes in Physics, Vol. 814 (Heidelberg and Berlin: Springer-Verlag, 2011); https://www.cbm.gsi.de/.

[5] Alexander Herler for the NUSTAR Collaboration, "The NUSTAR Program at FAIR: Overview and Present Status of the Project", *EPJ Web of Conferences* 71 (2014) 00064, https://fair-center.eu/user/experiments/nustar.

[6] PANDA Collaboration, "Physics Performance Report for PANDA: Strong Interaction Studies with Antiprotons", https://panda.gsi.de/oldwww/archive/public/panda_pb.pdf.

[7] https://www.gsi.de/fileadmin/oeffentlichkeitsarbeit/fair/FAIR-report_221025.pdf.

[8] https://www.gsi.de/en/work/organisation/ukraine.

Chapter 13

Astronomy

> "To give man ever more knowledge of the universe and to help him 'to learn humility and to know exaltation,' that is the mission of astronomy."
>
> — Robert G. Aitken

13.1 Introduction

As the sun, which provided light and warmth to ancient humans, went down the horizon, engulfing the world around him in darkness, stars and a luminous moon (its shape changing every night) lit the sky overhead. They must have venerated the dawn. This can be seen from the large number of prayers of exquisite literary beauty and profound veneration of Usha, the goddess of dawn, in one of the earliest known compositions of mankind — the Rig Veda. They must have been drawn by the lustre and radiance of golden and silver rays from the sun and moon, in turn. They must have looked in wonder and awe at the majestic "river of stars" (the Milky Way), the "shooting stars" (the meteorites), the stars that "retained" their positions, the "stars" that "wandered" (planets), the constellation of stars that rotated around a "fixed" star (Polaris, the North Star in the Northern Hemisphere), and the constellations of stars which with some imagination (pareidolia) looked like celestial "objects" and celestial "beings." The appearance of comets and the occurrence of solar and lunar eclipses must have filled ancient humans with fear and perhaps even foreboding. And then, over years, decades, and centuries, they must have come to use some of the stars for navigation across vast deserts or snow covered landscapes and oceans, to know the time, to devise a calendar to sow their crops, to make preparations for the approaching seasons, and to await the arrival of migratory birds and animals. This happened in all cultures, across the world.

By now, we know that the seafaring cultures and caravans crossing the deserts in the Northern Hemisphere used Polaris to mark the north and determine the latitude. In the Southern Hemisphere, without the trusty "North Star" in view, navigators such as the Pacific Islanders are known to have memorised the locations of other stars at different times of the night and year to steer their courses across the vast distances while sailing from island to island in their large canoes.

Even the birds, which migrate from north to south to escape harsh winters and then migrate back, have at least three different compasses at their disposal — two of which employ positional astronomy. The first one allows them to extract information from the position of the sun in the sky and the other one uses the patterns of the stars at night. The third is based on the Earth's ever-present magnetic field [2]. It has been suggested that some animals like harbour seals, which move over larger distances for food, follow some bright stars to maintain their orientation, while dung beetles in Africa follow the Milky Way to roll their food away from competitors [3].

Aitken — perhaps in response to the question "Why study astronomy?" — published a leaflet [1] in 1933, in which he declared the following:

> "We want to know! We want to know the relations of these objects that glorify the heavens to the Earth on which we dwell. We want to know how far away they are; and how large they are. We want to know their physical and chemical structure and composition, the laws that govern their motions and their evolution. We want to know why some stars are red, some yellow, others white, why some are bright, some faint; why some shine with a steady light and others vary in their brightness; why some have companion stars with which they revolve in closed orbits, and others do not. We want to know what nebulae are and how the stars are related to them. We want to know why the sun and the stars give out radiant energy we call light and heat and how it is that they can maintain their output of energy substantially without change through hundreds of millions of years. We want to know whether the stars form an orderly

system, and whether there are other stellar systems than our own; whether the universe is 'finite but unbounded' or infinite."

This quest has always had a significant impact on our worldview. Almost all ancient cultures identified celestial objects with the gods and even took their movements across the sky as prophecies of what was to come. Today, we call this aspect astrology. The discovery that the basic elements that we find in stars, and the gas and dust around them, are the same elements that make up our bodies has further deepened the connection between us and the cosmos and is immortalised in the phrase "*we are made of dust of stars*," commonly attributed to Carl Sagan (1934–1996 CE). This connection deeply touches our lives, and the awe it inspires is perhaps the reason why the beautiful images astronomy provides us are so popular in today's culture and have been one of the most effective attractions for students to science.

Research and developments in astronomy have led to far too many scientific and technological achievements [4] to be listed here. These have given a large competitive edge to the nations that made these discoveries. We have repeatedly seen during this monograph that science brings nations together, encourages collaboration, and creates a constant invigorating flow as researchers travel around the globe to work in international facilities.

"Astronomy is particularly well suited to international collaboration due to the need to have telescopes in different places around the world, to see the whole sky. At least as far back as 1887 — when astronomers from around the world pooled their telescope images and made the first map of the whole sky — there have been international collaborations in astronomy and in 1920, the International Astronomical Union became the first international scientific union [4]."

Even earlier, in 1761, astronomers from Britain, France, and Austria had travelled to destinations around the world, including Siberia, Newfoundland, Madagascar, and the Cape of Good Hope to observe the transit of Venus across the sun as predicted by Edmond Halley (1656–1742 CE). Even though they did not collaborate, they studied the same phenomenon from different vantage points.

Albert Einstein (1879–1955 CE) had already become quite famous following the publication of his special theory of relativity (1905) and general theory of relativity (1915). However, he attained cult/superstar status after Arthur Stanley Eddington (1882–1944 CE) organised expeditions to Sobral in northern Brazil and the island of Príncipe off the coast of West Africa to measure the deflection or bending of light by the curvature in space-time due to the sun's gravity, during a total solar eclipse in May 1919, exactly as predicted by Einstein. It was also found to be twice of that predicted using Newton's universal theory of gravitation and his corpuscular theory of light. There have been suggestions that, as it was a British experiment testing the ideas of a German theorist, perhaps people found in Einstein's triumph a symbol of some sort of reconciliation following four terrible years of death and destruction during World War I. It is a pity that this euphoria was soon drowned by the emergence of Nazis in Germany with terrible losses of men and materials across the world due to World War II.

In addition to the need to see the sky from different vantage points on the Earth, building astronomical observatories on the ground and in space is extremely expensive — financially, technically, and intellectually. Thus, most of the current and planned observatories are owned by several nations and all of these collaborations have thus far been peaceful and successful. We shall discuss some of these in the following to illustrate this point.

We emphasise that international collaborations in astronomy are as old as the subject itself, and exchanges about concepts and observations in astronomy among cultures and nations were always very enriching. This is a vast subject and beyond the scope of the present study.

13.2 Early Observations

Every civilisation from antiquity to the present time has been fascinated by astronomy, the study of the motion of the heavenly bodies across the sky. Today, astronomy has many tools of investigation: γ-ray, X-ray, ultraviolet, visible, infrared, microwave and radio wave radiations, and more recently even gravitational waves emanating from the (merging/colliding) heavenly bodies. This has opened up a horizon that was not available to the ancients. They only had the visible range of the spectrum. Their studies were thus restricted to what we now call positional astronomy and did not specifically address the nature of these bodies. However, extensive data about positional astronomy, collected through "naked eye" observations and simple instruments, exist in ancient Egyptian, Babylonian, Indian, and Chinese records [5].

The real sequence of developments in Indian astronomy is hotly debated. It is complicated due to the loss of texts, some of which exist only as "references" or "quotes" in other texts. Still, the Rig Veda, which perhaps reached its present form around 1500 BCE, is the earliest source of astronomy in India. The part of the Vedas known as Vedanga was compiled around 1200 BCE and is attributed to Lagadha. It described the movements of the sun and moon relative to the "nakshatras" (the 27 lunar "mansions") and stated categorically that positional astronomy's sole purpose was to determine the times for performing Vedic rituals, which required strict temporal order. It also included the Sulba Sutras which could be used to calculate the area of triangles, squares, and circles and prepare the ritual platform for sacrifices, and which also included what is now called the Pythagorean theorem [6].

The invasion of Alexander of the northwestern periphery of India in 323 BCE and Chandragupta's victory over Seleucus in 306 BCE are believed to have brought India into contact with Babylonian astronomy through the Greeks. There are, however, suggestions that the influence of Egyptian and Babylonian astronomy on Hindu astronomy was delayed till the advent of the Scythians in the region of Bactria and their victory over Parthians in 123 BCE [5].

It has also been suggested that the Scythians "grafted" the Babylonian system on Hindu astronomy, and by 400 CE, the so-called Siddhanta astronomy replaced the older Vedanga astronomy. Varahamihira (ca. 505–ca. 587 CE), writing in Pancha-siddhantika, mentions five Siddhantas, *viz.*, Pitamaha, Vasistha, Romaka, Paulisa, and Surya. The Romaka and Paulisa Siddhantas are known to show the influence of Babylonian and Egyptian astronomy.

Aryabhat (476–550 CE) was one of the greatest mathematician-astronomers from the classical age of Indian mathematics [6] and astronomy (Fig. 13.1). His works include the Aryabhaṭiya and the Arya-siddhanta. He was greatly influenced by Surya Siddhanta. Some of the notable findings of Aryabhat include an explanation for the lunar eclipse and solar eclipse, the rotation of the Earth on its axis, the reflection of light by the moon, sinusoidal functions, the solution of single-variable quadratic equations, the value of pi correct to four decimal places, and the diameter of the Earth. Brahmagupta (598–668 CE) in his epoch-making treatise "Brahmasphutasiddhanta," written in 628 CE, introduced (or extensively

Fig. 13.1. Statue of Aryabhat at the Inter-University Centre for Astronomy and Astrophysics, Pune (Credit: Wikipedia).

used an already existing) the seminal concept of zero as a number, which is so basic to all of mathematics, among other things. He also described gravity as an attractive force and used the term "gurutvākarṣaṇam" in Sanskrit to describe it.

The next important contribution to Hindu astronomy came from Bhaskaracharya II (ca. 1114–1185 CE), who authored Siddhanta Shiromani (consisting of Lilavati, Bijagaṇita, Grahagaṇita, and Goladhyaaya) and Karaṇa-Kautuhala, and developed algebra, arithmetic, and trigonometry. These are the first works with full and systematic use of the decimal number system, which provided solutions for quadratic equations, among other important results. Many historians have lamented that after Muhammad Ghouri's conquest of northern India at the end of the twelfth century CE, all the academic centres of excellence in northern India were completely destroyed and starved of financial support, which saw the end of ancient Hindu astronomy [5].

However, in the distant south India, in Kerala, the traditions of Hindu astronomy and mathematics were kept alive and reached their zenith during the fourteenth to sixteenth century. Some of the most notable contributions of this region were series expansions of trigonometric functions and what we now call differential calculus — the latter about two centuries before Europe. These mathematicians also derived a value of pi accurate to nine decimal places. A notable and interesting result of this school was obtained by Neelkantha Somayaji (1444–1545 CE), author of the famous treatise Tantrasamgraha. He developed a computational system for a partially heliocentric planetary model in which Mercury, Venus, Mars, Jupiter, and Saturn orbit the sun, which in turn, along with the moon, orbits the Earth, similar to the system later proposed by Tycho Brahe (1546–1601 CE) in the late sixteenth century. Most astronomers of the Kerala school who followed him, accepted this planetary model, as it was also easy to calculate the positions of planets and quite accurate.

By now, it is well accepted that as far as positional calculations of planets or eclipses are concerned, the methods developed by Hindu astronomers were far superior to any other civilisation at that time. These results could be accurately applied for about a couple of centuries before or after the astronomer writing his treatise. Beyond that, the precession of the Earth, which they did not account for, introduced noticeable errors. Most Indian astronomers, starting from Aryabhat, were aware of this limitation and always maintained that numerical calculations must be backed by observations and suitable corrections should be made. The numerical methods developed in the process were again ahead of their times.

Sawai Raja Jai Singh II (1688–1743 CE), ruler of the Kingdom of Amber in Rajasthan and a keen astronomer, was aware of this problem and also of the inadequacies of the astronomical tables prepared by Ulug Beg (1394–1449 CE) from observations in Samarkand some centuries earlier, which were in use during that period. As we have seen in an earlier chapter, under his supervision, five observatories were built at Delhi, Mathura, Benares, Ujjain, and Jaipur with specially designed large masonry structures (yantras or Jantar Mantars) to make accurate "naked eye" observations to predict eclipses, accurate times for religious festivals, and other astronomical observables. What remains a great surprise is that even though he was in contact with several Europeans, had several treatises from Europe translated, and was possibly aware of the discovery and use of the telescope by Galileo Galilei (1564–1642 CE), he seems to have not attempted to use it or incorporate it in his observatories.

A similar situation involved Mahamahopadhyaya Chandrasekhara Singha Harichandana Mahapatra Samanta or Pathani Samanta (1835–1934 CE), a self-taught astronomer. He lived in an isolated and hilly region of present-day Odisha, knew only Sanskrit and Odiya, and had access only to ancient books written in Sanskrit. He studied, mostly on his own, Surya Siddhanta and the other books mentioned earlier, and started making "naked eye" observations of heavenly bodies using some simple instruments made of wood and bamboo, which he himself had designed.

Finding his observations different from his calculations using the procedures discussed in the ancient Sanskrit treatises, Pathani Samanta prepared a meticulous table of these observations. He wrote a book Siddhanta Darpana in Sanskrit on palm leaves in 1869 in the Odiya script. It was published thirty years later in 1899 in the Devanagari script, with a long preface in English, and only then did his work receive some attention [7]. It is said that he caught the attention of British administrators by accurately predicting a solar eclipse, which was to be visible only in England.

He even proposed a model (Fig. 13.2) with both heliocentric and geocentric elements, where the Earth remained stationary, with both the moon and the sun moving around it in different orbits. In contrast, the inner as

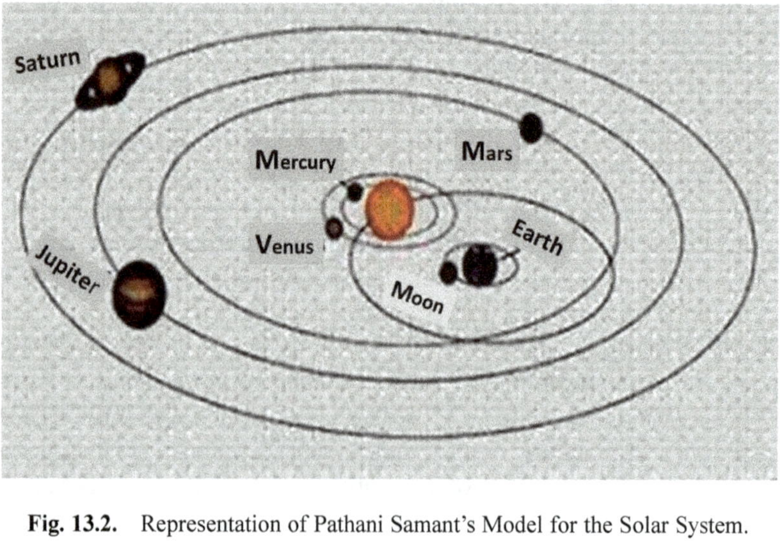

Fig. 13.2. Representation of Pathani Samant's Model for the Solar System.

well as outer planets moved around the sun. The model explained the retrograde motion of the moon, inner planets, equinoxes, solstices, eclipses, transit of planets, and more, and was similar to the models proposed by Neelkantha Somayaji and Tycho Brahe, of whose work he was not aware [8]. Needless to add, perhaps he was also not aware of the revolutions which had already been ushered in by the work of Nicolaus Copernicus (1473–1543 CE), Johannes Kepler (1571–1630 CE), and Isaac Newton (1642–1727 CE).

We have discussed the cases of Jai Singh II and Pathani Samanta in some detail, as they were operating at a time when astronomers elsewhere had made considerable progress. Pathani Samanta was handicapped due to his isolation, poverty, and even ill health. But Jai Singh II did not suffer any handicaps on these accounts. An international collaboration could perhaps have led them to make even more meaningful contributions of lasting importance. Perhaps their examples could have spawned a new generation of astronomers much earlier in the country.

Lamentably, the practice of writing down facts, observations, or ideas and preserving them for posterity, common in Europe, China, and the Middle East, was not so common in India. Thus, accounts of a supernova, whose remnants are now called the Crab Nebula, exist in historical records in China. These chronicles recorded the appearance of a "guest star" on 4 July 1054 CE, "which faded away after a year." There have also been indications from the rock carvings of the Native Americans and records of a physician in the Middle East that they had witnessed this spectacular event. However, no clear mention of this is found in Indian records [10]. Given this, the possible identification of a cave painting (Fig. 13.3) in Kashmir of the supernova HB9 provides the oldest depiction of a cosmic event in the world [9].

Fig. 13.3. Cave Painting in Kashmir (Left) Superimposed on Star Map (Right): possibly depicting Supernova HB9 [9] (Credit: Tata Institute of Fundamental Research).

13.3 International Terrestrial Observatories

Let us briefly discuss some of the major astronomical observatories which are an outcome of international collaboration. We emphasise that this list is only representative and not exhaustive.

ALMA: The Atacama Large Millimeter/Submillimeter Array (Fig. 13.4) is considered the crown jewel of international cooperation. It is located in Chile and is a radio interferometer with over 66 antennas spread across the high Atacama Desert to benefit from its high elevation (5000 m) and low humidity. These factors are crucial to reducing noise and decreasing signal attenuation due to the Earth's atmosphere. It has been built through a partnership amongst Europe, the United States, Canada, Japan, South Korea, Taiwan, and Chile. It observes the universe at millimetre and submillimetre wavelengths and provides information on cool gas, dust clouds, and the formation of stars and planets, and is part of the Event Horizon Telescope (see later). It was built at a cost of about 1.4 billion USD and has been in operation since 2013, though it had already started operating partially by 2011.

Its major findings include information about the formation of stars and planets within dusty nurseries, the observation of several new

Fig. 13.4. Two of ALMA's 12-m Antennas [11].

galaxies formed just 1.4 billion years after the Big Bang, and the identification of water vapour in distant galaxies, hinting at the potential for life.

VLT: The Very Large Telescope (VLT) was built by the European Southern Observatory (an intergovernmental organisation of 16 European member states) for ground-based astronomy. It is situated in the Cerro Paranal Region of Chile at a height of 2635m and was built at a cost of about 1.2 billion USD. It is one of the world's most advanced optical telescopes for deep-ultraviolet (300 nm) to mid-infrared (24 μm) wavelengths and consists of four unit telescopes with main mirrors of 8.2 m diameter and four movable 1.8-m-diameter auxiliary telescopes (Fig. 13.5). The unit telescopes have been given the Mapachu names Antu (sun), Kueyen (moon), Melipal (Southern Cross), and Yepun (now believed to be Venus).

Each telescope can detect objects that are roughly four billion times fainter than what can be seen with the naked eye. In the single-telescope mode, the angular resolution is about 0.05 arcsec. The telescopes can

Fig. 13.5. The Four Unit Telescopes That Form the VLT Together with the Four Auxiliary Telescopes (Credit: Wikipedia).

work together to achieve an angular resolution of approximately 0.002 arcsec. In this mode, they form the giant "ESO Very Large Telescope Interferometer." This mode of operation allows astronomers to pick up much finer details of the cosmos that would not be possible with the auxiliary and the unit telescopes alone. The VLT has pioneered cutting-edge astronomical technology, including adaptive optics to overcome atmospheric blurring and interferometric techniques, indicated earlier, to achieve even higher resolution. These have now been adopted by several other observatories.

The VLT has been providing a rich harvest of astronomical discoveries with several notable scientific firsts, including the first image of an extrasolar planet, tracking individual stars moving around the supermassive black hole at the centre of the Milky Way, and observing the afterglow of the farthest known Gamma-Ray Burst.

EHT: The Event Horizon Telescope (EHT) is a network of synchronised radio observatories around the world that combine to mimic a single telescope with an aperture the size of the Earth, which is used to observe sources of radio light associated with black holes. Event horizon represents the point at which not even light is fast enough to outrace the gravitational influence of a black hole and as such is the "deepest" we ever hope to see into a black hole. The EHT allows scientists to see the area of space immediately surrounding a black hole and the effects that the massive gravitational influence of the black hole has on this region.

The EHT observatories use a technique called Very-Long-Baseline Interferometry which synchronises telescope facilities (with the help of atomic clocks) around the world and exploits the rotation of our planet to form one huge Earth-sized telescope observing at a wavelength of 1.3 mm. It allows the EHT to achieve an angular resolution of 20 μ-arcsec — which can be enough to read a newspaper in New York from a sidewalk café in Paris!

The following are the participating telescopes:

(a) ALMA: Atacama Large Millimeter/submillimeter Array in Atacama Desert, Chile;
(b) APEX: Atacama Pathfinder Experiment in Atacama Desert, Chile;
(c) IRAM 30-m telescope: a radio telescope, located in the Sierra Nevada Mountain range, Spain, and operated by the Institute for Radio Astronomy in the Millimetre Range;

(d) JCMT: James Clerk Maxwell Telescope, a submillimetre-wavelength radio telescope at the Mauna Kea Observatory in Hawaii, USA;
(e) LMT: Large Millimetre Telescope, the world's largest single-dish steerable millimetre-wavelength telescope on the summit of Volcan Sierra Negra, Mexico at an altitude of 4600 m;
(f) SMA: Submillimetre Array, consisting of eight 6-m-diameter radio telescopes arranged as an interferometer for submillimetre-wavelength observations Mauna Kea, Hawaii, USA;
(g) SMT: Submillimetre Telescope, a submillimetre-wavelength radio telescope located on Mount Graham, Arizona, USA, where the dryness of the air is important for extremely high-frequency radio and far-infrared observations;
(h) SPT: South Pole Telescope, a 10-m-diameter microwave/millimetre/submillimetre-wavelength telescope located at the Amundsen–Scott South Pole Station in Antarctica. Currently, it is the best operational site on the Earth for mm-wave survey observations due to its stable, dry atmosphere.

Several other telescopes across the planet are likely to join this effort.

Petabytes of raw data from the telescopes are then combined by highly specialised supercomputers hosted by the Max Planck Institute for Radio Astronomy in Bonn and the MIT Haystack Observatory, near Boston, and analysed.

This mammoth effort resulted in the first-ever picture of a black hole (M87*) at the centre of the galaxy Messier 87, located 55 million light years away from the Earth.

The image (Fig. 13.6, released on 10 April 2019) showed a bright ring formed as light bends in the intense gravity around a black hole. These results were used to determine the mass of the black hole M87* as 6.5±0.7 billion solar masses and the diameter of its event horizon to be approximately 40 billion kilometres. This long-sought image provided the strongest evidence to date for the existence of supermassive black holes and opened a new window into the study of black holes, their event horizons, and gravity. Later, an image of the black hole M87* in polarised light was also released.

On 12 May 2022, astronomers unveiled the first image of a supermassive black hole at the centre of the Milky Way, Sagittarius A*. Details of other notable achievements — such as confirmation of general relativity by observing the behaviour of matter near the black hole, which matches Einstein's predictions and evidence of variability in the black hole's

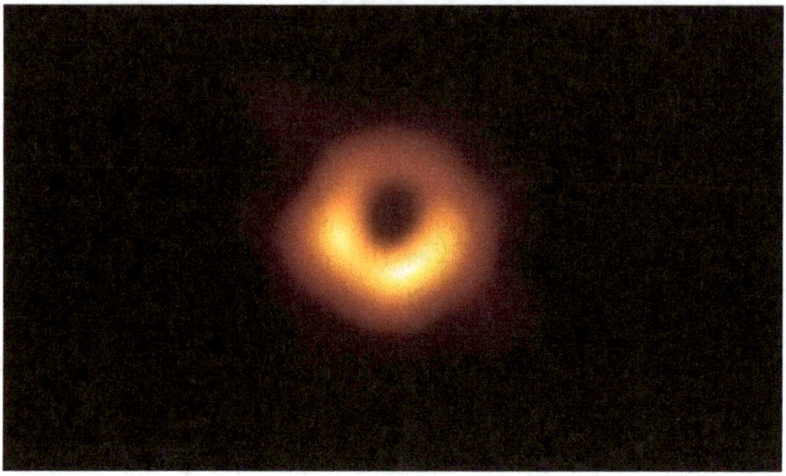

Fig. 13.6. The First Image of a Black Hole (Called M87*), Using Event Horizon Telescope Observations of the Centre of the Galaxy Messier 87 (Credit: Event Horizon Telescope Collaboration, 10 April 2019).

accretion disk, suggesting complex processes at play — can be seen in the continuously updated Ref. [12] and associated links.

SKAO: The Square Kilometre Array Observatory (SKAO) is an international radio telescope project aiming to be the largest scientific instrument ever built. It will collect radio waves from across the cosmos and offer incredibly sensitive and detailed observations about the universe.

The SKAO is being built in two phases, SKA1 and SKA2.

SKA1 will be spread across two locations — Murchison Shire in Western Australia (low-frequency antennas) and the Karoo region in South Africa (mid-frequency dishes). SKA2 is planned for later expansion, with sites likely in Africa and/or Australia.

Western Australia has vast stretches of unpopulated land with extremely low radio frequency noise, which is deemed perfect for studying the early universe and faint stellar nurseries, using low frequencies. The Karoo region's dry climate and radio-friendly environment make it perfect for observing celestial objects at mid-frequency radio wavelengths. The Southern Hemisphere also provides the best view of the Milky Way galaxy with unobstructed views of the galactic centre and southern sky and is expected to provide insights into our galaxy's formation and evolution.

Fig. 13.7. An Artist's Impression of SKA-Mid array [13].

Thus, the telescopes have been designed to cover two different frequency ranges and are named *SKA-Mid* and *SKA-Low* to reflect this.

SKA-Mid, an array of 197 parabolic dishes (Fig. 13.7) to collect radio waves of frequencies between 500 MHz and 3 GHz and a collecting area of 33,000 sq.m., is being built in South Africa's Karoo region. The existing MeerKAT radio telescope will form part of the array. Its 13.5-m-diameter dishes will be joined by the (slightly) larger 15-m-diameter SKA dishes, all integrated into one system.

From a central core measuring around 1 km across, the antennas will radiate out in three spiral arms covering a vast distance; the two furthest antennas in SKA-Mid will be 150 km apart. Research institutions in China, Australia, Canada, France, Germany, Italy, South Africa, Spain, the United Kingdom, Sweden, etc., contributed to the SKA-Mid design. Its

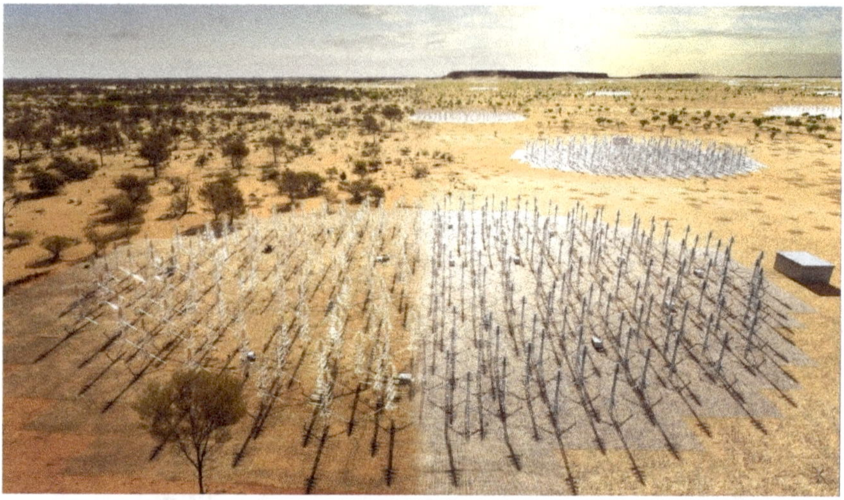

Fig. 13.8. An Artist's Impression of SKA-Low Array [13].

various components will be manufactured all over the globe before being shipped to South Africa for assembly.

SKA-Low, an array of 131,072 smaller tree-like antennas (Fig. 13.8), sensitive to lower frequencies between 50 MHz and 350 MHz and with a collecting area of 419,000 sq.m., is being built in Western Australia. The arrays will be spread across large distances, up to 65 km. Institutions in Australia, China, Italy, Malta, the Netherlands, and the UK contributed to the design of SKA-Low, and its many components will be manufactured all over the world before being shipped to Western Australia for construction at the Murchison Radio-astronomy Observatory.

The total collecting area of the array would be 1 square kilometre, which gave rise to the name Square Kilometre Array. The processing and analyses of these signals will involve vast resources of high-performance computing, and SKAO has agreed to work with CERN.

The SKAO is poised to revolutionise our understanding of the universe by providing opportunities to study the following:

- the formation and evolution of the first stars and galaxies by peering back to the cosmic dawn to unravel the mysteries of early universe formation and shed light on the birth of stars and galaxies;

- cosmic magnetism by mapping the distribution and evolution of magnetic fields throughout the universe, which is crucial for comprehending star formation, galaxy evolution, and black hole activity;
- exoplanets and their potential for habitability, thereby contributing to the quest for extraterrestrial life;
- the nature of dark matter and dark energy, where the SKAO's high sensitivity might unveil new insights into their properties and impact.

Organisations in 16 countries are currently taking part in the SKAO project, valued at about 3 billion AUD, at the government or national coordination level or are represented as observers: Australia, Canada, China, France, Germany, India, Japan, Italy, the Netherlands, Portugal, South Africa, South Korea, Spain, Sweden, Switzerland, and the United Kingdom. Eight African partner countries are involved in coordinated action to support the future expansion of the SKAO project in Africa. Many more countries are eager to join the collaboration.

Furthermore, around 100 organisations across about 20 countries (including India with its long experience of building and operating the Giant Metrewave Radio Telescope (GMRT), now upgraded as uGMRT) participated in the design and development of the SKAO project, including the detailed design of the telescopes over the last 10 years. Many of these organisations and their experts are now involved in construction activities.

The SKAO is a non-commercial, international scientific endeavour driven by a shared quest for knowledge, and not by profit motives. It operates under a unique governance structure that combines elements of an intergovernmental organisation with aspects of a research consortium, somewhat like CERN, Geneva.

LIGO: The Laser Interferometer Gravitational-Wave Observatory (LIGO) is a large-scale physics collaboration and international observatory. It has been designed to detect cosmic gravitational waves and develop gravitational wave observations as an astronomical tool.

The general theory of relativity predicts that when a massive object accelerates or undergoes intense gravitational interaction, it produces a ripple in space-time that we call a gravitational wave. The gravitational waves cause space to stretch in one direction and simultaneously compress in a perpendicular direction, then vice versa (i.e., the space in the

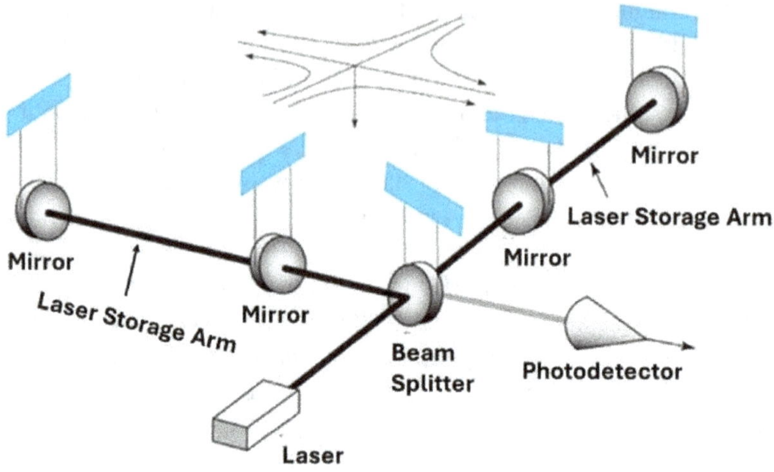

Fig. 13.9. Basic Schematic of LIGO's Interferometers with an Incoming Gravitational Wave Depicted as Arriving from Directly above the Detector (Credit: Adapted from Caltech/MIT/LIGO Lab).

first direction is compressed and the one in the perpendicular direction is stretched). This back and forth goes on as long as the wave is passing. The LIGO uses this occurrence to detect gravitational waves, as depicted in Fig. 13.9.

In the arrangement shown in Fig. 13.9, a laser beam is split into two beams perpendicular to each other. It travels an identical distance along the two arms and is merged again. The LIGO uses a laser beam with a (single) consistent frequency, which can lead to total constructive or destructive interference. The LIGO is deliberately designed to ensure a destructive interference in the merged beam when no gravitational waves are passing — no wave, no interference pattern!

When a gravitational wave from the top arrives (as shown in the figure), the lengths of the arms change, so too does the distance travelled by each laser beam. As a result, the beam in the shorter arm will return to the beam splitter before the beam in the longer arm. As the wave passes, each arm will oscillate between being the shorter arm and the longer arm. When the laser beams arrive back at the beam splitter (where they re-merge), the light waves would be out of phase and shift in and out of alignment for as long as the wave is passing. This shifting alignment would result in a flicker of light emerging from the interferometer and will signal the arrival of the gravitational wave. This flicker ("chirp") would last only for the duration of the passage of the gravitation wave — a fraction of a second for a binary black hole merger to tens of seconds for a binary neutron star merger.

The stretching and compression discussed here occur in specific planes relative to the direction of the wave. This "polarisation" can provide insights into the source of the wave.

The change in the length of the arms caused by a gravitational wave can be about 1/1000 the size of a proton or 10**(−18) m. Thus, disturbances that can shake the mirrors, like earthquakes or traffic on nearby roads, need to be filtered out before the flicker of light caused by a gravitational wave is detected.

The LIGO consists of two identical and widely separated (3002 km) interferometers situated in extremely isolated places: Hanford in southeastern Washington and Livingston in Baton Rouge, Louisiana (Fig. 13.10). As mentioned previously, the detectors are sensitive to the tiniest vibrations on the Earth from local, nearby sources to big events occurring hundreds or thousands of kilometres away (in the case of large earthquakes). Thus, earthquakes, moving trucks driving along nearby

Fig. 13.10. LIGO at Livingston (Credit: Caltech/MIT/LIGO Lab).

roads, farmers ploughing fields, heavy constructions, severe thunderstorms, and even local winds can cause disturbances that can mask or mimic a gravitational wave signal in each interferometer. The two widely separated interferometers are not likely to experience the same disturbance simultaneously.

Moreover, gravitational waves travel at the speed of light. Thus, the longest time difference between the times of arrival of the gravitational waves would be about 10 ms. Thus, similar signals detected at the two detectors which are farther apart than 10 ms are also ignored, as those could not have been caused by the passage of a gravitational wave.

The construction of the LIGO required extreme precision. The vacuum system consists of two 4-km-long beam tubes with a diameter of 1.2 m, which need to be evacuated and kept at a level of vacuum of a nano-Torr, or one-trillionth of an atmospheric pressure. This is necessary to ensure that the mirrors do not shift due to the impact of moving air molecules, the laser is not deflected by them, and dust does not get into the vessels and on the mirrors. The LIGO's mirrors are of the highest available quality. They are made of ultra-pure materials, laid down in nanometre-thick layers. Their reflective surface absorbs just one in 5 million photons. The LIGO uses an extremely stable (200 W), high-purity laser beam having a precise wavelength of 1064 nm. This is one of the most precise laser beams ever produced.

Since the LIGO's arms are long enough, over the 4 km length of each arm, the Earth curves away by nearly a metre! Thus, precision was required during concrete pouring on the path upon which the beam tube was to be installed to counteract this curvature and ensure that the laser beam travelling in a straight line is not lost.

The LIGO Laboratory Charter envisages that the LIGO's mission is to open the field of gravitational wave astrophysics through the direct detection of gravitational waves. The primary goal, direct detection of gravitational waves, was achieved on 14 September 2015, and it was announced on 11 February 2016, one hundred years after the prediction by Einstein. These observations matched calculations for the inward spiral and merger of a pair of black holes of around 36 and 29 solar masses and have been named GW150914. This merger took place 1.3 billion years ago (Fig. 13.11). Since then, the LIGO has made numerous other detections, offering insights into phenomena like merging black holes and colliding neutron stars.

Fig. 13.11. The Laser Interferometer Gravitational-Wave Observatory (LIGO) Discovered Ripples in Space-Time Created by Merging Black Holes (Credit: Caltech).

By observing gravitational waves from various sources, the LIGO is expected to help us probe the most extreme environments in the cosmos and throw light on stellar evolution, cosmic distances, and the formation of heavy elements.

The LIGO Scientific Collaboration started in 1997, and by now it has more than 1400 scientists from 19 countries spread over 127 institutes. To improve the sensitivity of (simultaneous) detection and accuracy in the location of the source of the gravitational waves, the LIGO-Virgo-KAGRA collaboration was formed. The VIRGO detector is an inerferometer with two arms of 3-km each, near PISA, Italy. It is operated by the European Gravitational Observatory and funded by Italy, France, and the Netherlands. The KAGRA observatory is an underground interferometre with two arms of 3-km each in Kamioka, Japan.

On 14 August 2017, the Virgo and the two advanced LIGO detectors coherently observed a transient gravitational wave signal as (GW170814) produced by the coalescence of two black holes having masses of about 30.5 and 25 solar masses and located about 1.8 billion light-years away. A network of three detectors improved the sky localisation of the source, reducing the likely region of the source to 60 square degrees, compared to 1160 square degrees, when only the two LIGO detectors were used [14].

GW170817 was the first event (17 August 2017) where gravitational waves, in addition to light from the spectacular collision of two neutron stars, were observed. The discovery was made using the LIGO detectors in the USA, the Virgo detector in Europe, and some 70 ground- and space-based observatories.

These studies are expected to get a considerable fillip when *LIGO India*, to be located near Aundha Nagnath, Hingoli, Maharashtra, starts operating. It is a collaboration between the LIGO Laboratory and three institutes in India: the Raja Ramanna Centre for Advanced Technology (Indore), the Institute for Plasma Research (Ahmedabad), and the Inter-University Centre for Astronomy and Astrophysics (Pune).

As indicated earlier, just the two LIGO detectors cannot effectively localise the source of the waves in the sky, nor can they reveal a wave's actual polarisation. The inclusion of the third detector — Virgo — has already led to considerable improvement in the localisation of the source. KAGRA in Japan is now operating, but it does not yet have the gravitational wave sensitivity of the LIGO and Virgo. Once LIGO India is operational, it will ensure that at least four comparable detectors are operating across the globe. This is needed to localise a source of gravitational waves anywhere in the sky.

A gigantic version of the LIGO observatory is planned in the form of a Laser Interferometer Space Antenna (LISA), consisting of three spacecraft forming an equilateral triangle, with sides precisely kept at 2.5 million km, which follow the Earth, about 20° behind on its journey around the sun. Laser interferometry is then used to measure the minute change in these lengths as gravitational waves pass through them. The sensitivity of the detection is assured as the three planes are located extremely precisely. Moreover, the sides of the triangle which the laser beams have to traverse in going from one plane to the other are very long. The project was adopted by the European Space Agency in January 2024 and is expected to be launched in 2034. It follows the successful test of the principle in the LISA Pathfinder experiment performed in collaboration with NASA.

The LISA (Fig. 13.12).will be humanity's first detector that will be capable of detecting space-time ripples from super-massive black holes and objects that fall into them, high-mass binary black hole companions to lower mass objects like stellar-mass black holes, or even neutron stars or possibly white dwarfs.

Fig. 13.12. LISA (Courtesy: NASA).

TMT: The Thirty-Meter Telescope (TMT) is a proposed astronomical observatory sensitive to near-ultraviolet to mid-infrared observations (0.31–28 μm wavelengths), with a 30-m aperture, which would help it collect more light and "see" farther, fainter objects. It was to be built on

Mauna Kea, a dormant 4,200-m-tall shield volcano in Hawaii. However, for Native Hawaiians, Mauna Kea is the first-born Son of the Sky Father and Earth Mother and a sacred site. This, along with concerns about environmental damage, has led to opposition to the setting up of the observatory there, and alternative sites, e.g., the Canary Islands, are being considered. It is a collaborative project involving several institutes from the USA, Canada, Japan, China, and India.

An India-based Neutrino Observatory (INO) has also been indefinitely delayed due to opposition and misgivings of some community leaders.

13.4 International Observatories in the Sky

Observatories in the sky or space-based observatories provide an unobstructed view of the sky, as they are above the Earth's atmosphere, which acts like a giant filter and blocks ultraviolet radiation, X-rays, and gamma rays. Space observatories can access the entire electromagnetic spectrum, leading to much clearer and more detailed observations. The Earth's atmosphere also causes turbulence, creating twinkling stars and distorted images, while observations from space provide sharper and crisper images, which enable us to see finer details and study fainter objects. Space-based observatories can observe targets continuously for extended periods, which is valuable for the study of long-term processes like stellar evolution and black hole activity.

The Hubble Space Telescope (HST) is a low-Earth-orbit (about 530 km) telescope, with a 2.4-m-diameter mirror capturing faint and distant light for high-resolution images of ultraviolet, visible, and near-infrared wavelengths. It is a collaboration between the USA, Canada, and the European Space Agency. It was developed for 1.5 billion USD and launched in 1990. It has since provided valuable and precise information about the following: the age of the universe (13.8 billion years) and its rate of expansion (Hubble Constant) with an accuracy to within 2%, down from a factor of two in the 1990s; the accelerated rate of expansion of the universe, which for want of better understanding is attributed to dark energy; the existence of dark matter; the prevalence of black holes in the centre of nearby galaxies; and breathtaking glimpses of the deep universe, capturing galaxies billions of light-years away and showcasing the vastness and history of the cosmos. These images (Fig. 13.13) inspired a public fascination with space. It also analysed the atmospheres of planets

Fig. 13.13. A "Rose" Made of Galaxies (Left) [15] and "Pillars of Creation" (Right) (Credit: NASA).

within our solar system and beyond, revealing diverse compositions and weather patterns.

One of the notable feats of space science and technology has been the numerous servicing missions conducted for the Hubble Space Telescope with the help of space shuttles.

The James Webb Space Telescope (JWST) is a collaboration of over 200 institutions from across the globe, led by NASA, the European Space Agency, and the Canadian Space Agency as major partners. It was conceived in the 1980s, with a launch in the late 1990s. However, technical difficulties and cost overruns led to a delay of over 30 years. It reached its operational orbit at Lagrange Point, L2, about a month after launch. At a development cost of about 10 billion USD, it is one of the most expensive scientific projects ever. It has been designed to observe primarily in the infrared spectrum, revealing objects obscured by dust and gases, like early galaxies and forming planets, using a 6.5-m primary mirror which collects 100 times more light than the Hubble Space Telescope and enhances sensitivity and resolution (Figs. 13.13 and 13.14).

The JWST has captured stunning images of galaxies that existed just a few hundred million years after the Big Bang. These provide a valuable insight into galaxy formation. It has revealed the detailed structures of

Fig. 13.14. Crab Nebula (NIRCam and MIRI Image, Credit: JWST [16]).

forming stars and their surrounding dust disks, providing clues about planet formation processes and unprecedented views of the complex structures and compositions of gas and dust within galaxies, shedding light on interstellar processes. Moreover, it has analysed the atmosphere of distant exoplanets, detected water vapour and other molecules, and hinted at potential habitability.

The JWST is still at the beginning of its mission, and a flood of groundbreaking discoveries is expected in the coming years. Its capabilities hold out the promise of transforming our understanding of the universe's evolution, the formation of stars and planets, and the potential for life beyond Earth.

India launched its satellite telescope *AstroSat* into a 650-km low-Earth orbit in September 2015. It carried highly sensitive equipment to simultaneously observe near and far UV, soft and hard X-rays, and a limited optical range to study black holes, neutron stars, galaxies, and other celestial objects; understand the formation and evolution of stars and galaxies; explore the interstellar medium and its interaction with stars; and discover new astronomical phenomena. Its important achievements include the detection of extreme UV light from a galaxy 9.3 billion light-years away, shedding light on the early universe, the observation of the faint "ghostly glow" of hot gas around black holes, and the discovery of new pulsars, remnants of massive stars. It is open to national and international researchers, fostering collaboration and scientific advancements.

Very recently, in January 2024, India put its *XpoSat* satellite into a 650-km low-Earth orbit, which is dedicated to the study of the polarisation of X-rays from celestial sources to understand the magnetic fields and physical processes around black holes, neutron stars, and other energetic objects. Its other objectives include probing the accretion processes in binary systems, the study of the nature of pulsars and their surroundings, and getting possible insights into the early universe and extreme environments.

A space observatory to provide an unobstructed view of and continuously monitor the sun, *Aditya-L1* was launched by India in September 2023. It reached Earth–Sun Lagrange point 1 (L1), about 1.5 million km from the Earth, on 6 January 2024, and was placed into a halo orbit around it. It is designed to provide high-resolution details of solar activity like flares, prominences, and coronal mass ejections. It is expected to provide insights into the origin and propagation of solar storms and their impact on Earth, and the dynamics of the sun's atmosphere, including heating mechanisms and plasma behaviour, and help predict space weather events to protect satellites and communication systems.

13.5 Conclusions

Sir Isaac Newton, whose contributions to mathematics, physics, optics, astronomy, etc., started the scientific revolution and the enlightenment that followed, famously remarked, "*I do not know what I may appear to the world, but to myself, I seem to have been only like a boy playing on the seashore and diverting myself in now and then finding a smoother pebble or a prettier shell than ordinary, whilst the great ocean of truth lay all undiscovered before me.*" Astronomy has been undergoing almost a continuous revolution, which started with the formulation of the general theory of relativity by Albert Einstein. The latest discoveries in multi-messenger astronomy and the wondrous images and insights that these have produced re-emphasise the message contained in the observation of Newton and that of Aitken mentioned at the beginning of the chapter.

The most beautiful images of galaxies, stars, supernovae, black holes, exoplanets, and other celestial objects revealed through multi-messenger astronomy have whetted the appetite of astronomers and even lay persons like never before, while the vast space-time evolution of the universe in its myriad ways, with scientific secrets to be unveiled, lies before us. The future is bound to add neutrinos to our arsenal of gravitational and electromagnetic waves to explore the mysteries of the universe and its laws.

One of the most delightful and philosophically humbling aspects of research in astronomy is that the further away in space we see, the further away in the past we peep. With the ability to observe the details of supernova explosions, neutron star collisions, accretions of companion stars, and planets in their immeasurable varieties, a detailed understanding of the evolution of the universe and its celestial bodies is bound to emerge. Astrobiology and astrochemistry, which we have not even touched upon, will surely decipher the mystery of the emergence of life.

A bright future with a vast store of knowledge lies ahead of us in astronomy. International collaboration will take us there much faster and more efficiently, making this journey of exploration exciting, exhilarating, and fulfilling. Looking at the developments in the past, it goes without saying that these will also give rise to new technologies that will make life on our planet more enjoyable. It is our fervent hope that international collaboration in research in astronomy will exemplify the futility of conflict among nations, ushering in an era of everlasting peace as envisaged by visionaries like Carl Sagan.

References

[1] R. G. Aitken, "The Use of Astronomy", *Leaflets of the Astronomical Society of the Pacific* 59 (1933) 33–36.

[2] P. J. Hore and H. Mouritsen, "The Quantum Nature of Bird Migration", *Scientific American* 326 (2022) 70.

[3] J. J. Foster, J. Smolka, D.-E. Nilsson, and M. Dacke, "How Animals Follow the Stars", *Proceedings of the Royal Society B: Biological Sciences* (24 January 2018), https://doi.org/10.1098/rspb.2017.2322.

[4] See, for example, https://www.iau.org/public/themes/astronomy_in_everyday_life/.

[5] P. Rudra, "A Short History of Hindu Astronomy and Ephemeris", *Physics Teacher* 48 (2006) 4, https://arxiv.org/abs/0903.1778.

[6] See e.g., B. Kamble, "The Imperishable Seed: How Hindu Mathematics Changed the World and Why This History was Erased" (Gurugram: Garuda Prakashan, 2022).
[7] P. C. Naik and L. Satpathy, "Samanta Chandra Shekhar: Life and Work", *Current Science*, 69 (1995) 705–710.
[8] M. Katti, "Pathani Chandrashekhar Samanta: Traditional Eye Scanning the Sky", *Bhavana* (A Publication of Bhavana Trust) 7(1) (January 2023).
[9] H. Joglekar, K. Gangal, M. N. Vahia, and A. Sule, "Oldest Sky-Chart with Supernova Record", *Puratattva* 41 (2011) 207.
[10] J. V. Narlikar and S. Bhate, "Search for Ancient Indian Records of the Sighting of Supernovae", *Current Science* 81 (2001) 701–705.
[11] Credit: Iztok Bončina/ESO, http://www.eso.org/public/images/potw1040a/.
[12] https://eventhorizontelescope.org/.
[13] https://www.skao.int/.
[14] B. P. Abbott *et al.* (LIGO Scientific Collaboration and Virgo Collaboration), "GW170814: A Three-Detector Observation of Gravitational Waves from a Binary Black Hole Coalescence", *Physical Review Letters* 119 (2017) 141101.
[15] https://hubblesite.org/.
[16] https://webbtelescope.org/.

Chapter 14

The Indian Ocean Tsunami Warning System

> "Whether it is a tsunami, or whether it is a hurricane, whether it's an earthquake — when we see these great fatal and natural acts, men and women of every ethnic persuasion come together and they just want to help."
>
> — Martin Luther King III

14.1 Introduction

The Indian Ocean Tsunami occurred on 26 December 2004. It was caused by an earthquake of magnitude 9.2 with an epicentre off the west coast of Sumatra, Indonesia. It was one of the strongest and deadliest tsunamis in recorded history in the Indian Ocean (Fig. 14.1). It devastated the coastal regions of surrounding countries: Aceh, Indonesia; Andaman and Nicobar Islands, Tamil Nadu, and Kerala, India; Sri Lanka; Khao Lak, Thailand; Maldives; and many other countries [1]. It even reached Somalia, Tanzania, and Seychelles, causing loss of life and property. It is estimated that about 230,000 people died, about 1.7 million people were displaced, and damage of more than 17 billion dollars occurred. To date, it was the most destructive natural disaster of the twenty-first century.

The tsunami was generated due to a massive undersea earthquake because of the under-thrusting of the India Plate below the Burma Plate. It was the third most powerful earthquake since 1900 CE, and it was

Fig. 14.1. A View of the Vast Destruction at Moratuwa, Sri Lanka, Caused by the 2004 Indian Ocean Tsunami (Credit: UN Photo).

caused by rupture along the fault between the Burma Plate and the Indian Plate for about 1600 km, with a slip of about 15 km. The energy released was estimated to be 1.1×10^{17} joules or 1500 times that of the Hiroshima atomic bomb. The earthquake lasted about 10 min, the longest duration ever recorded, and moved at a velocity of 2.1–2.8 km/s. Its effects were felt worldwide. It suddenly displaced about 30 km^3 of water and created tsunami waves as high as 30 m.

Table 14.1 provides the time taken for the tsunami to reach various locations along the route of the tsunami waves, while Fig. 14.2 shows the travel time contour with 30-min intervals for the 2004 Tsunami.

The extensive loss of life and property was due to the limited warning infrastructure, the (initial) underestimation of the earthquake's intensity, the severe limits of data communication, and the lack of public awareness [1]. As we can see, the tsunami reached the Sumatra coast within 15 min. However, Thailand, the South Indian Coast, Sri Lanka, and the East African Coast would have had sufficient time for an orderly evacuation of the population from the affected zones if an Indian Ocean tsunami warning system — like the one for the Pacific Ocean, operating for several decades and serving all the nations and islands along the Pacific — had been in place.

Table 14.1. Time Taken by the 2004 Tsunami to Reach Various Locations after the Earthquake.

Location	Time for the tsunami to reach after the earthquake
Sumatra, Indonesia	15 min
Andaman and Nicobar Islands	Within 15 min
Thailand (southwestern coast)	2 h
South Indian Coast	2–3 h
East African Coast	7–8 h

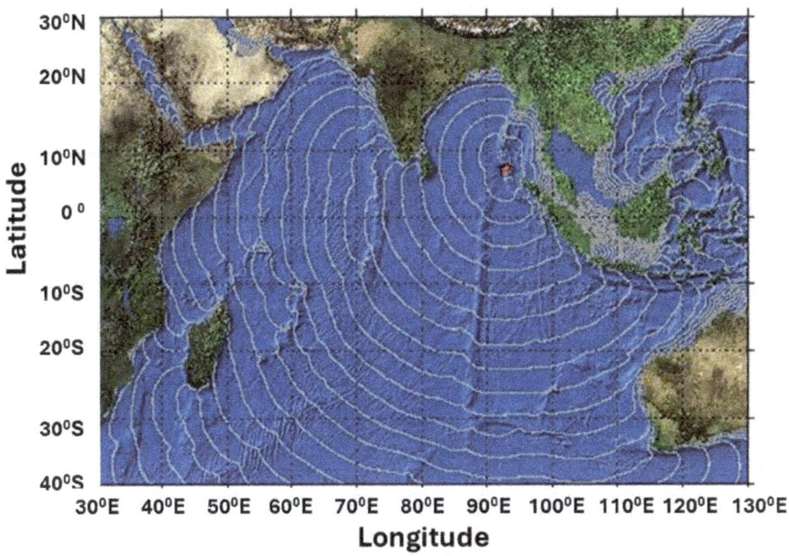

Fig. 14.2. Travel Time Contour for the 2004 Tsunami.

In the aftermath of the 2004 Indian Ocean Tsunami, in a remarkable achievement of international collaboration, the Intergovernmental Oceanographic Commission (IOC) of the United Nations Educational, Scientific and Cultural Organization (UNESCO) coordinated the establishment of regional tsunami early warning systems in the Indian Ocean and Caribbean and Mediterranean Seas. These regional systems, along with the Pacific Tsunami Warning System (PTWS), form a globally integrated system, built on three pillars: tsunami risk assessment and mitigation; tsunami detection, warning, and dissemination; and tsunami awareness and response.

India decided to set up a tsunami warning system to safeguard its population and infrastructure. It played a major role in setting up a Regional Tsunami Warning System for the Indian Ocean. The Indian Tsunami Early Warning Centre (ITEWC) was established in 2007 at the Earth System Science Organization (ESSO)-Indian National Centre for Ocean Information Services (INCOIS), Hyderabad, in collaboration with other scientific agencies and disaster management organisations in India [2]. Along with similar centres in Australia and Indonesia, the ITEWC also serves as one of the tsunami service providers (TSPs) in the IOC-coordinated Indian Ocean Tsunami Warning and Mitigation System (IOTWMS).

14.2 Types of Tsunamis

Tsunamis are large ocean waves that are usually triggered by an undersea earthquake (Fig. 14.3), a volcanic explosion, an asteroid impact, glacier calving, or submarine landslides. These are often a train of waves, one wave following the other. A tsunami travels very fast in the open ocean, having a velocity of several hundreds of kilometres per hour and a large wavelength but a wave height of only a few centimetres. When it approaches a coast, the speed reduces but wave height increases to the order of several metres, with disastrous consequences [1].

In the twenty-first century, tsunamis occurred near Chile (2010), Japan (2011), Palu (2018), and Sunda Strait (2018), and all of them caused disasters. A tsunami warning system essentially provides advisories to

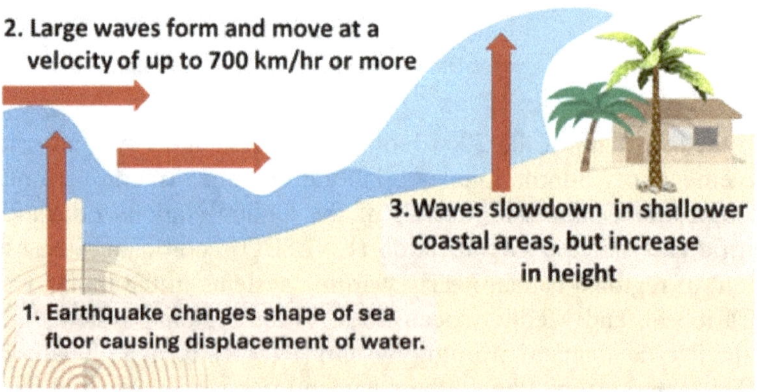

Fig. 14.3. Development of a Tsunami Due to an Undersea Earthquake [1].

disaster management authorities at the national, state, and local levels so that loss of life and damage to properties can be prevented.

Tsunamis are classified as local, regional, or distant based on the distance travelled from the source and the tsunami travel time according to the Intergovernmental Oceanographic Commission, UNESCO 2006. Local tsunamis originate from a nearby source and can affect areas within 100 km with a travel time of less than an hour. As the travel time is not significant, they are difficult to predict in time and thus cause major damage. Regional tsunamis travel up to 1000 km with a travel time between 1–3 h. Distant tsunamis are ocean-wide and originate from a source more than 1000 km away with a travel time of more than 3 h. Both regional and distant tsunamis can cause widespread destruction, but with the advancement of knowledge in observation, modelling, and geospatial and communication technologies, they can be predicted and advisories can be generated in time.

14.3 Status of Tsunami Warning Systems in the World

There are four regional tsunami warning systems: the Pacific Tsunami Warning Mitigating System (PTWS), Indian Ocean Tsunami Early Warning and Mitigation System (IOTWMS), Tsunami Early Warning and Mitigation System in the North-Eastern Atlantic, Mediterranean and Connected Seas (NEAMTWS), and Tsunami and other Coastal Hazards Warning System for the Caribbean and Adjacent Regions (CARIBE-EWS). These regional tsunami warning systems operate as "system of systems." The Indian Ocean tsunami warning system comprises the national systems of India, Australia, and Indonesia. These three systems are interoperable within themselves as well as with other systems. The tsunami bulletins include Estimated Wave Height(s) (EWH) and Estimated Time(s) of Arrival (ETA) for Coastal Forecast Points (CFPs) and threat status for Coastal Forecast Zones (CFZs) in both text and graphical formats.

14.3.1 *Key Components of the Indian Tsunami Warning System*

The key components of the tsunami warning system are as follows [2, 3]:

(i) Identification of hazard-prone areas.
(ii) Observational networks (seismic and sea level) for detection of earthquakes and evaluation of the potential for tsunami.

(iii) Modelling for generation and propagation of tsunami waves and inundation.
(iv) High-resolution database of bathymetry (the measurement of water depth in oceans or seas) and coastal topography.
(v) Communication facilities for real-time data acquisition and issuance of timely dissemination to stakeholders to facilitate the evacuation of coastal areas.
(vi) Computational infrastructure for real-time analysis and modelling.
(vii) A 24×7 operational facility housing the infrastructure, decision support system (DSS) based on standard operating procedures (SOP), and experienced staff for generation and dissemination of tsunami advisories.

The schematic layout of the Indian Tsunami Early Warning System developed by the Indian National Centre for Ocean Information Services (ESSO-INCOIS) is given in Fig. 14.4.

We have also provided a typical layout of the practices followed by the German–Indonesian Tsunami Early Warning System for the Indian Ocean in Fig. 14.5, which are essentially followed by most member states along the Indian Ocean.

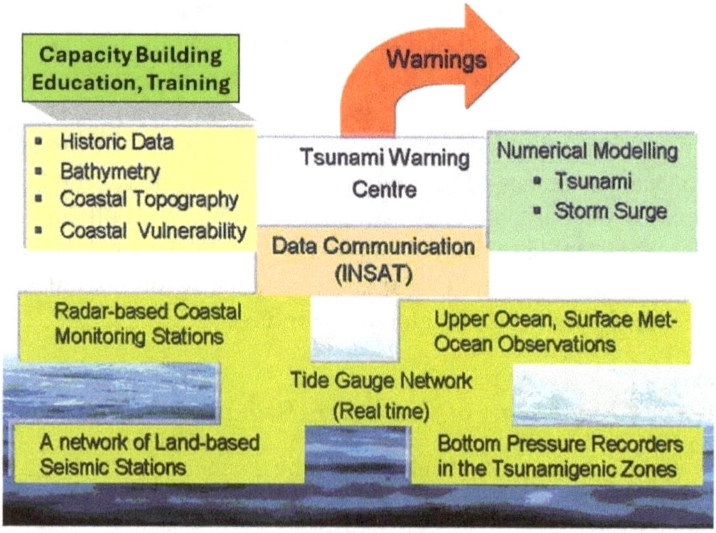

Fig. 14.4. Components of the Indian Tsunami Early Warning System [4].

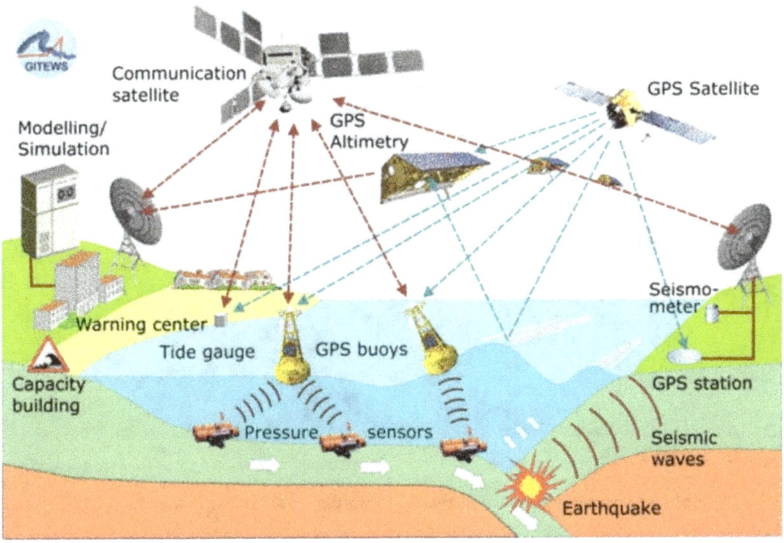

Fig. 14.5. Components of the German–Indonesian Tsunami Early Warning System [5].

14.3.2 *Identification of Hazard-Prone Areas*

A majority of tsunamigenic earthquakes occur at subduction zones along the Ring of Fire in the Pacific Ocean, the Andaman–Sumatra Arc in the Indian Ocean, the Mediterranean Sea, and the Caribbean Sea (see, Fig. 14.6). The Makran collision zone is another tsunamigenic area in the Indian Ocean. The Indian subcontinent can be affected by both these tsunami zones and both these areas are considered hazardous with observational systems planned accordingly.

14.4 Observation Network

14.4.1 *Seismic Network*

To locate and estimate the magnitude of potential tsunamigenic earthquakes, local and global seismographic networks transmitting seismograms in real time to continuously monitor seismicity are used. The timeliness of the data is crucial as the first bulletin is to be issued within 10 min of an earthquake event. The following networks provide seismic data in real time: IRIS Global Seismographic Network (GSN),

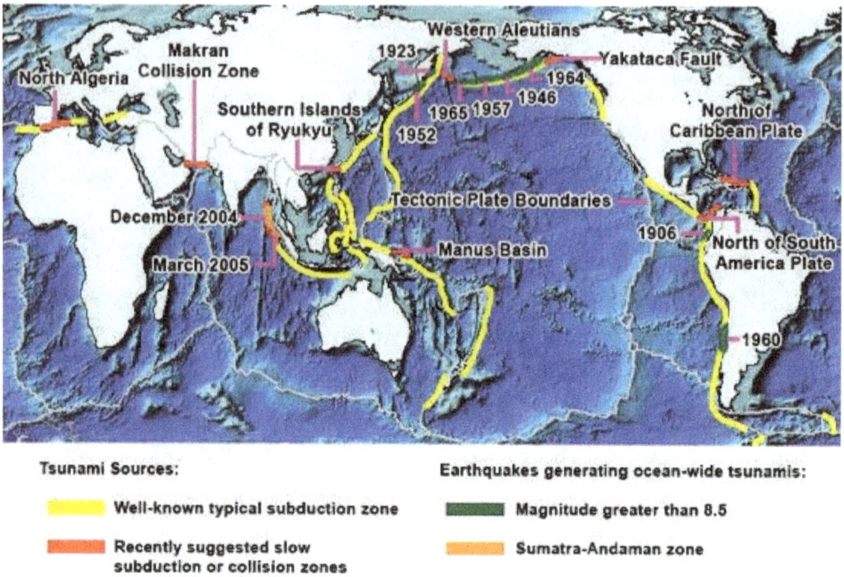

Fig. 14.6. Global Tsunami Source Zones (Credit: LDG-France).

Comprehensive Nuclear Test Ban Treaty Organization (CTBTO), the German Research Centre for Geosciences (GFZ) Geofon Extended Virtual Network, and stations from other national and regional networks (Fig. 14.7).

14.4.2 *Sea Level Network*

Tide gauges, bottom pressure recorders, and coastal ocean dynamics applications radar (CODAR) reporting in real time are used to measure changes in water level near a fault zone to confirm whether an earthquake has caused a tsunami and to gauge its severity. The Global Sea Level Observing System (GLOSS) is a network coordinated by the IOC of the UNESCO and World Meteorological Organization (WMO). The signature of tsunami waves at sea level gauges is generally quite distinctive and thus recognisable (see Fig. 14.8) and can be differentiated from a signal caused by a storm surge (UNESCO-IOC Manual and Guides, 2006).

Bottom pressure recorders or tsunami buoys measure wave heights on open oceans with a high accuracy to within a few millimetres and provide information on the generation of tsunamis. Tide gauge data are used to

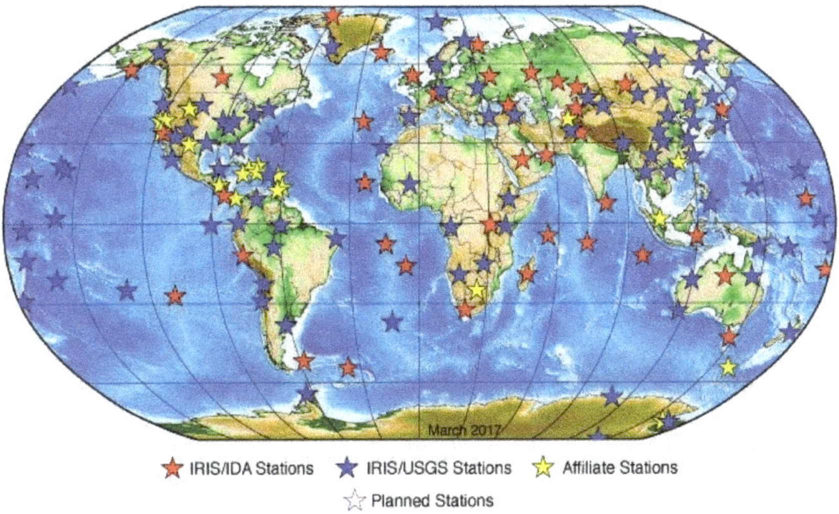

Fig. 14.7. Incorporated Research Institutions for Seismology (IRIS) Global Seismic Network.

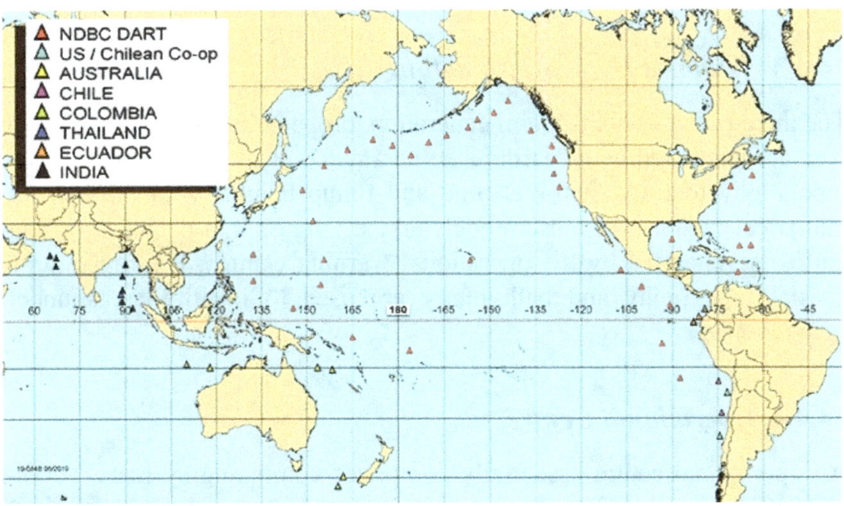

Fig. 14.8. Network of Tsunami Buoys Operated by Several Member States (Credit: National Data Buoy Centre, August 2019).

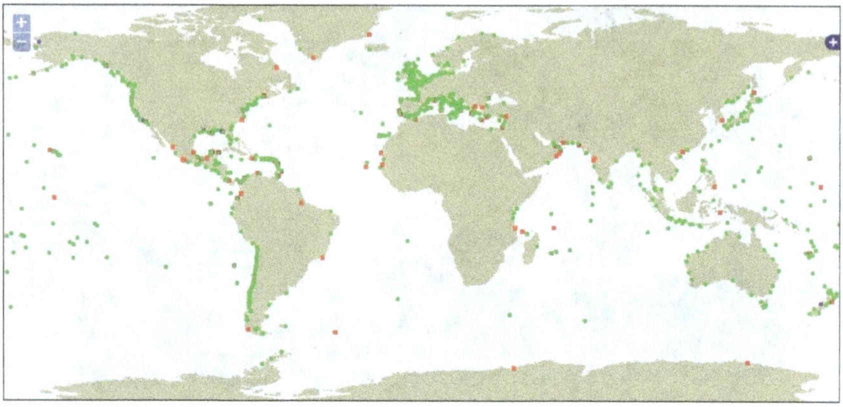

Fig. 14.9. Network of Tide Gauges Operated by Several Member States (Credit: IOC Sea Level Monitoring Facility, August 2019).

verify tsunami generation and threat-pass confirmation. Sea level networks also detect tsunamis that are triggered by volcanic eruptions or underwater landslides. Maps of currently available tsunami buoys and tide gauges are provided in Figs. 14.8 and 14.9, respectively.

14.4.3 *Tsunami Forecast Modelling*

The three processes of tsunami generation, propagation, and inundation of land are simulated using earthquake parameters, bathymetry, and coastal topography to forecast travel time and runup heights with the help of numerical models. As these models take a long time, a database of scenarios is generated by the operational warning centres. High-resolution coastal topography and bathymetry are used to identify areas under inundation.

14.4.4 *Operational Centre*

An operational centre essentially comprises computing systems, necessary models, a decision support system, and communication networks. A robust communication infrastructure is required to receive scientific data and disseminate advisories to all stakeholders. Geostationary communication satellites and internet services form the backbone of such infrastructure.

Advisories are delivered at a certain operational level. Currently, the IOTWS (UNESCO-IOC, 2007) classifies these levels into three groups, Service levels I, II, and III, based on incremental capacity development. At Service Level I, only earthquake information and qualitative tsunami threat evaluation are provided. At Service Level II, estimated wave height(s) (EWH) and estimated time(s) of arrival (ETA) for offshore forecast points, potential threat zones, advanced warning and notification products of threat status and zones in image format are provided. At Service Level III, the advisory products include local risk and hazard assessment using inundation models, which are more specifically directed towards communities at risk.

Tsunami warning systems constantly improve observation networks, analysis methods, computational capabilities, and communications.

14.5 Awareness and Response

The International Tsunami Information Centre (ITIC) was established by the IOC/UNESCO in November 1965 to mitigate the hazards associated with tsunamis in all Pacific Ocean nations and to develop relationships with scientific research institutes, academic organisations, and the general public. For each tsunami, the ITIC assists the World Data Service World Data System for Geophysics in soliciting and collecting all tsunami-related information such as earthquake parameters, tsunami wave arrival times, and tsunami wave heights together with supplementary data including tsunami damage assessments. Similarly, the Indian Ocean Tsunami Information Centre (IOTIC), the North-Eastern Atlantic & Mediterranean Tsunami Information Centre (NEAMTIC), and the Caribbean Tsunami Information Centre (CTIC) were established by the respective regional ICGs to promote tsunami awareness and response activities in the respective basins.

Mock exercises and drills are regularly conducted to improve the degree of preparedness of a community and to respond efficiently and promptly to a tsunami threat event.

"Tsunami Ready" is a voluntary programme of the National Oceanic and Atmospheric Administration (NOAA) in partnership with National Weather Service (NWS) with the aim of improving public safety during tsunami emergencies so that the public knows how to recognise the natural warning signs of a tsunami, such as a long duration earthquake, and understands the importance of moving to a safe location immediately. It also ensures that when a tsunami warning is issued, the public receives the warning, knows what to do about it, and takes appropriate action. The IOC-UNESCO is piloting the implementation of similar community preparedness programmes in the Caribbean and Indian Ocean.

14.6 Types of Tsunami Advisories

Based on earthquake information, tsunami modelling scenarios, and operational capability, different types of advisory products are issued to disaster management officials and the public. Each advisory has a distinct meaning and suggests the action to be taken. An earthquake information bulletin issued solely on earthquake information is a qualitative advisory while all others are quantitative advisories (Fig. 14.5) issued taking into consideration the modelling results and real-time sea-level observations, along with the earthquake information.

1. An earthquake information bulletin is issued to inform the concerned party about the occurrence of an earthquake and to advise on its potential to generate a tsunami based on its magnitude. The bulletin contains information about the earthquake's parameters (origin time, latitude, longitude, focal depth, and magnitude) and the name of the geographical area where it occurred.
2. A warning is the highest level of advisory where immediate action has to be taken. This advisory is issued when a potential tsunami with significant widespread inundation is imminent or expected. Warnings alert the public that widespread, dangerous coastal flooding accompanied by powerful currents is possible and may continue for several hours after the arrival of the initial wave. Warnings also alert emergency management officials to take action for the entire tsunami hazard zone. Warnings may be updated, adjusted geographically, downgraded, or cancelled.
3. A watch is issued to alert emergency management officials and the public of an event that may later impact the watch area. Immediate public evacuation is not required; however, local officials should be prepared for evacuation if it is upgraded to warning status. The advisory may be upgraded to a warning or cancelled based on updated information and analysis. Watches are normally issued based on

seismic information without confirmation that a destructive tsunami is underway.

4. An alert contains information about an earthquake and a qualitative tsunami evaluation message, indicating that a tsunami is expected. As conditions warrant, the advisory will either be continued, upgraded to a watch or warning, or cancelled. An alert implies that the public should avoid beaches since strong currents are expected. Local officials should be prepared for evacuation if it is upgraded to warning status.
5. Cancellation indicates the end of the threat of a tsunami. A cancellation is usually issued after an evaluation of sea-level data confirms that a destructive tsunami will not impact the warned area anymore. A cancellation will also be issued following a destructive tsunami when sea-level readings indicate that the tsunami is below destructive levels and is subsiding in most locations.

14.7 Conclusions

Tsunamis, from prehistoric times, have affected coastal regions with devastating effects, often completely obliterating cities and their populations, and even reshaping coastlines [1]. However, the relentless efforts of scientists from across the world have ensured that the understanding of the science of tsunamis has made rapid progress, especially in the areas of quick detection of earthquakes and forecasting the characteristics of tsunami waves. As all nations along oceans are susceptible, especially those in the so-called vulnerable zones, they must collaborate and cooperate to detect the earthquakes that can cause tsunamis and rapidly disseminate the warnings.

The challenges in forecasting tsunamis caused by submarine landslides, volcanoes, etc., remain. Similarly, capabilities in forecasting near-source tsunamis are yet to be developed. Educating coastal communities on appropriate responses to meet the challenges posed by near-source tsunamis must be made a priority. We have seen that international collaboration among all nations has played a significant role in providing an understanding of the mechanisms of tsunamis, developing a system that can provide sufficient warning to the people of regions likely to be affected, and helping them take necessary precautionary measures to save lives.

References

[1] V. S. Ramamurthy, D. K. Srivastava, and S. Nayak, "*Art and Science of Managing Public Risks*" (Singapore: World Scientific Publishing Company, 2022).

[2] S. Nayak and T. S. Kumar, "The First Tsunami Early Warning Centre in the Indian Ocean", *Tudor Rose: Risk Wise* (2008) 175–177.

[3] S. R. Nayak and S. T. Kumar, "Tsunami Watch and Warning Centres". In: H. K. Gupta (eds.) *Encyclopaedia of Solid Earth Geophysics*, Encyclopaedia of Earth Sciences Series (Dordrecht: Springer, 2011). https://doi.org/10.1007/978-90-481-8702-7_230.

[4] https://tsunami.incois.gov.in/TEWS/searlywarnings.jsp.

[5] "A New Approach in Tsunami-Early Warning", *Phys.Org* (10 November 2008), https://phys.org/news/2008-11-approach-tsunami-early.html; https://www.gitews.de/en/concept/; https://schmidtocean.org/cruise-log-post/tsunami-warning-systems/.

Chapter 15

Weather and Climate: Meteorology

> "I remain confident that despite many differences, the world can unite and rise to the challenge of the climate crisis. Multilateralism remains humanity's best hope. It is essential to come together around real, practical, and meaningful climate solutions that match the scale of the climate crisis."
>
> — António Guterres, *Secretary-General, United Nations, COP28, December 2023*

15.1 Introduction

For millennia, farmers and rulers across the world have depended upon astrology and the movement of planets through the zodiac signs; the behaviour of birds, insects, and animals; halos around the sun and the moon; the colour and shape of clouds, thunder, and lightning; and even the foliage, flowering, and fruiting of some trees to arrive at a guess about the weather [1]. Every region of the world also has its share of proverbs and ditties predicting the weather, handed down through generations. It is not uncommon, even now, to come across practices of prayers, quaint superstitious rituals, soothsayers, and fire/animal sacrifices to propitiate some gods or spirits to get relief from the vagaries of weather in some societies.

After all, the climate has shaped the people and the civilisations of the different regions of the world, and the regular seasons have triggered festivals to celebrate harvests. The day-to-day variations of weather are important for sailors, pilots, drivers, farmers, and firefighters, the launch of space missions, planning military missions and trade, and warning and preparing people for potentially dangerous or difficult weather conditions.

We add that while weather refers to short-term changes in the atmosphere, climate refers to atmospheric changes over longer periods, usually taken as 30 years or more. Climate is generally classified as tropical, arid, temperate, continental, or polar, which is further subdivided to emphasise finer details. On the other hand, weather is essentially local and described as sunny, partly cloudy, cloudy, rainy, hot/cold, dry/humid, having a clear or a cloudy night, or bringing thunder, storms, hail stones, blizzards, cloud bursts, cyclones, snow, etc. Thus, both are characterised by precipitation, speed and direction of the wind, humidity, and temperature, with the difference that climate is an average over a long period.

Climatic conditions exhibit significant spatial variation across the Earth's surface, which is also affected by the local terrain — which exerts a profound influence on the distribution, abundance, and composition of flora and fauna. Thus, for example, Antarctica exemplifies a polar climate characterised by temperatures going down to −45 degrees centigrade or lower, fierce winds, and extreme aridity. These harsh conditions have led to the evolution of highly adapted organisms, e.g., Emperor Penguins, which are specifically equipped to endure such an environment.

At the other end of the spectrum, the Amazon Rainforest offers a typical tropical climate. Here, consistently warm temperatures, high humidity, and abundant precipitation create a remarkably stable environment. This environmental stability has facilitated the development of exceptional species diversity, with a significant proportion of these being restricted geographically to this region.

Consider the Sahara Desert — covering over 9 million square kilometres, a brutally hot and dry place, with average temperatures exceeding 30 degrees centigrade throughout the year and remaining more than 40 degrees centigrade for months, and even reaching 58 degrees centigrade. Rainfall is extremely scarce. Plant life has adapted to the aridity, with drought-resistant shrubs, grasses, and succulents dotting the landscape. Animal life is sparser but surprisingly diverse. Its inhabitants include insects, reptiles like sand vipers and sand lizards, and mammals like fennec foxes, addax antelopes, and dromedary camels.

India's unique weather and climate are fashioned by the lofty Himalayas, whose high altitude, length, and direction allow it to intercept the summer monsoons from the surrounding Arabian Sea and the Bay of Bengal. The Himalayas, sort of, forces the monsoon winds to release their moisture content within the Indian subcontinent as rain or snow. Moreover, it stops the icy winds from Siberia from entering India.

Writers, poets, artists, and musicians have created works of exquisite beauty and lyrical charm describing the seasons, paintings changing landscapes, and creating fascinating and melodious ragas (recall the famous Ragamala paintings from India) and symphonies for various moods and times of different seasons. The climate and the seasons have also decided the outcome of wars, as indicated earlier.

Even the constantly wandering monks of India were expected to settle down at some place during the rains ("Chaaturmaas") and deliver sermons. The major discourses of Lord Buddha were given during such stays. The festival of Bali Yatra in Cuttack, Odisha, still celebrates the day when mariners would set sail for distant lands such as Bali, Java, Sumatra, Borneo in Indonesia, and Sri Lanka, to take advantage of relatively calm seas and favourable winds. It was a chance encounter with an Arab sailor, who told Vasco De Gama about the trade winds, that opened the sea route to India.

Meteorology has by now put the vastly important and interdisciplinary science of weather forecasting on a firm and quite robust scientific basis using observations of temperatures, humidity, wind speeds, pressure, clouds, atmospheric temperatures and currents, trade winds, ocean currents, ocean temperatures, polar ice extent and polar vortexes, oceanic currents, and convection (the process of warmer air rising and colder, denser air sinking and leading to a transfer of energy, heat, and moisture). There is mounting evidence that reduced Arctic Sea ice leads to more frequent extreme monsoon events in India, causing floods and droughts. The study of the corresponding effects of the extent of Antarctic Sea ice on monsoons in India is still in its infancy [3].

These and other extensive observations of the entire planet are now supplemented with measurements and observations from weather balloons, sounding rockets, weather radars, weather satellites, with modelling and computations using supercomputers offering real-time predictions [2]. The observations and studies to be made for these have been summarised in Table 15.1, taken from Elizabeth *et al.* [4].

Table 15.1. Challenges for Future Observations and International Collaborations [4].

Challenges in World Climate Projection	Monitoring Earth	Understanding Processes	Improving Projections
Clouds, Circulation, and Climate Sensitivity	Changes in Jet Stream, Expansion of Tropics, Aerosols	Impact of Stratus Clouds, Water Vapour in the Atmosphere	Survey of Cloud properties
Melting Ice and Global Consequences	Ice Volume, Ocean Salinity and Temperature	Sub-ice Ocean Temperatures, Ocean Salinity	Polynya Life Cycle
Understanding and Predicting Weather Extremes	Droughts, Floods, Cyclone Rapid Intensification	Rapid River Runoff, Ocean's Role in Droughts	Convective Initiation, Sub-ice Ocean Temperatures
Regional Sea Level Change and Coastal Impacts	Coastal Monitoring, Winds, Local Currents	Tidal Impacts, Erosion Protection Efforts	Smaller Glacier Impacts
Water for the Food Baskets of the World	Soil Moisture, Inland El Niño Impacts	El Niño Impacts to Heavy Precipitations, Ocean's Role in Droughts	Season Drought Forecasts, Storm Forecasts
Carbon Feedback in the Climate System	Global Methane Monitoring, Ocean Uptake	Wetland Emissions of Carbon	Identification of Vulnerable Coasts, Deep Ocean Convection and Currents
Near-Term Climate Prediction	Sub-surface Ocean Temperature, Upper Air Winds	Ocean Impacts on Seasonal Precipitation, Ocean Heat Content Anomalies	Seasonal Precipitation Forecasts, River Monitoring

A depiction of the international efforts towards enhanced weather observation systems in the 1970s [2] is shown in Fig. 15.1. Machine learning and artificial intelligence are being increasingly used for these studies now.

Various groups have developed detailed models to predict the day-to-day, place-specific, long-term, and large-area weather and climate, and it

Fig. 15.1. An Artist's Conception of the Enhanced Observing Systems for 1978/1979 Global Weather Experiment [2].

is always prudent to compare predictions from as many models as possible to assess the evolving weather and climate (as different models may use different simplifying assumptions, numerical techniques, or inputs). Of course, these weather predictions are getting more challenging with the rapidly progressing global warming, and the associated climate change, and as the frequency and ferocity of extreme events increase. For example, it has already been noted that global warming has intensified heavy rainfall events, such that even though the overall number of rainy days is low, the amount of rainfall on those days has increased significantly. This happens as a warmer atmosphere can hold more moisture and sustain strong, deep clouds for a long duration, leading to "rivers in the atmosphere," which leads to a sudden downpour or a heavy snowfall.

Over centuries, meteorology has emerged as a scientific field that requires and has prospered by global cooperation. The atmosphere's vastness and complexity and the universality of the fundamental laws of physics make weather and climate phenomena solvable only by international collaboration.

No single country or institution can observe, understand, predict, and protect against these issues in isolation. An excellent history of the development of international collaboration is given in Ref. [5]. As noted by the author, the foundations for this natural cooperation and collaboration started with the establishment of the non-governmental International Meteorological Organization (IMO) in 1873, which in 1950 transformed into the intergovernmental World Meteorological Organization (WMO). Meteorologists worked across the ideological divides of the Cold War to build the World Weather Watch and implement the Global Atmospheric Research Program (GARP) which transformed the 1960s and 1970s into the real golden age of international cooperation in meteorology. The benefits of these cooperations flowed immediately to almost every country and virtually every sector of society, which has continued to grow steadily.

The advantages and the necessity of international collaboration in meteorology were already realised by 1723, when the Royal Society took the initiative to assemble annual records from a network of volunteer observers across Europe, Asia, and North America. These studies got a further boost when Alexander von Humboldt introduced the concept of mapping climate variables. With the invention of Morse Code Telegraphy in 1843, it became possible to collect, plot, and analyse meteorological observations in real time [5]. Starting from these humble beginnings, international collaboration in meteorology has gone from strength to strength, and today it is possible to predict even such rapidly evolving phenomena as cyclones with remarkable accuracy [6].

In the following section, we discuss one of the most widely studied phenomena of monsoon in India to illustrate how international collaboration makes it possible to have an accurate prediction for this important phenomenon. Monsoon in India affects the livelihood of a large part of the world population and is a truly complex phenomenon with several competing forces at work. Next, we summarise the efforts of the Intergovernmental Panel on Climate Change to provide a solid scientific foundation for the mitigation of the potentially devastating effects of global warming.

15.2 Indian Monsoon

For millennia, the vast plains of India have experienced blazing hot sun, with temperatures ranging from 32 to 40 degrees centigrade across most of the country during summer (Fig. 15.2). The southern regions experience peak heat in March and April, while the north gets the worst of the heat in May.

This is followed by the rainy season (June to September, Fig. 15.3), bringing heavy rainfall and replenishing water supplies so important for agriculture and survival. The southwest monsoon winds (see later) bring moisture-laden clouds, causing widespread rain. The dry landscape transforms to lush green as plants revive with the rain and rivers swell, with some even flooding large areas.

The post-monsoon season or the fall (October to November) generally has mild temperatures and very mild rains, and the lush greenery from the rainy season gradually gives way to drier landscapes as temperatures rise before the winter (December to February) sets in. During this period, the northern plains experience the most significant drop in temperature,

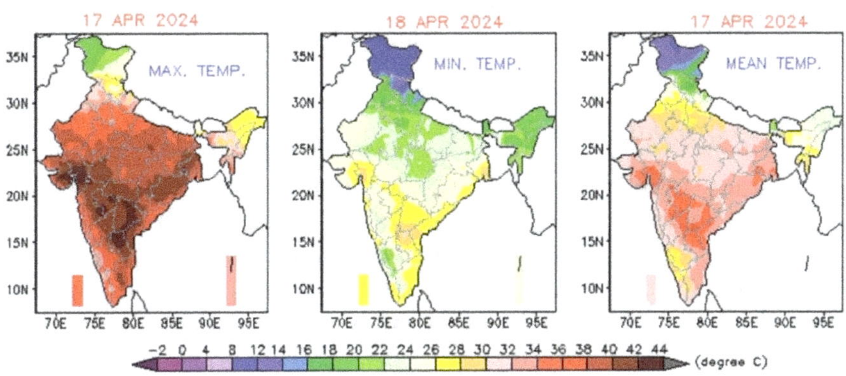

Fig. 15.2. Daily Temperature Map of India in April 2024 (Credit: Indian Meteorological Department [7]).

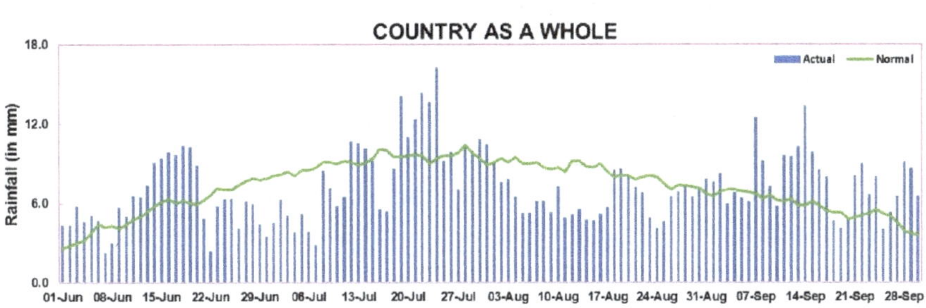

Fig. 15.3. Rainfall Distribution across India — Southwest Monsoon — 2021 [8].

Fig. 15.4. Monsoon Areas [9].

during nights, with minimal rainfall. This pattern with some shifting of dates and longer or shorter spells of the dry season is repeated across the plains. And it repeats year after year, on average, providing enough time to sow suitable crops, await their maturity, and harvest them.

15.2.1 *Monsoons of India*

Such clearly bounded dry and wet months are a definite hallmark of Indian and Southeast Asian monsoons. Monsoon derives from the Arabic word "Mausam" (season) and stands for seasonal winds. A large part of Africa and Southeast Asia (Fig. 15.4) experiences this definitive seasonal shift in wind from Africa to Southeast Asia, an area called the monsoonal region [9]. (See Ref. [10] for a more detailed discussion, including the effect of atmospheric currents in the upper atmosphere.)

Relentless heating by the sun, during the summer, heats both the sea and the land. The land's temperature rises more than that of the sea, and thus the air over it becomes lighter and moves up. The cooler and moisture-laden wind from the sea moves to the land, giving rise to gigantic sea-to-land breezes (Fig. 15.5). These are the monsoons.

These moisture-laden winds have been arriving with the regularity of clockwork for millennia and have shaped the agriculture, life, and civilisation of India. The progress of the monsoon and its abundance are eagerly awaited. The Indian Meteorological Department — which was established in 1875 following a disastrous tropical cyclone that struck Calcutta (now Kolkata) in 1864 and failures of the monsoon rains in 1866 and 1871, which led to severe famines — has by now developed very

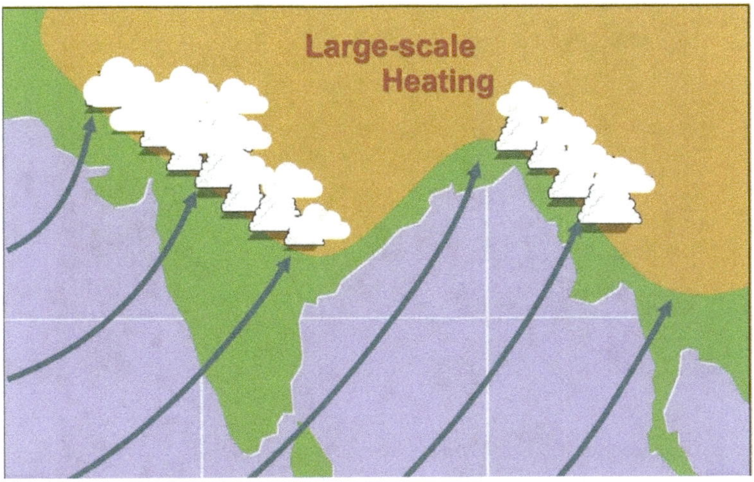

Fig. 15.5. Gigantic Sea-to-Land Breeze [9].

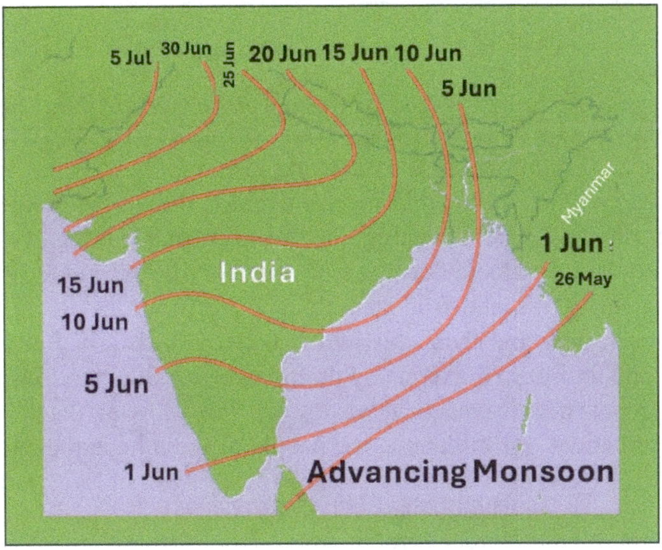

Fig. 15.6. Progress of Monsoon Across India [9].

commendable capability (and credibility) for these predictions, benefitting from extensive international collaborations and observational networks. Figure 15.6 shows the progress of the advancing monsoon across the Indian plane in an average year.

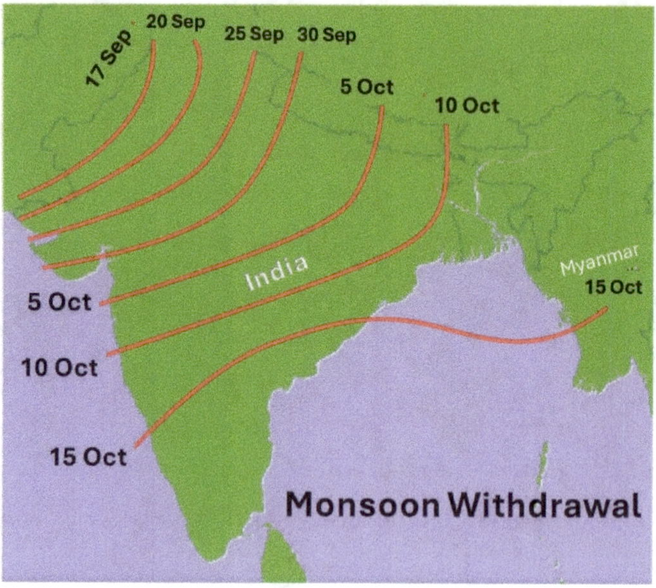

Fig. 15.7. The Withdrawing Monsoon [9].

As the sun gets milder, these moisture-laden winds retreat (see Fig. 15.7), as if returning after exhausting their supply of water, to return the following year. This regularity has decided the crops; their sowing, maturing, ripening, and harvesting; the flowering and fruiting of trees and vegetables; the arrival and departure of migratory birds; and the breeding of birds, insects, fish, and wild animals.

However, now and then, this idyllic pattern is disturbed at irregular intervals and to differing extents by droughts, excessive rains, unseasonal rains, cyclones, heat waves, cold waves, etc. Not knowing the reasons for these disturbances, for millennia, the people of India have prayed:

"काले वर्षतु पर्जन्यः। पृथिवी सस्यशालिनी।
देशो ऽयं क्षोभरहितः। ब्राह्मणास्सन्तु निर्भयाः।।"

"May the clouds provide rains on time, may the Earth be full of grains (give plenty of grains), may the country be free of calamities, may the intellectuals be without fear."

The relentless efforts over decades by thousands of meteorologists worldwide established that the weather of the entire planet is interconnected, and the heating/cooling of the oceans, the ocean–air interactions, and even volcanic eruptions in any part of the world affect the weather across the planet. Two phenomena in this respect are El Niño and La Niña Southern Oscillation, which strongly affect the rains in Africa, India, Indonesia, Australia, and southern United States. Similarly, the Indian Ocean Dipole also strongly affects the rains in several regions. These occur at irregular intervals and can occur together, aggravating or easing the distress.

15.2.2 *El Niño and La Niña*

The El Niño–Southern Oscillation (ENSO) cycle is a well-defined climate phenomenon that affects global weather patterns. It encompasses three distinct phases: El Niño (the warm phase), La Niña (the cool phase), and ENSO (neutral or normal conditions). Each phase exerts a distinct influence on trade winds, ocean temperatures, and ultimately rainfall patterns across the globe. During El Niño, chances for drought increase across India, Indonesia, Australia, and a large part of the Amazon, while southern USA tends to see more precipitation. During La Niña, the pattern is effectively reversed, with wetter conditions for India, Indonesia, Australia, and parts of the Amazon, with dry conditions in the southern tier of the USA in September. As is well known, the name El Niño (the Child Christ) was given to this weather pattern by the fishermen of Peru and neighbouring countries more than 300 years ago, as it peaked around Christmas. La Niña stands for "the girl child."

During *normal conditions*, trade winds function as steady conveyors, blowing consistently from west to east across the Pacific Ocean. These persistent winds act to push warm surface water westward, culminating in a warm water pool near Indonesia and Australia (Fig. 15.8). As a result, the eastern Pacific experiences cooler temperatures due to upwelling, a process where nutrient-rich cold water rises from the depths of the oceans.

Rainfall patterns tend to be relatively stable under normal conditions. The western Pacific, with its warm waters, experiences regular precipitation. The eastern Pacific, cooler and drier, sees less rainfall.

Upwelling in the eastern Pacific brings a wealth of nutrients to the surface, fostering a thriving ecosystem for fish populations. This translates to robust fisheries, particularly for species like anchovies that flourish on the abundance of plankton.

El Niño is triggered by a combination of ocean–atmosphere interactions, including weakening trade winds, changes in upwelling, alterations

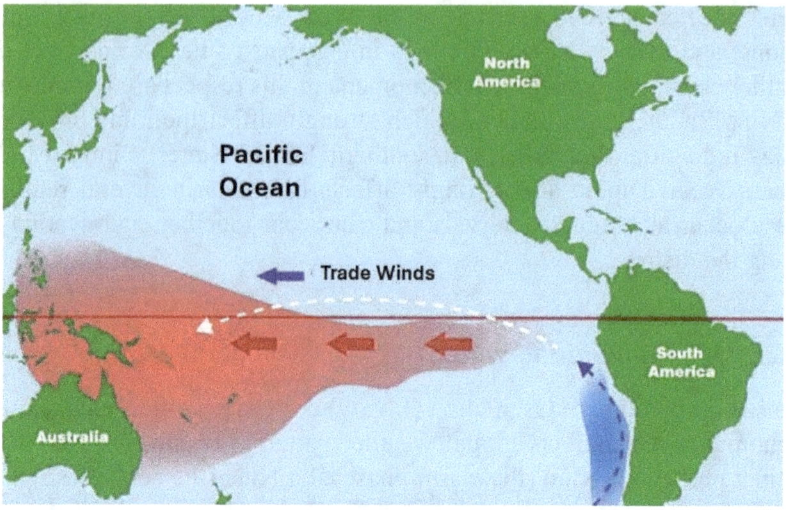

Fig. 15.8. Normal Conditions (Credit: NOAA).

in atmospheric circulation patterns, and feedback loops between the ocean and the atmosphere. Moreover, during an El Niño event, the surface waters in the central and eastern Pacific Ocean become significantly warmer than usual. These factors collectively contribute to the onset and development of El Niño events in the equatorial Pacific region (Fig. 15.9).

El Niño disrupts the established trade wind pattern. These winds weaken or even reverse direction, causing a dramatic shift in ocean currents. Warm surface water, previously confined to the west, sloshes eastward across the Pacific Ocean during this period, typically from December to April.

El Niño brings a dramatic shift in rainfall patterns. The western Pacific, generally wet, experiences drier conditions as the warm water pool diminishes.

In contrast, the eastern Pacific, usually dry, experiences increased rainfall and potential flooding from December to February. Additionally, El Niño can disrupt monsoonal patterns in India, particularly during the June-to-September window, leading to droughts.

The disruption caused by El Niño poses a significant challenge for fisheries. Upwelling in the eastern Pacific ceases, leading to a decline in nutrient availability and a subsequent decline in fish populations. This can have severe economic consequences for fishing communities. For example, in April 2016, nearly 8,000 tonnes of sardines died and washed up

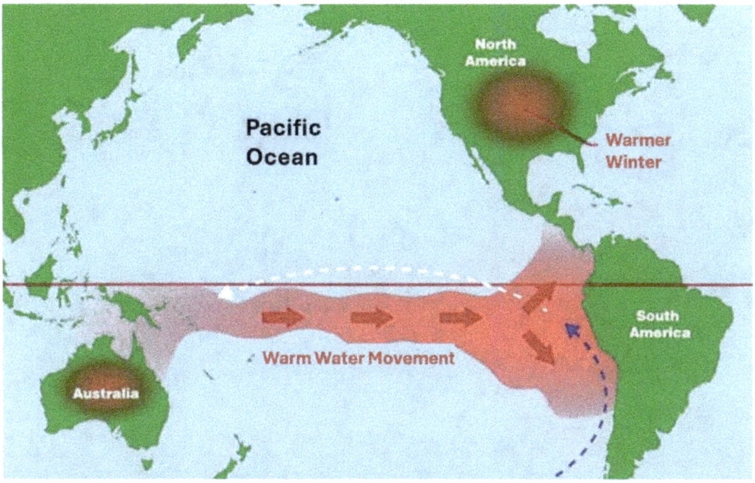

Fig. 15.9. El Niño (Credit: NOAA).

along the coast of Chile, believed to be a result of El Niño-related changes in the ocean. With the decrease in fish populations, sea birds can also die in large numbers due to a lack of food.

La Niña acts as the counterpoint to El Niño and often follows it. During a La Niña event, trade winds strengthen significantly, pushing even warmer surface water westward towards Asia, typically from June to August. The eastern Pacific experiences colder than average sea surface temperatures due to enhanced upwelling (Fig. 15.10).

La Niña flips the story of rainfall patterns. The western Pacific, including Australia, experiences increased rainfall and potential flooding from December to February.

Conversely, the eastern Pacific, particularly California and Peru, enjoys drier conditions. La Niña typically strengthens Indian monsoons, bringing more rain to the region from June to September.

La Niña brings a period of prosperity to fisheries in the eastern Pacific. The intensified upwelling creates a nutrient surge, leading to a population explosion for fish, like anchovies. This period can be a time of abundance for fishing communities.

International Cooperation in Understanding El Niño

The impacts of El Niño and La Niña are complex and can vary depending on the strength and duration of the event. However, understanding the

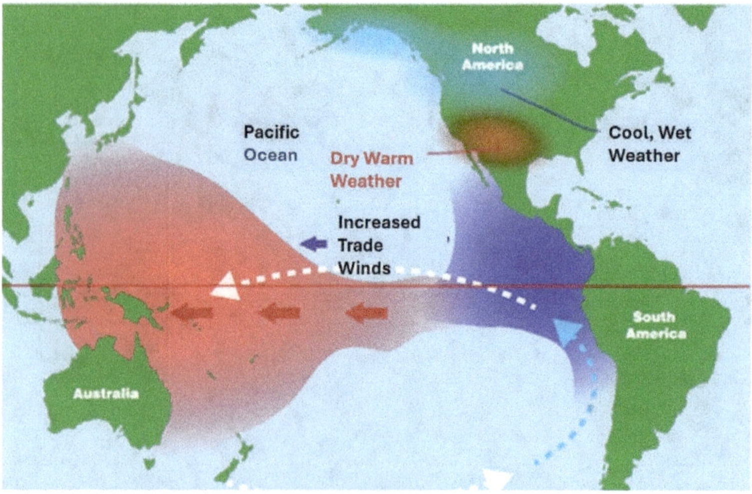

Fig. 15.10. La Niña (Credit: NOAA).

fundamental dynamics of ENSO allows us to better predict and prepare for potential disruptions in weather patterns, rainfall, possible crop failure, and even fisheries across the globe.

Since these events occur at irregular intervals of two-to-seven years, and their effect is not uniform, we must keep track of the sea surface temperature, particularly across the Pacific Ocean. This is done by using buoys across the Pacific Ocean and satellites. Figure 15.11 gives the satellite sea surface temperature departure in the Pacific Ocean for October 2015. The darker orange-red colours are above normal temperatures and are indicative of El Niño. These can be additionally supplemented with cloud cover to get valuable insight. We have seen how informative the pictures taken by the International Space Station have been.

15.2.3 *Indian Ocean Dipole*

The Indian Ocean experiences temperature swings, known as the Indian Niño or Indian Ocean Dipole (IOD). These swings cause the western Indian Ocean to get alternately warmer (positive phase) and cooler (negative phase) compared to the eastern side. Like El Niño and La Niña, these also occur at irregular intervals.

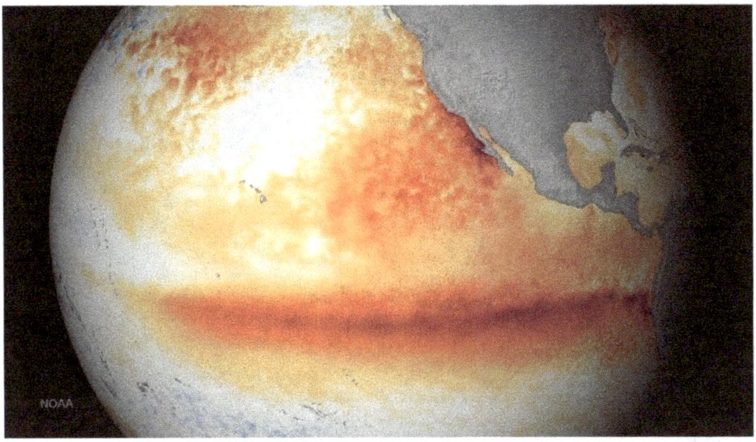

Fig. 15.11. Satellite Sea Surface Temperature departure in the Pacific Ocean for October 2015, Which Led to El Niño (Credit: NOAA).

During a positive IOD event, the waters near Indonesia become cooler than usual, while the western Indian Ocean gets much warmer. Like a seesaw, this temperature difference disrupts wind, temperature, and rainfall patterns across the region.

Typically, a positive IOD (Fig. 15.12) brings floods to eastern Africa and drier conditions to eastern Asia and Australia; the impact includes floods in the east African countries, stronger-than-normal Indian summer monsoon rainfall, and droughts in Indonesia and southeastern Australia.

On the other hand, during a negative phase of the Indian Ocean Dipole, the water near Indonesia gets unusually warm, while the western Indian Ocean cools down. This disrupts the usual air circulation (Fig. 15.13), and westerly winds, which normally carry moisture towards India, become weaker, leading to less rainfall over Africa and India and heavier rainfall over Indonesia and Australia.

Finally, in its neutral phase, the Indian Ocean Dipole leads to neither a positive nor a negative effect (Fig. 15.14).

We would like to re-emphasise that the development of the Indian Ocean Dipole requires the atmosphere and ocean to interact, as in the case of El Niño Southern Oscillation. The complexity arises from further interaction of the El Niño Southern Oscillation and the Indian Ocean Dipole,

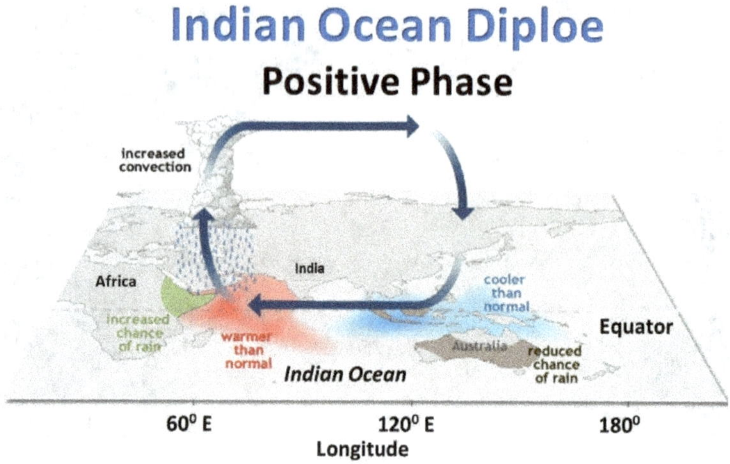

Fig. 15.12. Positive Indian Ocean Dipole (Credit: NOAA).

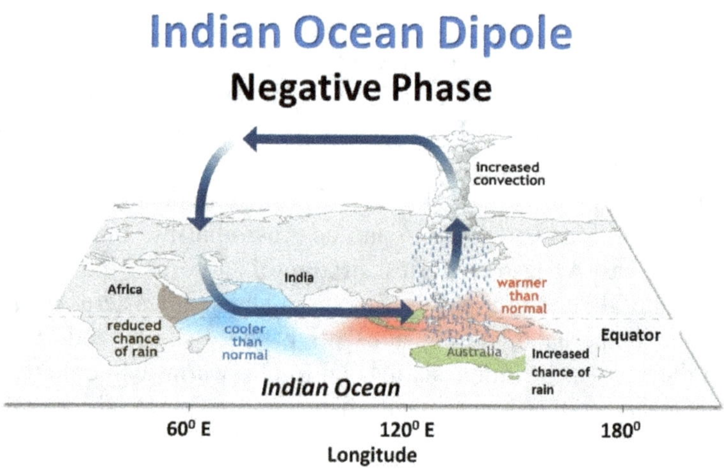

Fig. 15.13. Negative Phase of Indian Ocean Dipole (Credit: NOAA).

which can reduce or aggravate the outcome of El Niño or La Niña, which is only now being understood in detail.

15.2.4 *Cyclones*

Cyclones are swirling storm systems with low pressure at their centre. They can be enormous and bring strong winds, torrential rains, and storm

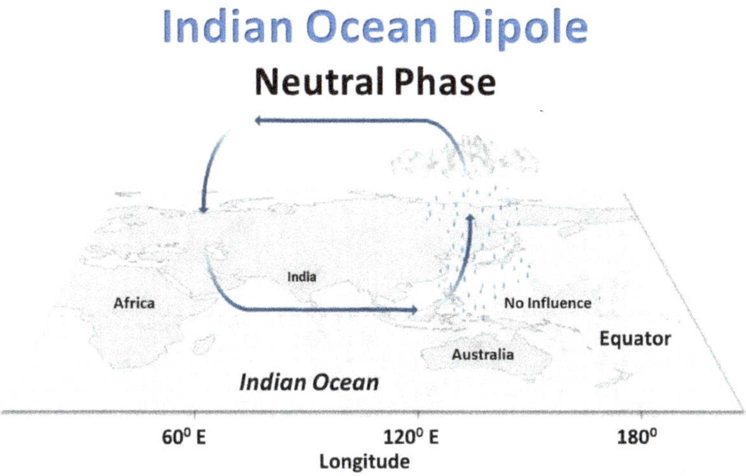

Fig. 15.14. Neutral Phase of Indian Ocean Dipole (Credit: NOAA).

surges (Fig. 15.15). The accuracy of the prediction of their power, movement over the sea and land, landfall, quantum and duration of rainfall, and their progress is one of the most notable achievements of modern meteorology. Cyclones are classified according to wind speed. Every year, many lives are saved by issuing timely warnings in cyclone-prone areas [6].

The energy for the formation of cyclones is mostly provided by the warm, moist air over oceans. As the warm air rises and cools, it condenses, releasing energy that fuels the storm's growth. The Coriolis effect, caused by the Earth's rotation, influences the wind direction to create the cyclone's characteristic spin: counterclockwise in the Northern Hemisphere and clockwise in the Southern Hemisphere.

Tropical cyclones are formed over warm tropical waters. They are further subdivided depending on wind speed [6]. Cyclones are called hurricanes in the Atlantic and eastern Pacific and typhoons in the western Pacific. As indicated earlier, cyclones raise powerful winds that can cause loss of life and massive damage to crops, infrastructure, and trees. They cause heavy rainfall, lead to flooding, and generate storm surges (the abnormal rise in sea level caused by the low pressure and strong winds of the cyclone which can inundate coastal areas).

An efficient early warning system providing accurate and timely forecasts is very critical to ensure relief, rescue, and shelter to communities. The improvement in the prediction of tracking, landfall, and intensity has been possible due to a better understanding of the physics through

Fig. 15.15. Severe Cyclonic Storm Aila, May 2009, North Indian Ocean (Credit: NASA).

enhanced global ocean and atmosphere observations, including satellite observations and their easy availability and accessibility [6]. The assimilation of data from satellite observations, Doppler weather radars, moored and drifting buoys, Argo floats (Fig. 15.16) [11], and automatic weather stations into 12-km-resolution global numerical models and mesoscale (3-km) models has led to a major improvement in the prediction of cyclones and helped save thousands of lives. This is a remarkable achievement of global collaboration. However, there are many countries which still need an effective response mechanism to respond to cyclone advisories.

The World Meteorological Organization (WMO) has set up Regional Specialized Meteorological Centres (RSMC) for tracking tropical cyclones and providing advisories. In India, the Earth System Science Organization (ESSO)–India Meteorological Department (IMD), has been identified as an RSMC by the WMO and provides cyclone information to all countries in the north Indian Ocean. The WMO has a Tropical Cyclone Programme

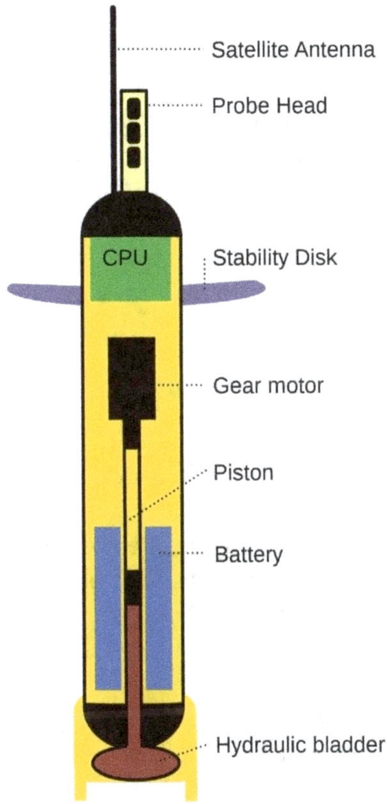

Fig. 15.16. Argo Floats Record the Vertical Profile of the Ocean by Taking Measurements of Temperature and Salinity at Different Depths as They Ascend Vertically, from a Depth of 2000 m [11] to the Surface.

(TCP) which coordinates national and regional systems and ensures that the loss of life and damage caused by cyclones are minimal.

There is a need to build a resilient response mechanism by building effective collaboration between countries to establish observational systems and developing standards for data exchange and networking of services so that lives can be saved [6].

Extratropical cyclones or mid-latitude cyclones are another category of cyclones which are low-pressure weather systems. These develop in cooler regions, typically between 30°- and 60° latitude due to collisions of warm and cold air masses. As before, the warm air rises, cools, condenses, and releases energy, which fuels the storm's growth. Again, the Coriolis effect causes the air to spin counterclockwise in the Northern Hemisphere and clockwise in the Southern Hemisphere. These can produce mild breezes to powerful gales, produce rain and snow, and lead to rain, sleet, or blizzards, depending on the temperature. These can also trigger thunderstorms, tornadoes, and coastal flooding due to high waves and storm surges.

We have already seen that the frequency and ferocity of cyclones are increasing due to global warming and climate change [6].

15.2.5 *Gulf Stream and Jet Stream*

We shall discuss the oceanic current Gulf Stream and the atmospheric current Jet Streams (only briefly) to illustrate how the weather and climate are affected by these currents.

The Gulf Stream is a powerful ocean current that carries warm water from the Caribbean up the eastern coast of North America and across the Atlantic Ocean to Europe. It brings warm, tropical waters to cooler areas in the north Atlantic, making places like Western Europe much milder than they would otherwise be.

However, climate change and the accompanying global warming are leading to the Gulf Stream warming faster than the rest of the ocean. Moreover, melting glaciers are adding fresh water to the North Atlantic, and it is feared that these could disrupt the circulation that drives the Gulf Stream.

A weakened Gulf Stream could lead to significantly more volatile swings in temperature, with both colder winters and hotter summers in Western Europe that could seriously impact agriculture and ecosystems. There are also apprehensions that changes in the Gulf Stream could lead to more extreme weather events like storms and floods.

Jet Streams are fast-moving (160 km/h), narrow bands of currents high up in Earth's atmosphere. They typically flow from west to east and are most concentrated around 11–19 km above the surface, though they meander around quite a bit.

These strong winds act like atmospheric highways, directing storm systems and weather fronts along their path, whose strength and locations are decided by the difference in the temperatures of their north and south ends.

Because of these issues, it is important to understand Jet Streams to forecast weather patterns and predict storm tracks. However, climate change and global warming are expected to impact Jet Streams in complex ways by reducing the temperature difference between the north and the south and weakening them, so that they start meandering much more viciously.

15.3 Monsoon Mission

The Indian Summer Monsoon Rainfall (ISMR) accounts for about 80% of the total rainfall in India (Fig. 15.3). It is critical for agriculture, drinking water, and energy production and affects socio-economic conditions significantly. Thus, the rainfall, the food production, and the Gross Domestic Product of India are strongly correlated.

Due to this, accurate predictions are needed at different temporal scales: short (3 days) and medium (3–10 days) terms, extended range (up to 3 weeks), and seasonal. In 2012, the Earth System Science Organization (ESSO), Ministry of Earth Sciences, Government of India, launched the Monsoon Mission to develop and implement a numerical prediction system using coupled general circulation models (GCMs) to improve ISMR predictions on different temporal scales [12]. It was a unique concept to conduct research, demonstrate improvement in the forecast skill, and integrate these into the operational modelling framework. High-performance computing facilities and a training school were set up for capacity building in modelling weather and climate.

This gigantic task was achieved by collaborating with national and international research and academic institutions. The ocean–atmosphere coupled model Climate Forecast System version 2 (CFSv2) was adopted from the National Centers for Environmental Prediction (NCEP), USA, for short-range, extended-range, and seasonal predictions. The Unified Model (UM) of the United Kingdom Meteorological Office (UKMO) was adopted to improve short- to medium-range forecasts. Indian researchers worked closely with partners from the UKMO; Korea Meteorological Administration; the Bureau of Meteorology/Commonwealth Scientific and Industrial Research Organisation, Australia; and the National Institute of Water and Atmospheric Research, New Zealand. India funded 40 projects to improve the prediction and predictability of these two chosen models and build capacity in the country. Out of these, funds for 20 projects were provided to institutions in the USA (12), United Kingdom

(3), Australia (1), Canada (1), France (1), Japan (1), and the United Arab Emirates (1). This was probably the first time that India supported research in weather and climate predictions in the developed world. At the same time, the development of the Earth System Model (ESM) was also initiated for climate predictions.

The Monsoon Mission also undertook an understanding of the physical processes through observational programmes of both the ocean and atmosphere. The "Ocean Mixing and Monsoon" project focused on fine-scale observations in the near surface in the Bay of Bengal and was implemented in partnership with Indian and US institutions. Other observational projects were supported as a part of the Indo-UK Collaboration between the MoES and the Natural Environmental Research Council (NERC) to understand factors influencing variability in the Indian monsoon.

The focused and concerted efforts of the Indian scientists led to the development of a high-resolution model for seasonal predictions with high skill, a coupled model for extended range, and a very high-resolution short-range prediction system. The skill of these forecasts is comparable to climate research centres such as the NCEP, the European Centre for Medium-Range Weather Forecasting, the Japan Meteorological Agency, and the UKMO. The Monsoon Mission improved its skill (defined as the correlation between observation and predictions) to 0.71, much higher than the earlier estimated potential skill of 0.65. The statical models that were used earlier for seasonal prediction had the skill of only 0.36. The ESSO–IMD uses these models for operational forecasts.

Considering the efforts made by India in monsoon research, both nationally and internationally, the International Monsoon Project Office (IMPO) was established at the ESSO–Indian Institute of Tropical Meteorology (IITM) under an agreement between the ESSO/MoES and the World Meteorological Organization (WMO). The IMPO provides support to global monsoon research from weather to climate under the WMO's World Weather Research Programme. The ESSO–IITM collaborates with many countries, including the USA, Canada, the UK, France, Germany, Italy, Switzerland, Norway, Russia, the UAE, Japan, Australia, and New Zealand to advance monsoon research.

15.4 Intergovernmental Panel on Climate Change

The Seventy-Fifth General Assembly of the United Nations held in September 2020, let out a cry (or rather a scream) of anguish and pain and identified climate change as the defining crisis of our time. It also noted that it is happening even more quickly than we feared, but also expressed the conviction that even though "the climate emergency is a race we are losing, it is a race we can win." It went on to state, before discussing the sad predictions about climate change, which are being enacted right now, that "We must realise that no corner of the globe is immune from the devastating consequences of climate change. Rising temperatures are fuelling environmental degradation, natural disasters, weather extremes, food and water insecurity, economic disruption, conflict, and terrorism. Sea levels are rising, the Arctic is melting, coral reefs are dying, oceans are acidifying, and forests are burning. We also know that billions of tonnes of CO2 are released into the atmosphere every year because of coal, oil, and gas production and human activity is producing greenhouse gas emissions at a record high, with no signs of slowing down [13]."

The platform for these assertions was created by reports of the Intergovernmental Panel on Climate Change (IPCC), detailing broad, robust, and independent scientific results, in one of the most impressive instances of international scientific collaboration in response to a growing global concern about climate change, to which thousands of scientists from 195 countries contributed.

Before embarking on a discussion of the IPCC, we recall the story of the "ozone hole" and its successful mitigation to illustrate how international concern and determined international collaboration can solve such problems for good.

Scientists at the British Antarctic Survey had been monitoring ozone levels over Antarctica since the late 1970s. In 1985, their study revealed a significant and unexpected decline in ozone values during springtime and confirmed a massive hole in the ozone layer, specifically over Antarctica. It raised serious concerns about increased ultraviolet radiation reaching the Earth's surface, leading to potential health problems like skin cancer. The situation's urgency led to an international collaboration under the umbrella of the United Nations Environment Programme (UNEP). In 1987, the *Montreal Protocol* was established. This international treaty aimed to freeze the production and consumption of ozone-depleting substances, mainly chlorofluorocarbons, commonly used in refrigerants and aerosols. Subsequent amendments strengthened the Montreal Protocol, leading to a complete phase-out of these chemicals by 2010.

This phase-out of ozone-depleting substances has significantly slowed down the depletion of the ozone layer. While the hole over

Antarctica persists, it is slowly recovering, demonstrating the effectiveness of global action.

The predecessor of the IPCC, the Advisory Group on Greenhouse Gases, was established in 1986 by the World Meteorological Organization (WMO), the United Nations Environment Programme, and the International Council of Scientific Unions. The scientists involved in this group, from across the world, reviewed research on greenhouse gases and rising concentrations, laying the groundwork for a more comprehensive approach.

Once the complexity of climate science and the need for a broader response were recognised, the WMO and UNEP established the IPCC in 1988. Its creation was officially endorsed by the UN General Assembly, later that year, which marked a shift towards a more intergovernmental approach, with scientists collaborating with governments to assess climate change.

The initial task and the core mission of the IPCC were ambitious: to provide a comprehensive assessment of climate science, its social and economic impacts, and potential response strategies. This scientific knowledge was expected to help develop international agreements to address climate change.

Since its inception, the IPCC has produced a series of Assessment Reports, which are considered the most comprehensive scientific reports on climate change globally.

These reports are created through rigorous processes:

- *Assessment by Working Groups*: Three working groups within the IPCC assess different aspects of climate change: the physical science basis; impacts, adaptation, vulnerability, and mitigation of climate change.
- *Peer-reviewed Science*: The IPCC relies on thousands of scientists who contribute their expertise by reviewing and synthesising existing scientific literature.
- *Government Consensus*: For a report to be adopted, all member governments of the IPCC must reach a consensus, reflecting the intergovernmental nature of the organisation.

The Assessment Reports of the IPCC, released every 6–7 years, have played a crucial role in shaping international climate policy:

The First Assessment Report (AR1) published in 1990 highlighted the urgency of climate change and its global consequences and paved the way for the creation of the United Nations Framework Convention on Climate Change (UNFCCC) in 1992. This report confirmed the scientific evidence for climate change and provided the basis for negotiations on the climate change convention. The Second Assessment Report (AR2) was published in 1995. The Third Assessment Report (AR3) was published in 2001. The Fourth Assessment Report (AR4) was published in 2007. The Fifth Assessment Report (AR5) was published in 2013–2014.

The most recent and comprehensive assessment, the Sixth Assessment Report (AR6), was published in stages between 2021 and 2023:

- AR6 Climate Change 2021: The Physical Science Basis (August 2021).
- AR6 Climate Change 2022: Impacts, Adaptation and Vulnerability (February 2022).
- AR6 Climate Change 2022: Mitigation of Climate Change (April 2022).
- AR6 Synthesis Report (March 2023).

The history of the IPCC underscores its commitment to providing policymakers with the best available science to address the complex challenge of climate change. Its work continues to be a cornerstone of international efforts to mitigate and adapt to a changing climate [13].

An intergovernmental platform, the Conference of the Parties (COP), was set up to make policy decisions based on the knowledge provided by the IPCC. This is an annual meeting organised by the UNFCCC. COPs are essentially gatherings of countries ("parties") to discuss and act on climate change. COP1 was held in 1995 in Berlin, Germany, and it marked a significant step towards international collaboration on climate change. The *Berlin Mandate* required countries to start talks on reducing emissions beyond 2000 with specific deadlines and quantitative objectives, agreed on a programme of work and a budget for the secretariat, and adopted decisions on guidance for the Global Environment Facility. Unfortunately, the agreement was not implemented, which continued to aggravate the situation. However, parties agreed that the Conference's commitments were "inadequate" and decided to establish a process to negotiate stronger commitments for developed countries, which was later adopted by COP3.

As of today, 28 COPs have been held. COP 28 was held in December 2023 in Dubai, United Arab Emirates, with over 85,000 participants.

We note the important outcomes of COP28 in the following:

- For the first time, the COP included language acknowledging the need to move away from fossil fuels. This is a significant step towards a global commitment to reducing greenhouse gas emissions.
- Discussions continued for setting a new collective financial goal to support developing countries in their climate efforts.
- It highlighted the importance of protecting nature and biodiversity in tackling climate change. This reflects a growing understanding of the interconnectedness of these environmental challenges.
- The review process assessed the progress made by countries in achieving the goals of the Paris Agreement.

While COPs do not have direct enforcement power, they play a critical role in setting international goals and fostering cooperation to tackle climate change.

Two of the most important agreements (protocols) to arise from COP meetings were the Kyoto (1997) and the Paris (2015) agreements.

The *Kyoto Agreement* was adopted at COP3 in Kyoto, Japan. It was the first international agreement with binding emission reduction targets for developed countries. These targets varied by country but overall aimed for a reduction of greenhouse gas emissions by an average of 5.2% below 1990 levels by 2008–2012 (the first commitment period). It was a stepping stone, paving the way for the more comprehensive Paris Agreement.

The *Paris Agreement*, adopted at COP21 in Paris, aimed to limit global warming to well below 2°C, ideally 1.5°C, compared to pre-industrial levels. It focused on a framework for countries to set their own emission reduction targets and tighten them over time. Developed countries also committed to providing financial assistance to developing countries for climate action.

The non-binding agreement reached at COP15 in *Copenhagen* (2009) aimed to limit warming below 2°C and established a framework for developed countries to provide financial aid to developing countries. While it did not have the legal force of the Kyoto Protocol, it kept the momentum going on climate action.

India has consistently acknowledged the seriousness of the issue of climate change but has also sought to balance it with the country's development needs. India is also one of the most vulnerable countries to climate change, facing threats like rising sea levels, extreme weather events, and water scarcity. Yet, India emphasises the need for a development path that addresses climate change without hindering economic growth and poverty reduction.

Even with these constraints, India announced a pledge (at COP26) to achieve *Net Zero Emissions by 2070*. It also pledged to lower projected carbon emissions by one billion tonnes by 2030, reach 50% of installed electric power capacity from non-fossil fuels by 2030, and reduce the emission intensity of GDP by 45% by 2030 (compared to 2005 levels). India is vigorously pursuing renewable energy production, with a target of 500 GW of non-fossil fuel energy capacity by 2030.

India has also proposed and is pursuing an ambitious *International Solar Alliance* under its vision of *One Sun, One World, One Grid*, to escape the variability of solar and wind power and minimise the requirement of expensive batteries.

15.5 Tragedy of Commons

We pause here to look back on history and recall that global warming is rightly called a tragedy of commons, a concept that has been around for a long time among economists; thinkers like Aristotle even discussed this concept in ancient times. However, the term itself and the modern economic theory we know today are credited to William Forster Lloyd (1794–1852 CE), a British writer, who created it in 1833. Though Lloyd laid the groundwork, Garrett Hardin (1915–2003 CE) popularised the concept in 1968 with his article "The Tragedy of the Commons" [14]. Simply stated, the tragedy of the commons is a situation where everyone acting in his/her self-interest ruins a shared resource. Consider a field everyone can graze their cows on. Each person adds a few more cows because it benefits them in the short term. But eventually, the field becomes overgrazed and unusable for everyone.

Global warming is seen as a classic example of the tragedy of the commons because the atmosphere is a shared resource. Each country or company polluting it to benefit itself (e.g., by using cheap fossil fuels) contributes a bit to loading it with greenhouse gases like CO_2 and warming the planet. It is likely that while the economic benefit of using cheap fossil fuel accrues to the polluting nation or company, the disadvantage or the suffering caused is distributed to all, and the share for the polluters is marginal. The effects, like rising sea levels and extreme weather, hurt everyone eventually. One can take this analogy further and imagine all of us travelling in a giant ship, where the greenhouse gas emission of each country acts like a leak. The safety of this ship is guaranteed only if every country, big or small, plugs its leak. The ship would surely sink even if just one country left its leak unplugged

We can further elaborate on this by noting that the Earth's atmosphere is a shared resource that can handle a certain amount of pollution, like greenhouse gases from burning fossil fuels (which we are on the verge of overshooting). Every country, company, and individual using energy contributes to this pollution. We note that fossil fuels are cheap and convenient. Reducing emissions often requires upfront costs or lifestyle changes. So, it is tempting for everyone to keep using them. Each bit of extra pollution adds to the overall problem. Even small individual contributions, when multiplied by billions of people, cause the atmosphere to become overloaded. We have argued again and again and yet again that the resulting climate change is causing rising sea levels, extreme weather events, and disruptions to ecosystems, hurting everyone, even those who did not contribute as much. Thus, while developed countries have historically emitted more greenhouse gases (about 75%), they are also best positioned to protect themselves from damage. On the other hand, developing countries like India and African nations tend to have low per capita emissions, need economic development, and are more vulnerable to climate change impacts.

Thus, as repeatedly emphasised at the IPCC and COP platforms (mentioned previously),

- Countries must set limits on emissions and work towards shared goals.
- Countries must invest in renewable energy and cleaner technologies to reduce our reliance on fossil fuels.
- All of us can and must participate and contribute our bit by minimising our energy usage.

By acknowledging the shared nature of the atmosphere and working together, we can avoid a climate catastrophe and ensure a healthy planet.

The same considerations apply to fishing, pollution/utilisation of water bodies (lakes, rivers, oceans), and the overexploitation of groundwater for agriculture and industry.

Otherwise, unfortunately, climate change (as well as pollution and exploitation of water bodies) will unfold its tragic consequences like the gruesome and tragic prophesies of classical Greek dramas.

15.6 Discussions

It is time for us to take a sober look and remember that Earth is the only known planet in our solar system that has an abundance of liquid water, along with a life-sustaining atmosphere. This makes our blue planet a truly unique and extraordinary place in the universe. Its axis, around which it rotates- giving day and night, is tilted at about 23.5° from the plane of its orbit around the Sun, which gives us seasons, with their myriad variations of weather and climate. The tilt changes between 22.1° and 24.5° in a cycle of about 40,000 years, which can exaggerate the seasons. The axis also experiences a precession in a cycle of about 26,000 years, which shifts the seasons in relation to months. The gravitational pull of Saturn and Jupiter causes the eccentricity (which is a measure of the deviation of the orbit from a circle; it is zero for a circular orbit and larger, though less than one, for an elliptic orbit) of the Earth's orbit to oscillate between 0.0034 and 0.058, in a cycle of about 100,000 years, which triggers (triggered) the onset and the end of ice ages, following the Milankovitch Cycles [15, 16]. While the latter two effects are very long-term effects, the first one relating to seasons shapes our lives. However, these are completely beyond our control.

We have seen, from the example of the monsoons of India, how the weather and seasons, rainfall, wind, and temperature at any place on the Earth are affected by conditions across the planet.

Our immediate concern is the onset of climate change and, as a consequence, rapidly changing (and greatly fluctuating) weather. This is due to the human activities of using fossil fuels which are loading our atmosphere with more than 30 billion tonnes of carbon dioxide every year. It has converted our planet into a greenhouse that traps a large part of the heat from the sun, which is continuously raising the temperature of the air, land, and oceans. In addition to melting glaciers, and rising sea levels, this temperature increase is disrupting ecosystems, harming wildlife, and putting stress on food and water resources [6, 15]. Left unchallenged and unattended, it is likely to lead to social unrest and human suffering of gigantic proportions and produce millions of climate refugees, the beginnings of which we are already witnessing.

This is also providing a formidable challenge to meteorologists across the world who have made tremendous progress towards understanding and predicting the weather and climate by leveraging international collaborations and the most modern observation tools and supercomputing.

The challenge arises as global warming heats our planet and leads to more extreme weather events like heat waves, droughts, floods, and intense storms. Even if any further release of greenhouse gases is stopped by some magic wand today, it will take several decades for the weather and the climate to stabilise, and we can only speculate with dread and horror the suffering of our future generations, if we do not immediately address this onrushing calamity.

Zillman [5] invoked the compulsive and persuasive considerations of the famous problem called the prisoner's dilemma from game theory to make a case for international cooperation in meteorology and climate change. As the problem of the prisoner's dilemma perfectly captures the dilemma of international collaboration on climate change, let us discuss it in some detail.

The prisoner's dilemma imagines two criminals (prisoners) who can either stay silent (cooperate) or confess (betray each other). Cooperation benefits both in the long run (and leads to lighter sentences). Betrayal may seem better individually (lighter sentence if the other stays silent), but if both betray each other, they both end up worse off. The challenge is the lack of trust and fear of being betrayed. This scenario reflects situations where cooperation is best for everyone, but acting selfishly can seem more tempting in the short term.

Now, consider countries as the "prisoners" in this game. Each country must choose to reduce emissions (cooperate) or continue polluting (betray). If all countries reduce emissions (cooperate), everyone benefits from a healthier planet with fewer extreme weather events and a stable climate. If one country pollutes (betrays) while others reduce emissions, the polluting country enjoys short-term economic benefits. However, everyone suffers the consequences of a worsening climate. If all countries prioritise short-term gain and continue polluting (mutual betrayal), everyone will suffer the consequences of devastating climate change.

In the above-mentioned discussion, the fear of being cheated can discourage action. A country might hesitate to reduce emissions (cooperate) if it believes that others will not follow suit (betray). This can lead to a situation where everyone prioritises short-term gain, resulting in a disastrous outcome for the planet (mutual betrayal).

This dilemma highlights the difficulty of achieving international cooperation on climate change. Countries might be hesitant to take on the economic burden of reducing emissions if they believe others will not do the same. This can lead to an unfortunate situation where no country takes significant action and the climate crisis worsens.

However, the alarms raised by the IPCC discussions held at COPs and agreements like the Kyoto and the Paris agreements do hold out a glimmer of hope. Firstly, the cost of renewable energy is decreasing, making it a more attractive option for countries. Secondly, and more importantly, as the effects of climate change become more evident, countries have started to realise the essential role of prioritising long-term sustainability.

Verily, the prisoner's dilemma not only reminds us of the challenges but also highlights the importance of international cooperation and innovative solutions to tackle climate change effectively.

International collaboration in long- and short-term meteorology has never been as crucial for the survival of life on Earth as today. Fortunately, enough progress has been made in this regard. It is important to fully understand the complex interactions between atmospheric deep convection and the equatorial modes of ocean variability, which lie behind events like El Niño. It has also been argued that reliable regional climate models are vitally required so societies can adapt to climate change, assess the urgency of reaching net zero, or implement geoengineering solutions if things get bad.

Climate change leads to frequent and ferocious extreme weather events. It will help our efforts to adapt to these changes, if we know whether droughts, or heat waves, or storms, or floods will cause a greater havoc at a given place- as preparing for all eventualities may sadly be even impossible.

It is well known that before the 1980s, poorly forecast tropical cyclones often killed hundreds of thousands of people in vulnerable societies. However, with improved model resolution (and the ability to communicate the forecasts), excellent week-ahead predictions are made. It is rare for more than a few tens of people to be killed by extreme weather [6]. Do we need to say more?

References

[1] See e.g., V. B. Vaidya, V. Pandey, and S. Dhable, "Ancient science of weather forecasting in India with special reference to rainfall prediction", *Journal of Agrometeorology* 25 (2023) 477–490, https://doi.org/10.54386/jam.v25i4.2422.

[2] G. O. P. Obasi, "A Decade of Progress: The World Meteorological Organisation in the 1990s and the New Century", WMO no. 956 (Geneva: Secretariat of the World Meteorological Organisation, 2003).

[3] S. S. A. Azhar *et al.* "The Linkages between Antarctic Sea Ice Extent and Indian Summer Monsoon Rainfall", *Polar Science* 25 (2020) 100537, and references therein.
[4] "Guide to Meteorological Instruments and Methods of Observation", WMO-No. 8 (Geneva: World Meteorological Organisation, 2008). See e.g., E. C. Weatherhead *et al.*, "Designing the Climate Observing System of the Future", *Earth's Future* 5 (2017), https://doi.org/10.1002/2017EF000627.
[5] J. W. Zillman, "International Cooperation in Meteorology, Part 1: Origin and Early Years", *Weather* 73 (2018) 295, https://doi.org/10.1002/wea.3217; "International Cooperation in Meteorology, Part 2: The Golden Years and Their Legacy", *Weather* 73 (2018) 341, https://doi.org/10.1002/wea.3219.
[6] V. S. Ramamurthy, D. K. Srivastava, and S. Nayak, "*Art and Science of Managing Public Risks*" (Singapore: World Scientific Publishing Company, 2022); S. Nayak "Addressing Resilience to Cyclone — A Perspective", *Guest Editorial. Current Science* 120 (2021) 247–248.
[7] https://mausam.imd.gov.in/imd_latest/contents/climate_services_daily_temeprature_maps.php.
[8] B. P. Yadav *et al.*, "Rainfall Statistics of India — 2021" (Indian Meteorological Department). Report No. MoES/IMD/HS/Rainfall Report/02(2022)/60.
[9] See e.g., https://www.e-education.psu.edu/meteo3/node/2271.
[10] S. Gadgil, "Physics of the Monsoon", *Resonance* (May 2007), and references therein, https://www.researchgate.net/publication/263251955.
[11] https://www.climate.gov/news-features/features/argo-revolution.
[12] S. A. Rao *et al.,* "Monsoon Mission — A Targeted Activity to Improve Monsoon Prediction Across Scales", *Bulletin of the American Meteorological Society* 100(12) (2019) 2509–2532, https://doi.org/10.1175/BAMS-D-17-0330.1.
[13] https://www.un.org/en/un75/climate-crisis-race-we-can-win.
[14] G. Hardin, "The Tragedy of the Commons", *Science* 162 (1968) 3859, https://www.science.org/doi/10.1126/science.162.3859.1243.
[15] D. K. Srivastava and V. S. Ramamurthy, "*Climate Change and Energy Options for a Sustainable Future*" (Singapore: World Scientific Publishing Company, 2021).
[16] https://science.nasa.gov/science-research/earth-science/milankovitch-orbital-cycles-and-their-role-in-earths-climate/.

Chapter 16

The Arctic and Antarctica

"The world cannot live without the Arctic; it affects every living thing on Earth and acts as a virtual thermostat, reflecting sunlight and cooling the planet."

— Philippe Cousteau, Jr.

"If Antarctica were music, it would be Mozart. Art, and it would be Michelangelo. Literature, and it would be Shakespeare. And yet it is something even greater; the only place on Earth that is still as it should be. May we never tame it."

— Andrew Denton

16.1 Introduction

Polar regions — the Arctic and Antarctica — are remote but have spectacular landscapes. These are a tangible heritage for humankind. Our knowledge about polar regions has advanced manifold during the last two centuries. We now know that snow and ice play an important role in modulating the Earth system. The changes in the polar environment can modify the planetary energy budget, regulate the temperature gradient between the equator and poles and variability in atmospheric circulation, and modulate global weather and climate. The importance of polar research has increased due to this improved understanding of the connections that the polar regions have with physical and biological processes as well as sun–Earth interactions and space weather.

A comprehensive observation, and analysis, of the roles played by the polar atmosphere, ocean, and cryosphere, both present and past, is important to understand changes in climate and future projections. Thus, for example, major changes in the sea level during the last thousands of years were mainly controlled by the melting of the Antarctic ice sheet. The analysis of the historical ice in Antarctica provided an accurate correspondence between the temperature and carbon dioxide and methane (see later).

The *Arctic* encompasses the North Pole and stretches across the northernmost reaches of North America and Eurasia, covering them in a vast expanse of frozen ocean and tundra. It has its own stark beauty and delicate balance, with clear air and the ethereal and mesmerising charm of the Aurora Borealis caused by the ionisation of the atmosphere by the passage of very high-energy cosmic rays, often during solar storms dancing across the sky (Fig. 16.1). The area around the southern pole has its own Aurora Australis. Will their locations change as the magnetic north and south poles shift?

Fig. 16.1. Aurora Borealis (Credit: Wikipedia).

During winters, the temperatures drop to −30 to −40 degrees centigrade. Even during the brief summer, the temperature stays below 10 degrees centigrade. While the interior regions have meagre precipitation, the coastal regions of Alaska, Iceland, northern Russia, and Scandinavia see snowfall and rainfall accumulating between 600 and 1250 mm. The wind speeds are generally low, except during storms. Polar vortexes are fascinating formations of low pressure and extremely cold air trapped near the poles. Their southward expansion can occasionally reach mid-latitude regions, bringing outbreaks of frigid air to areas in North America, Europe, and Asia. Polar bears, walruses, and Arctic foxes have adjusted to the climate over the millennia. A vibrant ecosystem of fish and marine mammals teems below the ice.

Indigenous communities like the Inuit, Saami, Yupik, Aleut, Evenki, and Chukchi with distinct languages, traditions, and social structures have lived in the Arctic region for millennia. Their life and economy are marked by subsistence hunting, fishing, and herding (e.g., reindeer) and a deep respect for the environment — leading to their unique culture. They have adapted to the harsh climate, long winters, and limited resources. Their traditional knowledge and practices provide valuable insights (also to the Arctic Council, see later) for sustainable living in a fragile ecosystem.

The Arctic environment is changing and impacts global weather, climate, and ecosystems. The possibility of accessing newer energy resources, critical and strategic minerals, fishing grounds, and opening shipping routes has brought global attention to the Arctic for scientific, economic, and geopolitical reasons. These developmental activities are likely to lead to oil pollution and additional release of methane from permafrost (which is already being released due to global warming). These will also affect the environment and lead to extreme weather conditions in developing countries.

Unfortunately, global warming is seriously affecting the Arctic ice extent, endangering the biodiversity, and affecting the weather across the globe via complex ocean–air couplings and the rotation of the Earth.

Antarctica, Arabia, Australia, India, Madagascar, South America, and Zealandia were parts of the Gondwanaland supercontinent, which started breaking up around 200 million years ago. Antarctica currently encompasses the South Pole, and unlike the Arctic, which is an ocean surrounded by land, Antarctica is a continent encased by the Southern Ocean. It has a profoundly cold and dry environment, with the average annual temperature hovering around −60 degrees centigrade at the high interior plateau, which rises to a relatively balmy −10 degrees centigrade on the coast. The coastal regions experience a brief summer thaw, when temperatures can even reach above freezing, though for just a short period. The precipitation in Antarctica is normally as snow, with the coastal areas receiving the most, while the interiors receive very little snowfall due to the extremely cold and dry air, converting it into what is technically a desert.

Antarctica is the windiest continent on Earth. Katabatic winds — powerful downslope winds created by the continent's high elevation and vast ice sheet — roar across the landscape, with average wind speeds reaching around 100 km/h and gusts exceeding 300 km/h.

The Southern Ocean surrounding Antarctica teems with life. Krill — the shrimp-like creatures — form massive swarms and serve as food for whales, seals, penguins, and fish. Blue whales migrate to the Southern Ocean during the summer months to feed on the abundant krill. Elephant seals, Weddell seals, and crabeater seals are just a few of the species that live here.

The most famous residents of Antarctica are Emperor Penguins, Adélie penguins, chinstrap penguins, and gentoo penguins, which waddle, swim, and feed around the continent's shores. As the cold becomes unbearable and food scarce, penguins travel north, covering distances up to several hundreds of kilometres to feed and breed, in one of the most fascinating migrations in the animal kingdom. Adélie penguins living in the Ross Sea region in Antarctica migrate an average of about 13,000 km each year [1].

Antarctica and its environment are critical for our planet's future. Its vast ice sheets hold a significant amount of freshwater and their stability is important for maintaining sea levels. It is also important to study and monitor penguin populations, krill abundance, and the impact of climate change on these ecosystems.

16.2 The Science of the Arctic

The Arctic region consists of parts of Northern Norway, northernmost Sweden, Northern Finland, Northern Russia, Northern Alaska, Northern Canada, Greenland (part of the Danish Realm), and Northern Iceland, along with the Arctic Ocean and adjacent seas, 66° and 30 min north (Fig. 16.2). These nations also constitute the Arctic Council (see later). According to the UN Convention on the Law of the Sea (UNCLOS), these

Fig. 16.2. The Arctic Region (Marked as Blue Circle) (Credit: Wikipedia).

nations have exclusive economic rights up to 200 nautical miles and over the continental shelf, up to 350 nautical miles.

The Arctic's unique location, extreme weather, nights lasting up to 40 days (during the winter), and the sun staying above the horizon for months (during the summer) have fascinated mankind for a long time. The Dutch and English are known to have explored the Arctic in the sixteenth century to trade with the Far East. Russians discussed exploring the Arctic for economic reasons as far back as 1525 CE. The early explorers went to the Arctic to reach an unknown land, while some explorers looked for shorter sailing routes to and from the Far East to Europe. Other explorers went to hunt whales and seals for profit.

Recent explorers have sought to gather scientific information about the polar regions and study their effect on global weather and climate (change). For example, it has been reported that since 1979, the Arctic has warmed almost four times faster than the rest of the world in a phenomenon called Arctic amplification, defined as the ratio of Arctic warming to the global mean warming [2], while other studies had earlier reported a value between two and three for this. Some studies have reported that the Barents Sea north of Russia and Norway is warming at an even more alarming rate — seven times faster than the global average.

Detailed studies have provided valuable insights into Arctic amplification, which is mainly driven by the ice–albedo feedback, lapse rate feedback, water vapour feedback, and ocean heat transport. Sea ice and snow have high albedo, which implies that they reflect most of the solar radiation as opposed to water and land. Global warming is leading to diminishing sea ice (Fig. 16.3).

As the sea ice melts, the Arctic Ocean absorbs more of the solar radiation, which further drives the heating and melting of sea ice. Like sea ice, melting snow exposes darker land surfaces, enhancing surface warming. Moreover, warmer air can hold more moisture, which acts as a greenhouse gas, further amplifying warming. We know that the heat radiated by the surface of the Earth gets trapped by greenhouse gases. This trapping effect is stronger at lower altitudes, which warms lower layers much faster than higher altitudes. This again aids the warming, through a complex interplay of trapping of heat by greenhouse gases and the radiation from the upper layers of the atmosphere, in a process called lapse rate feedback.

A further cause for concern is that not only is the extent of Arctic Sea ice decreasing but it is also getting much thinner, and scientists have observed a transition to younger ice.

The reduced ice cover adversely affects the habitat of polar bears (Fig. 16.4), who depend on sea ice to travel, hunt, and mate, and in some areas to den. They now need to travel greater distances and swim more than previously for food, which is particularly difficult and life-threatening for young cubs.

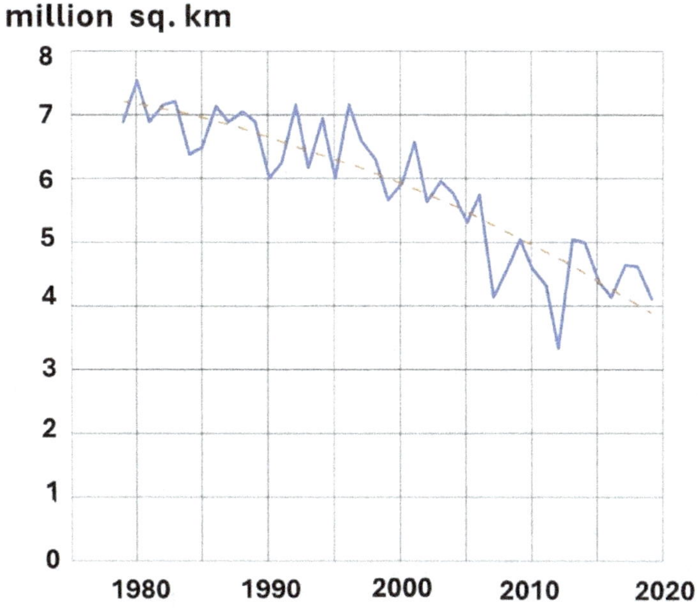

Fig. 16.3. Decreasing Arctic Ice Extent (Credit: UNEP and NOAA).

Fig. 16.4. Polar Bears Depend on Sea Ice for Dens, Food, and Mating. The Loss of Sea Ice is Affecting the Health and the Population of Polar Bears. (Credit: NOAA).

Yet another cause for concern, if not managed judiciously and sustainably, is the fact that the Arctic holds vast resources. Scientific and responsible resource management practices are very urgently needed if we are not to mess up this pristine part of our planet, which will have disastrous effects across the globe.

16.3 The Arctic Council and Other International Collaborations

The vastness and complex effects of the Arctic on the weather and the climate necessitate international collaboration. The Ottawa Declaration of 1996 formally established the Arctic Council. It is a high-level intergovernmental forum to provide a means for promoting cooperation, coordination, and interaction among the Arctic states, with the involvement of the Arctic indigenous communities and other Arctic inhabitants on common Arctic issues. The issues of sustainable development and environmental protection in the Arctic receive special attention from the Council.

The Council has Canada, the Kingdom of Denmark, Finland, Iceland, Norway, the Russian Federation, Sweden, and the United States as members. The Arctic has about 4 million inhabitants, out of which about 500,000 are indigenous peoples. These indigenous peoples represented by the Aleut International Association, Arctic Athabaskan Council, Gwich'in Council International, Inuit Circumpolar Council, Russian Association of Indigenous Peoples of the North, and Saami Council, are Permanent Invitees to the Arctic Council. In addition, France. Germany, Italy, Japan, the Netherlands, China, Poland, India, South Korea, Singapore, Spain, Switzerland, and the UK, along with several intergovernmental and inter-parliamentary organisations as well as non-governmental organisations, are Observers of the Council.

The Council's work is conducted by six working groups: Arctic Contaminants Action Programme; Arctic Monitoring and Assessment Programme; Conservation of Arctic Flora and Fauna; Emergency Prevention, Preparedness and Response; Protection of the Arctic Marine Environment; and Sustainable Development Working Group. More details about the workings of the Council and the Working Groups can be found in Ref. [3].

Even before the platform of the Arctic Council came into being, International Polar Years (IPYs) were considered a prime example of international collaboration in scientific research. They coordinated essentially intensive research periods focused on the Arctic and Antarctic regions, where multiple countries came together to share resources, expertise, and findings. The first IPY with 11 countries was held during 1881–1884. The last one held during 2007–2008 had 67 countries participating in it, oversaw 200 joint projects on oceanography, atmospheric studies, glaciology, etc., and led to more robust research designs and findings. The fifth IPY is to be held during 2032–2033.

The Arctic Circle was set up by the former president of Iceland, Olafur Ragnar Grimsson, and others in 2013 and provides governments, research organisations, corporations, universities, think tanks, and civil societies with a platform to conduct a dialogue about the future of the Arctic and our planet.

The University of Arctic (UArctic), based in Finland, is a network of universities and academic and research institutions engaged in research and education that promotes and strengthens collective resources and infrastructure in member organisations.

The Asian Forum on Polar Sciences (AFPOS) comprising China, India, Japan, South Korea, Malaysia, and Thailand is a non-governmental organisation to advance polar sciences, both Arctic and Antarctic, through collaboration and cooperation.

The International Arctic Science Committee (IASC) is a non-governmental international scientific organisation with 24 members countries. The IASC promotes interdisciplinary research for scientific understanding of the Arctic environment and its role in the Earth system. The IASC has the following five working groups addressing scientific issues: atmosphere, cryosphere, marine, social and human, and terrestrial. Though the working groups are formed according to a specific discipline, they address scientific questions across disciplines.

Thus, we see that the Arctic has a long history of scientific cooperation, and the focus should be on the Earth system and the development of appropriate environmental policies. The question is how the Arctic should be governed given the availability of energy and mineral resources and the role it plays in influencing global weather and climate [4].

One of the most important aspects of these collaborations is data, under which countries standardise data units and collection methods to ensure compatibility. Additionally, large centralised data repositories are established for researchers worldwide to access and utilise information from all participating nations. This fosters transparency and allows for comprehensive analysis across borders.

Major scientific organisations like the International Council for Science (ICSU) and the World Meteorological Organization (WMO) have provided the framework for collaboration, facilitated communication between nations, and set overarching research goals for each IPY. It has been suggested that more advanced technologies like telepresence and advanced communication platforms will allow for more real-time collaboration despite the physical distance of the polar regions.

The association of India with the Arctic is more than 100 years old. The Svalbard Treaty signed in 1920, recognised the sovereignty of Norway over the Svalbard region. The treaty allowed India (as a British Overseas Dominion) to engage in commercial activities, mainly exploiting coal reserves in Svalbard. The first scientific expedition was undertaken by the late Prof. Mahendra Nath Bose (1925–2011), in association with Norwegian scientists, to collect plant fossils.

Himadri, the first Indian research station was established in Ny-Ålesund, Svalbard, Norway, in 2008. It has focused on a broad range of scientific disciplines like atmospheric sciences, glaciology, oceanography, marine biology, and environmental chemistry as well as climate change and its impact on the Arctic, with a particular interest in understanding the link between the Arctic and monsoon patterns in India, as well as a focus on the Arctic flora and fauna to compare them with those in the Himalayas. Since 2018, India has been conducting winter expeditions, which allow year-round research.

We reiterate that India has a major stake in the Arctic for several reasons. Changes in the Arctic affect our monsoon. Moreover, Arctic may prove to be an alternate source of oil and gas, provide access to critical metals and rare earth elements, and shipping routes may develop to and via Arctic to access these resources [5].

International collaboration in scientific research in the Arctic is proof of the power of global partnership in advancing scientific knowledge, and it continues to shape our understanding of the planet and guide efforts to address global challenges.

16.4 The Science of Antarctica

Antarctica is a unique territory that has no sovereign control, and yet it is free from any conflict or discord. It is a natural laboratory to conduct research in glaciology, geoscience, paleoclimate, climate change, and biological and atmospheric studies.

Sailors had started reporting sightings of Antarctica in the seventeenth century, and Captain James Cook (1728–1779 CE) provided the first concrete evidence of a vast landmass south of the polar circle in the late 1770s. The early expeditions in the nineteenth century mapped parts of the Antarctic coastline and documented encounters with penguins, seals, and other wildlife. A Norwegian expedition led by Roald Amundsen (1872–ca. 1928 CE), followed by a tragic British expedition led by Robert Falcon Scott (1868–ca. 1912 CE), was the first to reach the South Pole in 1911.

By the mid-20th century, scientific research in Antarctica started in earnest with the establishment of permanent research stations, and international collaboration increased significantly. These were facilitated by technological advancements like aircraft, icebreakers, and sophisticated instruments, and studies were conducted in meteorology, glaciology, geology, biology, physics, and oceanography.

One of the most interesting results to come out of these studies has been the climate history trapped in ice cores, giving the extent of carbon dioxide (and methane) in the atmosphere, and the temperature of the last 800,000 years (Fig. 16.5) obtained by analysing the air bubbles trapped in the ice in the continuous ice core which was drilled [6]. This is also one of the most important results in support of the argument on the increase of carbon dioxide in the atmosphere (due to the use of fossil fuels) which is leading to global warming and climate change. The carbon dioxide content, which never crossed the value of 300 ppm (part per million) during the last 800,000 years, has touched 423 ppm now and is showing no signs of stabilising or decreasing yet.

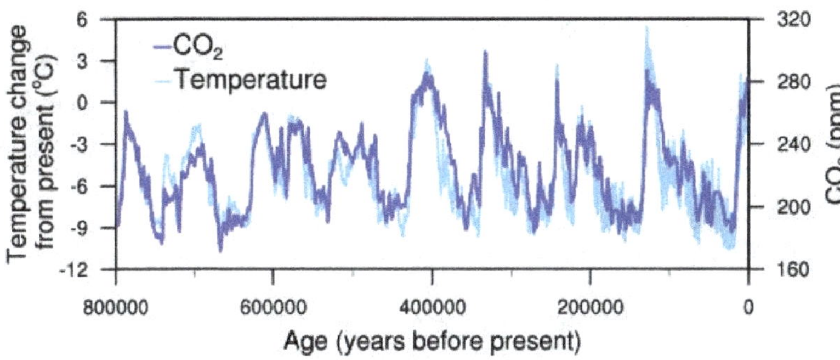

Fig. 16.5. Correspondence between Temperature and the Concentration of Carbon Dioxide in the Atmosphere, Observed during the Glacial Cycles of the Past Several Hundred Thousand Years (Credit: NOAA).

These collaborations have also studied and documented how the unique ecosystems of Antarctica have adapted to extreme conditions, analysed Antarctica's role in regulating ocean circulation and climate patterns, and explored the geological formations, revealing past connections with other continents, as mentioned earlier. Such collaboration is continuing with an emphasis on understanding the impact of climate change on Antarctica, including melting glaciers, rising sea levels, and potential disruptions to marine ecosystems.

The South Pole IceCube Neutrino Observatory, located close to the South Pole, is the biggest neutrino detector ever built, using 1 km^3 (one billion tonnes) of the purest ice. It is an international collaboration of about 350 researchers from 58 institutes in 14 countries and has been operating since 2011. It observed the first high-energy neutrinos from the Milky Way (2023), offered the first indication of high-energy neutrinos from outside the solar system (2010–2012), gave the first link between high-energy neutrinos and a known source outside the galaxy — a blazar — a supermassive black hole that shoots out particles at nearly the speed of light (2017), and provided the first glimpse into the inner depths of an active galaxy. The next generation of IceCube will have 8 km^3 of ice and several other unique features [7].

16.5 The Antarctic Treaty

The Antarctic Treaty, initially signed by 12 nations in 1959, ushered in a new era of peaceful scientific cooperation and ensured that Antarctica is used for peaceful purposes only and prioritises scientific research. Today, over 50 countries have acceded to the treaty, reflecting its global importance. The treaty established the Scientific Committee on Antarctic Research (SCAR) to facilitate international collaboration on Antarctic science by doing the following:

- Coordinating research programs across disciplines.
- Standardising data collection methods for better comparability.
- Encouraging the exchange of scientific personnel and expertise.
- Providing scientific advice to the Antarctic Treaty System.

The Antarctica Treaty Consultative Meetings (ATCMs) are held annually, where 56 Contracting Parties discuss issues relating to science, policy, governance, management, and preservation. The ATCMs are administered through the Antarctic Treaty Secretariat headquartered in Argentina.

The Protocol on Environmental Protection to the Antarctica Treaty (the Madrid Protocol 1991) recognising Antarctica as "a natural reserve devoted to peace and science" was the first legal instrument adopted under the Antarctica Treaty. It came into effect in 1998 and provides a mechanism to protect the Antarctic environment and ecosystems and coordinates the saving of the protected areas, environmental impact assessment, waste management, marine pollution, etc. The demarcation of areas, such as the Antarctic System Management Area (ASMA) and Antarctic Specially Protected Areas (ASPAs), and the inspection of research stations are under the purview of this protocol. Many countries have exercised their intentions in this regard. The Committee for Environmental Protection (CEP) has been established under the Madrid Protocol and advises the ATCM on environmental protection and conservation in Antarctica.

The marine living resources are governed by the Convention on Conservation of Antarctic Marine Living Resources (CCAMLR). The marine harvesting of krill and toothfish in the Antarctic, biological prospecting related to intellectual property rights, etc., are regulated under this convention.

The Council of Managers of National Antarctic Programs (CONMAP) is an international association that supports scientific research in Antarctica on behalf of the member governments. It provides expertise in air and marine operations and safety, science facilitation, environment protection, human biology and medicine, and education, outreach, and training.

The Scientific Committee on Antarctic Research (SCAR) provides scientific support to the Antarctic Treaty System (ATS). The SCAR 2011–2016 Strategic Plan called to institute the Antarctic and Southern Ocean Science Horizon Scan. It was established to identify the highest-priority science questions [8]. An International Steering Committee (ISC) representing fourteen countries was appointed by SCAR to oversee the selection processes. An open call was made to solicit questions from the entire Antarctic scientific community. More than 800 questions were received after this call. A meeting of scientists, national programme managers, policymakers, and early career scientists was organised to define 80 questions across various disciplines. The selection of participants was done by the ISC. A systematic transparent system was adopted during the meeting to prioritise questions. After intense discussions, questions were selected in the following disciplines:

- Antarctic atmosphere and global connections;
- Southern Ocean and sea ice in a warming world;
- Antarctic ice sheet and sea level;
- Dynamic Earth — probing beneath Antarctic ice;
- Antarctic life on the precipice;
- Near-Earth space and beyond — eyes on the sky;
- Human presence in Antarctica.

The Horizon Scan activity was concluded in 2014.

Most countries have aligned their national programmes to address the 80 questions. This is one of the best examples of cooperation among scientists to identify priorities and work towards them. The technological challenges and logistical requirements to meet these objectives are also being addressed.

International collaboration is the backbone of modern Antarctic research, and countries have pooled resources to establish and maintain joint research stations, e.g., Amundsen-Scott South Pole Station (US), Concordia Station (France and Italy), and Maitri Station (India), leveraging each other's strengths in logistics, technology, and scientific expertise. Operating in Antarctica's harsh environment requires specialised skills and equipment. Countries share best practices in areas like icebreaker technology, survival training, and environmental protection protocols. The Antarctic Treaty System emphasises open access to scientific data, and centralised data repositories ensure that researchers worldwide can access and utilise information from all participating nations, promoting transparency and collaboration.

The Antarctic Treaty offers a robust framework for continued cooperation, ensuring that scientific discovery and responsible stewardship go hand in hand in this critical region.

16.6 India at Antarctica

Indian scientists and explorers participated in various international Antarctic expeditions during the pre-independence era, primarily as part of British expeditions. Once India recognised the importance of polar research and the need to establish a scientific presence in Antarctica, it formally initiated its Antarctic programme in 1981, organising its first Antarctic expedition. It established its first permanent research station in

Fig. 16.6. The Dakshin Gangotri Research Station of India, Antarctica, before It Sank into Ice.

Antarctica, named "Dakshin Gangotri" in 1983 (Fig. 16.6). Remarkably, Indian women scientists (e.g., Aditi Pant and Sudipta Sengupta) were part of these expeditions from the very beginning. The station is currently being used as a supply base and transit camp after it sank into ice.

The station served as India's base for scientific research and exploration in Antarctica for several years until it was replaced by the Maitri Research Station in 1989 (Fig. 16.7).

Maitri is situated in the rocky mountainous region called Schirmacher Oasis. It is just 5 km from the Russian Novolazarevskaya Station and close to Lake Priyadarshini (named after India's Prime Minister, Mrs. Indira Gandhi), which supplies it with fresh water. India's early Antarctica expeditions were marked by the grit and determination demonstrated by the researchers due to the hardships of the extremely difficult conditions for living, working, and communication. It was possible only due to international cooperation and goodwill among the different stations from the different countries. One can find an interesting narrative [1] of the thrilling experiences of the researchers in the early days and the Eleventh Expedition (1991–1993) which was led by Dr Sharadindu Mukerjee [1].

Fig. 16.7. An Aerial View of the Indian Station Maitri, Antarctica.

During these expeditions, Indian researchers visited and mapped several new areas for the first time, collected samples from the remotest and most difficult places, provided ample proof that the ozone hole is recovering at a slower rate than earlier anticipated, tracked the Earth's magnetic field in the region of the South Pole, and conducted research in glaciology, human physiology in extreme environments, geology, atmospheric sciences, marine biology, and microbiology. Indian scientists have played a key role in discovering new extremophile microbes, organisms that thrive in harsh conditions, and are credited with finding 30 out of 120 new microbes discovered in Antarctica by 2005. We do add that one of the greatest fears of Antarctic warming is the release of unknown viruses buried in its ancient ice over millennia.

India inaugurated its second permanent research station in Antarctica, named "Bharati," in 2012, in the region of the Larsemann Hills in East Antarctica. The Bharati Station further enhanced India's research capabilities in the region. India expanded its research activities in Antarctica, covering various scientific disciplines such as glaciology, geology, oceanography, meteorology, and biology.

India actively collaborates with other countries involved in Antarctic research. It participates in joint scientific expeditions, shares data and resources, and contributes to international scientific initiatives aimed at understanding climate change and environmental conservation.

These activities are coordinated by the National Centre for Polar and Ocean Research, Goa.

16.7 Conclusions

The term "poles apart" is often used to describe very diverse and opposing views or conditions. The North Pole and the South Pole are very different — one is a frozen ocean of ice and the other is a frozen continent. Both have their fauna as well as flora. Yet, global warming and the associated climate change are affecting them seriously, with the warming of the Arctic region being even more rapid (Arctic amplification) than that of the rest of the planet. The melting ice is affecting the habitat of both regions and altering it far too rapidly, leaving very little time for polar bears in the north and penguins in the south to adjust and "rewire" their lives adapted to extreme weather over millennia. The interconnectedness of the atmosphere, oceans, various currents, glaciers, the melting of ice sheets, rising sea levels, the fragility of the young ice, and their effects on climate and weather, especially the life-sustaining monsoons across Africa and Asia, make it imperative that we study these systems urgently. Fortunately, the Arctic Council in the North and the Antarctic Treaty in the South provide a very robust framework to facilitate international collaborations for the study of the science of the polar regions. We only need to equitably strengthen these organisations and implement their recommendations. These organisations would of course face their biggest challenge when some countries decide to start mining the Arctic and the Antarctic regions for minerals, gas, and oil.

There is a need to have greater engagement with international organisations such as the UNFCC, IPCC, WMO, and World Climate Research Programme (WCRP) to address issues related to climate change and its impact on the Arctic and Antarctic regions and vice versa.

References

[1] S. Mukerji and D. Srivastava, "*The Edge of The Earth*" (Allahabad: Redgrab Books, 2017).

[2] M. Rantanen, *et al.*, "The Arctic Has Warmed Nearly Four Times Faster than the Globe since 1979". *(Nature) Communications Earth and Environment* 3 (2022) 168, https://doi.org/10.1038/s43247-022-00498-3, [3] https://arctic-council.org/.

[4] S. Nayak, "*Issue Brief III: India and China in the Arctic*" (Chennai Centre for China Studies, 2023). https://www.c3sindia.org/post/issue-brief-iii-india-and-china-in-the-arctic-by-dr-shailesh-nayak.

[5] S. Nayak and D. Suba Chandran, "Arctic: Why India Should Pursue the North Pole from a Science and Technology Perspective?" *Current Science* 119 (2020) 901–904.

[6] D. K. Srivastava and V. S. Ramamurthy, "*Climate Change and Energy Options for a Sustainable Future*" (World Scientific Publishing Company, 2021).

[7] https://icecube.wisc.edu/, https://icecube-gen2.wisc.edu/.

[8] M. C. Kennicutt II, *et al.* "A Roadmap for Antarctic and Southern Ocean Science for Next Two Decades and Beyond", *Antarctic Science* 27 (2015) 3–18. DOI: 10.1017/S0954102014000674.

Chapter 17

Collaboration Among Animals and Birds

"Organisms are inherently competitive, yet cooperation is widespread. Genes cooperate in genomes; cells cooperate in tissues; individuals cooperate in societies. Animal societies, in which collective action emerges from cooperation among individuals, represent extreme social complexity. Such societies are not only common in insects, mammals, and birds, but exist even in simple species like amoebas."

— D. Rubenstein and J. Kealey, *Nature Education Knowledge*, 3 (2010) 78.

17.1 Introduction

Collaboration and cooperation are not unique to humans. The kingdom of animals and birds, and even bacteria, is replete with examples of collaboration. These creatures cooperate in activities such as hunting, migrating, protecting members, and securing territories. Collaboration for them is often a matter of life and death.

17.2 Emperor Penguin, Fish, Whales, and Beavers

Consider *Emperor Penguins*. These flightless birds have developed remarkable adaptations to survive in one of the Earth's most extreme environments. During the Antarctic winter, temperatures plummet to −45 degrees centigrade or lower, while wind speeds can reach 100–300 km/h.

Additionally, the sun stays below the horizon for several weeks in a year. By congregating in large groups and huddling together (Fig. 17.1), penguins share their body heat, thereby maintaining their core body temperature and mitigating the harsh effects of the Antarctic winter.

Consider a school of small fish, e.g., *sardines*. When threatened by a predator, they form a dense cluster (Fig. 17.2) called a bait ball, which

Fig. 17.1. Emperor Penguins Huddle in a Blizzard at Auster Rookery (Credit: Frederique Olivier, Australia Antarctica Programme).

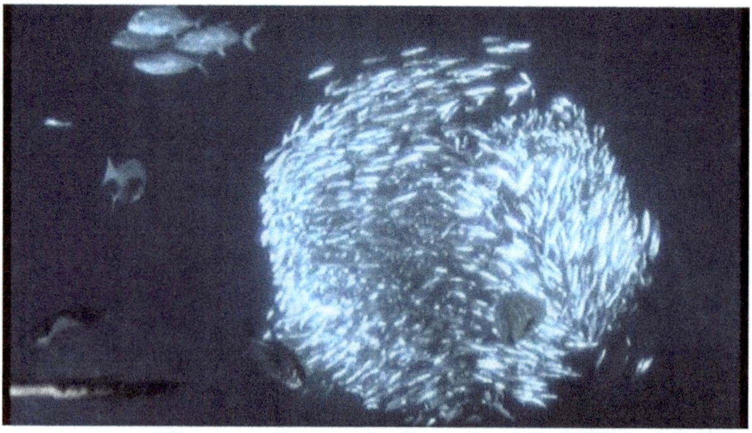

Fig. 17.2. Sardine Bait Ball (Credit: Wikipedia).

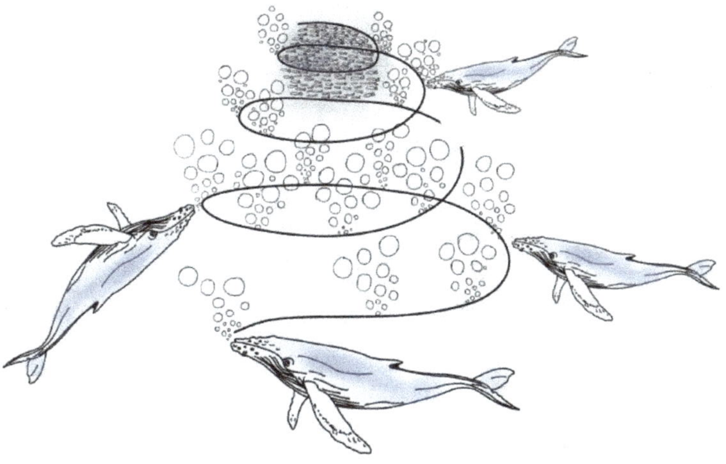

Fig. 17.3. A Group of Whales Engaged in Bubble Net Fishing.

makes it difficult for a single predator to pick out an individual fish. This tactic of forming a dense cluster is also adopted by several herding animals like zebras, buffaloes, and wildebeest, who additionally move the young to the interior of the cluster to protect them. Predators, on the other hand, often cooperate to isolate a particular member and drive them away from the safety provided by these numbers.

Now consider *humpback whales*. They are known to operate in groups of ten or 12 individuals, which surround a school of krill and herd them into an ever-shrinking (and thus dense) bait ball by blowing air bubbles around them, which the krill are unable to cross, and then feed upon them (Fig. 17.3). The study of this behaviour has given rise to the whale optimisation algorithm.

Beavers are some of the most impressive engineers of the animal kingdom, who leverage their instincts and teamwork to create impressive and functional homes within their environment. They construct dams (Fig. 17.4), using branches, logs, and mud, stretching to hundreds of metres, controlling water flow and creating a calm, deep pond, perfect for their lodge. Their multi-level lodges have multiple underwater entrances with dry chambers above the water line. These are elaborate structures made using sticks and branches plastered with mud, keeping the beavers warm during winter and safe from predators. They constantly monitor and repair their lodges. It is known that they even built canals to transport

Fig. 17.4. A Dam Constructed by Beavers (Credit: Wikipedia).

Fig. 17.5. Pack of African Wild Dogs, Kruger National Park (Credit: Wikipedia).

heavy building materials to their building sites! And they do all this in good time before the winter sets in and the water in the dams freezes!

17.3 African Wild Dogs and Elephants

African wild dogs are among the most efficient hunters (Fig. 17.5). It has been estimated that a whopping 90% of their hunts lead to a kill. They doggedly pursue their prey within a short distance of at most 2 km for an

Fig. 17.6. A Small Herd of African Elephants.

hour or so to wear it out. The stamina of these dogs and their ability to work as a well-knit pack help them succeed so very often.

The cheetahs and lions of Africa are also known to hunt in packs and thus maximise their chances of success.

African elephants (Fig. 17.6) are highly social creatures that depend on collaboration for survival and success. They live in herds led by a wise old female, the matriarch, who uses her vast experience to guide the herd to food and water sources, remembers migration routes, and leads them to safe areas during danger. The matriarch is also a living repository of knowledge for the herd. The herd members work together to raise young calves and communicate through rumbles, trumpets, and infrasonic calls.

The matriarch's leadership and the collaborative spirit of the herd are what keep these magnificent creatures thriving in the African savanna.

Several animals, e.g., meerkats or prairie dogs, take turns keeping watch for predators while others forage for food. Langurs and mynahs are the well-known sentries of Indian forests [1], who warn the entire jungle of approaching predators.

17.4 Birds

Birds migrating over long distances often fly in a V formation (Fig. 17.7), which is a wonderful study in cooperation, as the birds at the back take advantage of the upwash generated by those in front and spend less energy

Fig. 17.7. Migrating Birds Flying in a V Formation.

flying. As the bird in the front gets tired, it moves back, and the one at the back takes its place. The V formation of fighter jets is believed to have copied this practice to save up to 20% of their fuel.

Animals, birds, fish, and insects have evolved these cooperative systems over generations for their survival and propagation.

Who is not fascinated by the breathtaking and mesmerising *murmuration* of a large flock of starlings (Fig. 17.8), where they fly in close formation, creating a giant, fluid mass that appears to shimmer in the fading light? The flock constantly changes shape, twisting, turning, swooping, and swirling like a dark, living cloud. Despite their vast numbers, the starlings seem to move as one, their changes in direction happening in perfect unison. While the cause of their murmuration is not yet precisely known, it is believed that by constantly changing shape and direction, the starlings confuse predators like peregrine falcons and make it difficult for them to pick out a single bird. The study of their behaviour has given rise to the fascinating and vibrant field of swarm intelligence, which studies the collective motion of a large number of self-propelled entities and their emergent (self-organised) behaviour without any central authority [2].

Fig. 17.8. Murmuration of Starlings.

17.5 Ants Collaborate too, but Humans Are Different

Before ending this chapter, we would like to discuss one more aspect, which has not been emphasised sufficiently in this monograph. All of us have seen ever-busy ants moving as swiftly as their little legs can carry them, often in a line — which can teach a lesson or two to troops marching in a ceremonial parade — towards a food source, an insect, grains of cereal or sugar, or crumbs of bread. We have also seen hundreds of arts pushing, pulling, lifting, and working together to carry a large food item, skirting obstacles along the way. Even singly, ants are known to be capable of carrying things that could weigh as much as 50 times their body weight. Ants are known to form a bridge or even a raft across (water) gaps using their bodies so that members of their nest can cross to the other side (Fig. 17.9).

The secret of how ants find the shortest route from their nest to a food source and their collaborative effort to manoeuvre and move food items have fascinated mankind since time immemorial. A better understanding of ants' behaviour was used by Marco Dorigo to develop the ant colony optimisation algorithm in 1991. It is now used to find the shortest route through a maze, design efficient delivery routes, schedule tasks in a

Fig. 17.9. Ants Making a Bridge (Credit: Wikipedia).

complex project, optimise traffic flow in a city, design telecommunication networks, predict results of protein folding, and solve all such problems that can be reduced to the classical problem of a travelling salesman. The study of ants has also helped us design better robots.

The ant nests are an architectural marvel (Fig. 17.10), with sophisticated ventilation systems that regulate temperature and humidity, formed by strategically placing openings and chambers of different sizes to create air circulation patterns along with an efficient waste disposal system. Despite being built from tiny grains of soil, these nests are strong and durable. Certain species, like leafcutter ants, have developed farming practices, cultivating fungus gardens as their primary food source.

An ant colony is collective intelligence in action, where each ant plays a small role. Still, the coordinated effort of the entire colony leads to the creation of complex, efficient, and functional structures and society. At some stage during their development over millions of years, there was an emergence of sterile worker ants dedicated to the queen and brood (eusociality organisation), which is critical for larger colonies and efficient resource acquisition.

Ants have a highly developed and sophisticated communication system, using chemical signals (pheromones), which allows complex tasks

Ant Nest/Hill **Ant Trail**

Fig. 17.10. (Left) Ant Nest/Hill and (Right) Ant Trail (Credit: Wikipedia).

like trail-laying, food sharing, and colony defence. The same signals are also used to decide to feed the larvae differing amounts of nutrients to make them into the queen, drones, or workers by giving expression to different aspects of their genes, even though they (mostly) share the same DNA, like in the case of honeybees of a beehive.

Honeybees are like ants. It is also important that we understand the behaviour of honeybees thoroughly as they are one of the most important pollinators for flowers, fruits, and vegetables as they transfer pollen between the male and female parts, allowing plants to grow seeds and fruits [3]. They have even developed a language of dance to precisely tell other honeybees of the location of a source of food.

Ants, honeybees, and termites are examples of highly organised societies where individuals work together for the colony's collective benefit. This is seen in their behaviours of foraging, nest-building, and caring for the young, which require cooperation and coordination among colony members.

The collaboration among ants is phenomenal and the envy of any group of people collaborating to solve a problem and find a solution. Yet, ants of different colonies do not collaborate; at best (and only in some species), they may share food from the same source. As their communication is via pheromones, which are specific to the queen and her nest, ants from different nests cannot communicate with each other! Any alien ant entering a nest is treated as an enemy, an alarm is raised, and they fight the alien ant and in general kill it. One can only speculate on how ants would have fared if different ant colonies could collaborate — they could have brought down even large animals and people/civilisations. After all, it is estimated that there are about 2.5 million ants per person on our planet!

17.6 Conclusions

We see from the above-mentioned examples that animals, birds, and insects have developed very effective strategies of cooperation, which is essential for their survival.

Many species engage in symbiotic relationships where they rely on each other for survival, e.g., birds that latch on to large animals like rhinos and buffaloes and remove the ticks and other parasites and obtain food. Cleaner fish remove parasites from larger fish, benefiting both species. The large fish recognise the cleaner fish, do not devour them, and allow free passage to them into their mouths and gill chambers for the cleaner fish to operate. Similarly, Egyptian plover birds clean crocodiles' teeth and protect them from infections by eating the meat stuck there.

The most important example of symbiotic relationships is the domestication of animals and plants, which has helped humans, plants, and animals.

Of course, we also have examples of parasites, e.g., tapeworms, that live in the intestines of humans and animals and absorb nutrients from the food the host eats, making them weak and causing other health issues. Our bodies host many bacteria, both beneficial and harmful. We also have examples of parasite plants like mistletoe and Cucusta or dodders (note the yellow string-like creepers without leaves, Fig. 17.11) which are completely dependent on their hosts for their sustenance and may even cause their death. It has been suggested that trees in a forest communicate and collaborate using the system of the "Wood Wide Web," or the underground networking of their roots, and even share nutrients and concerns!

Fig. 17.11. Cucusta or Dodder Parasitic Creepers.

References

[1] V. S. Ramamurthy, D. K. Srivastava, and S. Nayak, "*Art and Science of Managing Public Risks*" (Singapore: World Scientific Publishing Company, 2022).

[2] H. Satz, "*The Rules of the Flock: Self-Organization and Swarm Structure in Animal Societies*" (Oxford: Oxford University Press, 2020).

[3] D. K. Srivastava and V. S. Ramamurthy, "Apiculture: The Sweet Trail to Prosperity", *Science Reporter* 58 (2021) 44–48.

Chapter 18

Basic Sciences: Competition, Cooperation, Collaboration, Conflict, and Communication

"The only thing which will redeem mankind is cooperation."

— Bertrand Russel, Scientist, Mathematician, and Philosopher

18.1 Introduction

Our understanding of the world around us and the laws of nature has progressed on the wheels of competition, cooperation, collaboration, and communication — the last one creating the ecosystem which provides support to scientists and scientific endeavours. But let us not forget that these had their humble beginnings in barter or exchange.

Long before the emergence of formal nations and currencies, barter — the direct exchange of goods and services between clans and regions — laid the groundwork for interaction. It is still common in isolated communities. Today, barter in sciences manifests as the exchange of knowledge, datasets, or expertise. Researchers from different institutions often agree to share their findings or technical capabilities in exchange for access to specialised equipment or unique research samples. This fundamental act of exchange ultimately evolves into a foundational level of international collaboration. One of the finest examples is Tycho Brahe's (1546–1601 CE) extremely accurate observations which were used by

Johannes Keplar (1571–1630 CE) to arrive at his laws (though he could access them fully only after Tycho Brahe's death, as the latter guarded his data zealously), which laid the foundation for Newton's laws of gravitation.

Competition, cooperation, and collaboration are three stages of the vehicle that propels basic sciences to heights beyond our wildest imaginations, till it encounters conflicts, which have the potential to withhold or derail rapid progress.

While competition can drive individual ambition and rapid advancements, collaboration and cooperation unlock a collective scientific power that transcends national borders.

18.2 Competition

Competition is the adrenaline that spurs scientific enquiry and motivates researchers to push boundaries, refine hypotheses, and strive for groundbreaking discoveries. The reward is a breakthrough and a surfeit of dopamine, as recognition follows. We know that scientific discoveries are the cultural heritage of mankind. Competition also fosters a healthy scepticism. It drives researchers to rigorously critique each other's work, which ultimately leads to a more robust understanding of scientific phenomena and reliable measurements.

Thus, for example, completely unaware of the pioneering developments of the Kerala School, Isaac Newton and Gottfried Wilhelm Leibniz independently developed their versions of calculus around the same time in the late seventeenth century, which laid the groundwork for the mathematical framework we use today. The quarrel between Newton and Leibniz about who developed calculus first remains one of the uglier episodes of Newton's life, even though now it is accepted that both developed it independently, while its discovery by the Kerala School preceded both by a few hundred years.

The iconic double helix structure of DNA, unveiled in 1953 by James Watson and Francis Crick, was a race with a photo finish. Rosalind Franklin, Maurice Wilkins, and their colleagues were also working on X-ray crystallography of DNA and had crucial data. The pressure to be the first to publish likely fuelled Watson and Crick's intense work, ultimately leading to their groundbreaking paper, even though Franklin's X-ray crystallography data were critical to their success. Although she was not formally credited in their paper, her work played a vital role.

The theory of evolution by natural selection is another case where independent thinkers reached similar conclusions almost at the same time. Both Charles Darwin, through his meticulous observations during the voyage of the Beagle, and Alfred Russel Wallace, working independently in Southeast Asia, arrived at the concept of natural selection as the driving force behind evolution. While Darwin published his work first, Wallace's independent discovery highlights the power of competition in scientific thought. It is said that Darwin hurried to publish when he realised that Wallace was preparing to publish similar findings. It is now believed that Alexander von Humboldt had earlier started thinking along the lines of the theory of evolution.

In more recent times, we have seen that the competition between the United States and the Soviet Union during the Cold War accelerated advancements in space technology, including the development of satellite technology, human spaceflight, the moon landing, and exploration of planets. The race to sequence the human genome in the 1990s between the publicly funded Human Genome Project and a private company spurred rapid advancements in sequencing technology and bioinformatics. Competition between pharmaceutical companies drives innovation in drug development, the vaccines for COVID-19 being a very recent example. The competition between research institutes and technology companies has led to rapid advances in the field of artificial intelligence in machine learning algorithms and natural language processing. The race to develop sustainable energy sources in response to climate change has accelerated research in solar, wind, and battery technologies, with an increase in efficiency and cost-effectiveness of the technology. Advancements in techniques such as functional magnetic resonance imaging and optogenetics owe their origins to the competition between research groups to map the human brain and understand its functions. Development of genetically modified crops, precision agriculture technologies, and sustainable farming practices are driven by the competition between agricultural companies. The race to develop stronger, lighter, and more versatile materials has led to discoveries such as graphene, carbon nanotubes, and advanced ceramics. Competition between climate research institutions and international collaborations has accelerated understanding of climate change, leading to better predictive models and mitigation strategies.

Thus, in some cases, the competitive environment helped accelerate discoveries. In others, it might have created unnecessary rivalries that slowed progress.

However, unbridled and unhealthy competition can have detrimental effects. The pressure to be first can lead to hasty publications with questionable methodologies, potentially hindering scientific progress. It can also discourage researchers from sharing preliminary data, hindering collaboration and slowing down the overall pace of discovery. A very strong emphasis on being the first, which puts excessive emphasis on winning, can unfortunately take away the pleasure of research and learning, as players, athletes, students, and researchers have learned over millennia. This can only be avoided by a strong faith in human values and morality, several examples of which can be found in the history of competitive sports and athletics too.

It is also our feeling that when several people work on a particularly intractable problem, there can be near-simultaneous breakthroughs at several places, for the simple reason that scientific laws are universal. In technology, however, the breakthroughs may lead to the same solution, which may differ in detail.

18.3 Cooperation

Cooperation is a step beyond mere competition. It involves researchers acknowledging the limitations of their efforts and seeking to combine expertise and resources for mutual benefit. This can take many forms, such as sharing datasets, jointly designing experiments, or co-authoring research papers. Cooperation allows researchers to leverage the strengths of diverse perspectives and approaches, leading to richer scientific insights.

For instance, large-scale experiments requiring sophisticated equipment, vast datasets, or supercomputers often necessitate cooperation between research teams with complementary expertise. International cooperation plays a vital role here as it allows researchers from different countries with unique resources to combine forces for groundbreaking research.

India received valuable help with the design and construction of the Variable Energy Cyclotron, the Superconducting Cyclotron, and the Radioactive Ion Beam Facility at Kolkata from several countries. The Pelletron Facilities at the Tata Institute of Fundamental Research, Mumbai and the Inter University Acceleration Centre, New Delhi benefited from international collaborations, especially during the design and the fabrication of their Superconducting Linear Accelerators. The synchrotron radiation source facilities and the superconducting RF cavities developed at the Raja Ramanna Centre for Advanced Technology, Indore, similarly benefited from international collaborations. These accelerators have helped create a very vibrant community of researchers in atomic, nuclear, condensed matter, and accelerator physics, which has led to valuable contributions to these fields of work with these researchers proceeding to participate in international collaborations at CERN, FAIR, ITER, LIGO, etc.

We saw examples of international cooperative efforts involving researchers from dozens of countries when they successfully mapped the entire human genome. This groundbreaking project paved the way for personalised medicine and advancements in understanding a wide range of diseases. We also saw that the International Space Station is a shining example of cooperation in space exploration, where scientists from several countries collaborate on conducting experiments in microgravity, furthering our understanding of space science and paving the way for future human space missions.

However, this occasionally leads to a situation where the people with more specialised facilities leverage their positions to lead the effort. Many third-world researchers have expressed their frustration that their contribution is forgotten, for example, after their initial contributions to the sample collection. Again, this calls for a more equitable sharing of credit and continuous engagement.

In brief, cooperation is like a relay race, where individuals work with each other to achieve their individual goals. There is a common understanding that helping others benefits everyone in the long run. Thus, the outcome focuses on the completion of individual tasks or goals, even though they contribute to a larger objective. Cooperation also involves a less intense level of involvement, so that people might aid or provide resources to each other without necessarily needing to be in constant communication or deeply involved in each other's work. This, as we shall see, is different in the case of collaboration.

18.4 Collaboration

Collaboration goes a step further than cooperation. It builds on a shared vision and long-term commitment. Collaborative projects involve researchers working together as a unified team, often with a formalised structure and joint leadership. This fosters a sense of shared responsibility and ownership, which leads to highly impactful discoveries.

International collaborations in areas like particle physics, where giant multi-national facilities like the Large Hadron Collider are central to research, or astronomy exemplify the power of collaboration. Yet another example is the Human Microbiome Project (HMP) taken up by the National Institutes of Health (NIH), USA, in 2007, which aimed to characterise the human microbiome — the vast and diverse community of microorganisms, which number trillions and reside in and on our bodies. These microbes, primarily bacteria but also archaea, fungi, and viruses, outnumber human cells by ten to one and play a crucial role in our health and well-being. The HMP, while spearheaded by the NIH, involved researchers from various countries. This international collaboration was crucial for sharing expertise, data, and resources, accelerating progress in understanding the human microbiome.

There is also a rise in regional collaboration, especially in the emerging science powerhouses of non-western countries. Recall the tuberculosis network of Brazil, Russia, India, China, and South Africa established in 2017, which is making significant advancements in basic and applied research into the disease.

Collaborative efforts not only benefit research but can also foster international understanding and goodwill. By working side by side on scientific challenges, researchers develop a shared sense of purpose, fostering collaboration in other areas as well.

Thus, collaboration is the art of co-creating a painting or an orchestra, where the final product is a result of the combined vision and effort of everyone involved, everyone has a stake in the outcome, and the success is collective. Collaboration is a very intensive process, and people need to actively exchange ideas, brainstorm solutions, and yet work interdependently. It also involves a high level of interaction and joint decision-making. The outcome is a new, unified creation or solution jointly developed by everyone involved. A collaborative effort fosters a sense of shared ownership and achievement.

A good collaboration is best described as a well-knit team. Mahendra Singh Dhoni, one of the most successful and finest captains of the Indian cricket team, described his team in the following terms:

"Here we're, a team in which everyone has always tried to give his best and backed each other to succeed. Everyone's success is everyone else's and everyone celebrates everyone else's success. Individuals have done their best and shown their class at the right time. It's all been subsumed into a collective whole, as it were, turning the team into a match-winning combination. ..."

This embodies all that a great collaboration, at an international level, requires to succeed.

Recall the excitement and sense of achievement engulfing the entire world when man first landed on the moon, when the Large Hadron Collider started working, and when the discoveries of the Higgs boson and the gravitational waves were announced, and as the giant roar of applause reverberated across the planet. Those will remain some of the most delightful moments in the history of mankind.

18.5 Conflicts

Political, ideological, or personal conflicts have severely impeded scientific progress throughout history.

The burning of the Library of Alexandria (48 BCE), either accidentally or intentionally during Julius Caesar's siege of the city, is believed to have resulted in the loss of countless ancient texts and scientific knowledge, setting back the progress of various fields such as mathematics, astronomy, and medicine. Unfortunately, this act was not an isolated event, and it was repeated at several places worldwide, including at Bagdad, Nalanda, Vikramashila, Odantapuri, and Jhansi, resulting in the loss of valuable knowledge and setting advancement back by centuries.

Galileo Galilei's conflict with the Catholic Church over his support for the heliocentric model of the solar system is well known. His trial and condemnation by the Inquisition led to the suppression of his ideas and hindered the advancement of astronomy and physics in Europe for some time. Earlier, Giordano Bruno (1548–1600 CE) was burnt alive for his support of the Copernican system of planetary motion. Centuries later, S. Chandrashekhar (1910–1995 CE), a young researcher, was severely (brutally) criticised for his (correct) work on the theory of stars in the 1930s by the much more senior and well-established Arthur Eddington (1882–1944 CE). Chandrashekhar's work later won him the Nobel Prize (1983).

Trofim Lysenko of the USSR rejected Mendelian genetics in favour of Lamarckian inheritance (which suggests that an organism can pass on to its offspring physical characteristics that the parent organism acquired through use or disuse during its lifetime). He used his political influence and power to suppress dissenting opinions and discredit, marginalise, and imprison his critics. This ideological conflict stifled genetics research and caused significant setbacks to biological sciences in the Soviet Union. The persecution of Jewish scientists during the rise of Nazism in Germany is well known due to which many Jewish scientists fled the country.

It deprived Germany of invaluable scientific talent and impeded progress in various fields. There was a time when Einstein's theories were ridiculed as Jewish science.

In more recent times, Abdus Salam, despite his wonderful contribution to particle physics in general and the progress of science in Pakistan in particular by nurturing a large number of talented researchers, faced extreme hostility as the community to which he belonged was declared heretic and un-Islamic by Pakistan. Opposition to the teaching of Darwin's theory of evolution in several parts of the world is still not uncommon.

Once the importance of the discovery of nuclear fission (1938) in making an immensely powerful weapon was realised, research related to nuclear physics was labelled "top secret," making it inaccessible to the public and even many scientists not directly involved in the war effort. International scientific collaboration, a hallmark of scientific progress, came to a halt and physicists who might have shared their findings freely in peacetime were now restricted from doing so, and scientific progress suffered.

The anti-communist witch hunts led by Senator Joseph McCarthy during the Cold War resulted in the persecution and blacklisting of many scientists and intellectuals suspected of communist sympathies in the USA, and the climate of fear stifled scientific inquiry and inhibited collaboration among researchers.

The Cold War arms race (1940–1990s) diverted resources away from civilian research and towards military technologies, as repeatedly commented upon by Abdus Salam. It slowed progress in areas such as space exploration, environmental science, and healthcare.

And yet, despite diplomatic nonrecognition, vast political differences, a long-standing trade embargo, and strict limitations on travel, US–Cuba scientific collaboration in oceanography, fisheries, climatology, etc., was jointly undertaken by scientists from the two countries. The United States and Cuba began operating a newly installed Global Positioning Satellite system in Camagüey, Cuba, designed to measure the amount of water vapour in the atmosphere above the region to be used as input to produce more accurate weather forecasts [1].

More than three centuries earlier, one of the most famous Dutch scientists, Christian Huygens (1629–1695 CE), helped establish the French Academy of Sciences in 1666. The Academy played a great role in accelerating the progress of science in France, like the Royal Society of

London on which it was modelled. And Huygens did it at a time when the French were at war with the Dutch.

We have indicated the delays triggered by the Ukraine war of several international projects. This has led to the near-global cessation of collaboration with Russian scholars — in everything from humanities research to climate science in the Arctic. Scientific collaboration between China and the US also keeps blowing hot and cold, especially in fields like microelectronics and quantum computing because of national security concerns on both sides.

There is also the question of sanctions, which have been imposed on different countries from time to time due to political differences or conflicts. While one may somehow justify the ban on sharing of dual-use technologies, which could be used by the country that is banned, the ban on exchange and collaboration in fields like basic sciences, vaccines, diseases, and climate change becomes difficult to justify.

International collaboration is not without its challenges. Intellectual property disputes, competition for funding, and differing research priorities can lead to conflict. Additionally, national security concerns and political tensions can create barriers to collaboration. For example, restrictions on data sharing or limitations on researcher mobility due to political disagreements can significantly hinder scientific progress.

The ideal state for international basic science collaboration lies not in the dominance of any single force, but in a dynamic equilibrium between them. Healthy competition can drive ambition, while cooperation and collaboration pave the way for groundbreaking discoveries. Recognising the potential for conflict is crucial, and effective conflict resolution mechanisms are vital for navigating these challenges.

By promoting a culture of openness, respect, exchange (of ideas, expertise, and resources), and shared scientific pursuit, the international basic science community can harness the power of all four forces, propelling scientific progress towards a future filled with exciting discoveries and advancements.

18.6 Communication

Effective communication, clear agreements on authorship and intellectual property, and a commitment to mutual respect are crucial for navigating conflicts. International scientific bodies and national science academies can play a vital role in fostering dialogue and establishing frameworks that encourage collaboration despite political or cultural differences.

Effective and honest communication is the key to avoid conflict and inculcate a scientific temper and appreciation of sciences, which translates into support for science and scientists. Effective science communication also helps create a scientifically literate populace and scientific temper, which are so important for the progress of science [2, 3]. Scientists across the world should and must develop a sincere approach to science outreach, as their survival is guaranteed by a society that is alert to scientific research. It is our experience that scientists have often expressed regret for their failure in such attempts. We add the following two comments in connection to this.

A very popular and legendary poet of the Persian and Urdu languages, Mirza Ghalib (1797–1869 CE), while lamenting about his "beloved" not understanding and reciprocating his feelings, prayed, "Dear God, if you cannot give more of 'heart' (*ability to comprehend*) to my beloved, give more of language (*skills of expression*) to me."

Recall the famous comments of Joseph Pulitzer (1847–1911 CE), the reputed newspaper publisher and editor, which have remained a guiding light for generations of authors:

> "Put it before them briefly so they will read it, clearly so they will appreciate it, picturesquely so they will remember it, and above all, accurately so they will be guided by its light."

Reaching out to the masses is essential for sustained and enthusiastic support for scientific research, and communication is the only means at our disposal for this.

18.7 Conclusions

Given the discussions in the preceding chapters, we conclude that science being universal benefits immensely from collaborations beyond national boundaries, within which mankind has divided itself due to historical reasons. Not only are the laws of science universal but the language of science is also universal, and scientists of two countries can still communicate when all other avenues of communication and discourse break down due to a misunderstanding or a conflict. Mankind must keep this channel alive and active.

References

[1] See e.g., G. Machlis, *et al.*, "US-Cuba Scientific Collaboration: Emerging Issues and Opportunities in Marine and Related Environmental Sciences", *Oceanography* (Special Issue on Internal Waves) 25 (2012) 227–231; G. A. Ronda-Pupo, "Cuba — U.S. Scientific Collaboration: Beyond the Embargo", *PLoS One* 16(7) (2021) e0255106. DOI: 10.1371/journal.pone.0255106; R. Anthes *et al.*, "Cooperation on GPS Meteorology between the United States and Cuba", *Bulletin of American Meteorological Society* 96 (2015) 1079–1088, https://doi.org/10.1175/BAMS-D-14-00171.1.

[2] V. S. Ramamurthy, D. K. Srivastava, and S. Nayak, "*Art and Science of Managing Public Risks*" (Singapore: World Scientific Publishing Company, 2022).

[3] S. Mohanti, "Nehru's Vision of Scientific Temper", *Journal of Scientific Temper* 4 (2016) 154–166.

Chapter 19

Science Diplomacy: Emerging World Order, Impending Tech War, and Challenges to Science Diplomacy

19.1 Introduction

The discussion in previous chapters has been with the wishful assumption/scenario of a near-ideal world, disturbed perhaps only occasionally by local or widespread conflicts between countries. Here, we proceed to have a reality check and a hard look at how science diplomacy can shape international relations and how (or rather how far) science cooperation can tackle global challenges and build bridges between countries. We have seen that the foundations for international collaboration on science and technology projects like the International Space Station (ISS) and the International Thermonuclear Experimental Reactor (ITER) originated in science diplomacy.

First of all, the idea of science diplomacy, which witnessed a forward push during the first two decades of the twenty-first century, is likely to face new challenges in the post-Russia–Ukraine War period. The ongoing war in Ukraine is already changing the nature of the world order; it is expected to change further in the 2020s. The Ukraine War and the change in the world order are likely to have implications for science diplomacy. We discussed this briefly in connection with the Facility for Antiproton and Ion Research (FAIR).

Second, during the last decade, there has been a "Tech War" brewing at the global level. During the Cold War from the 1960s to the 1980s and the Trade War that followed since the 1990s, science diplomacy continued. Countries collaborated during the peak of political differences. The next phase of the world order is likely to have a "Tech War," with implications for science diplomacy.

Third, as part of the emerging international equation, the US, Europe, and China are likely to influence other countries interested in using science and technology (S&T) to develop their economies and infrastructure. This will, in turn, impact the "Tech War."

We proceed to discuss these three important issues in this chapter.

19.2 The Idea of Science Diplomacy During the Cold War

According to an essay published in 2023, science diplomacy became a global phenomenon in the 1970s due to the "globalising impulses" of the decade [1]. According to the essay, "The growing number of newly independent countries represented in the United Nations (UN), the emergence of regional geopolitical powers and alliances, and the increasingly important role of science and technology in tying distant economies and societies together have all profoundly shaped science diplomacy, just as they shaped changes in the global economy, popular culture, and geopolitics [1]." Undoubtedly, during the early phase of globalisation, as a part of the phenomenon, these multiple domains witnessed increasing interaction, including the S&T field. However, the idea of science diplomacy came about much earlier, though it was not pursued as "Science Diplomacy."

Some authors have been linking the origin of the idea of science diplomacy to the address made by Dwight Eisenhower, the then US President, to the 470th Plenary Meeting of the United Nations General Assembly on 8 December 1953. Addressing Vijaya Lakshmi Pandit from India, who was then the President of the UN General Assembly, President Eisenhower said, "The United States knows that if the fearful trend of atomic military build-up can be reversed, this greatest of destructive forces can be developed into a great boon, for the benefit of all mankind. The United States knows that peaceful power from atomic energy is no dream of the future. The capability, already proved, is here today. Who can doubt that, if the entire body of the world's scientists and engineers had adequate amounts of fissionable material with which to test and develop their ideas, this capability would rapidly be transformed into universal, efficient and economic usage? [2]"

In this speech, Eisenhower proposed, "The governments principally involved, to the extent permitted by elementary prudence, should begin now and continue to make joint contributions from their stockpiles of normal uranium and fissionable materials to an international atomic energy agency. We would expect that such an agency would be set up under the aegis of the United Nations. The ratios of contributions, the procedures and other details would properly be within the scope of the "private conversations" I referred to earlier." Making a pitch for the creation of an atomic energy agency at the international level, President Eisenhower said in the speech, "The atomic energy agency could be made responsible for the impounding, storage and protection of the contributed fissionable and other materials. The ingenuity of our scientists will provide special safe conditions under which such a bank of fissionable material can be made essentially immune to surprise seizure [3]."

This led to the birth of the International Atomic Energy Agency (IAEA), which could be seen as the first international collaboration on S&T. In one of its notes, the International Atomic Energy Agency says, "The Agency was set up as the world's "Atoms for Peace" organisation within the United Nations family. From the beginning, it was given the mandate to work with its Member States and multiple partners worldwide to promote safe, secure and peaceful nuclear technologies."

According to its statute, "The Agency shall seek to accelerate and enlarge the contribution of atomic energy to peace, health and prosperity throughout the world. It shall ensure, so far as it is able, that assistance provided by it or at its request or under its supervision or control is not used in such a way as to further any military purpose [4]." Article III of the IAEA statute, referring to its functions, refers to fostering "the exchange of scientific and technical information on peaceful uses of atomic energy" and "to encourage the exchange and training of scientists and experts in the field of peaceful uses of atomic energy [5]."

While President Eisenhower's reasons for suggesting an international atomic energy agency may have been strategic and specific to American interests, the idea sowed the seeds of science diplomacy in the early 1950s. During the Cold War, despite their differences, the US and the former Soviet Union collaborated on specific areas related to S&T. It could very well be a part of Cold War diplomacy, but it gave space for the emergence of S&T collaboration subsequently. Olga Krasnyak in an essay published in 2019 in *American Diplomacy* argues, "Cold War diplomacy produced critical armament control treaties — the Non-Proliferation Treaty (NPT), the Anti-Ballistic Missile Treaty (ABM), and the Intermediate-Range Nuclear Forces Treaty (INF) to name but a few. In all these cases, scientific and technical people-to-people exchanges were critical for enabling communication and dialogue." She also says, "People-to-people exchanges began with the 'Lacy-Zarubin Agreement' signed in 1957. The agreement reinforced scientific contacts in space and medicine and laid out the terms for exchanges between the two parties in what evolved into an Inter-academy Scientific Exchanges agreement, renewable every two to three years. In the following decades, the agreement remained unbroken. The latest agreement was prepared in 1984 but finalised two years later. The exchanges were symmetrical. If an American delegation visited the Soviet Union for a joint conference, then a Soviet delegation visited the US for a similar joint conference [6]."

19.3 End of Cold War and the Emergence of Tech War

Adverse relations between the US and the Soviet Union, after World War II, especially during the 1950–1980s, were described as part of the "Cold War." The focus on describing the world order through the Cold War prism declined during the 1990s. Following the disintegration of the Soviet Union, the 1990s witnessed a discussion on the world order being steered away from the Cold War prism. Defined as a unipolar moment, with the US being the sole superpower, there was an expectation that international relations would not be adversarial.

In the 1990s, there was also a widespread expectation that there would be increased cooperation at the global level, including on issues relating to S&T. Developments relating to "The World Academy of Sciences" would highlight this expectation. Originally founded in 1985 as the "Third World Academy of Sciences," it became "The Academy of Sciences for the Developing World" in 2004, and then later "The World Academy of Sciences for the Advancement of Science in Developing Countries." Based in Trieste, The World Academy of Sciences (TWAS) functions as a part of the United Nations Educational, Scientific and Cultural Organization. According to TWAS, in 1991, UNESCO became responsible for "administering TWAS funds and personnel [7]." Now referred to more as "The World Academy of Sciences," TWAS was expected to play a larger role during the 2000s and 2010s in increasing international cooperation. According to TWAS, its objectives were the following [7]: Recognise, support, and promote excellence in scientific research in the developing world; respond to the needs of young scientists in countries that are still developing in science and technology; promote South–South and South–North cooperation in science, technology, and innovation; and encourage scientific research and sharing of experiences in solving major challenges facing developing countries. TWAS established regional partners in five countries — Egypt, Brazil, China, South Africa, and India.

Thus began a new era of structured science collaboration after the 1990s.

However, this period was short-lived, as a new equation entered global politics with China's rise during the 2000s. While the US and Europe looked at the post-Soviet Union world order as a unipolar moment in global politics, it was short-lived. While a section expected China's emergence to be a "peaceful rise," another section started pushing a thesis referred to as the "Thucydides Trap." Almost two decades ago, in 2005, the Chinese scholar Zheng Bijian projected that the growth of China to a great power status would be a "Peaceful Rise [8]." According to Bijian, "China does not seek hegemony or predominance in world affairs. It advocates a new international political and economic order, one that can be achieved through incremental reforms and the democratisation of international relations. China's development depends on world peace — a peace that its development will in turn reinforce."

On the contrary, the Thucydides Trap thesis refers to an assertion made by General Thucydides (ca. 460–ca. 400 BCE) during the Peloponnesian War (431–404 BCE) between Athens and Sparta. Thucydides predicted a conflictual relationship between Sparta and Athens, stating that the former as a rising power would challenge the latter. A recent thesis authored by Graham Allison predicts, "… as China challenges America's predominance, misunderstandings about each other's actions and intentions could lead them into a deadly trap…The past 500 years have seen 16 cases in which a rising power threatened to displace a ruling one. Twelve of these ended in war [9]." The Graham Allison thesis about the Thucydides Trap is being debated across the world.

Where does the China–US relationship stand in the 2020s, and how would it evolve during the next decade and shape the global order? The idea of the "peaceful rise" of China has been increasingly challenged during the last two decades. China's recent responses to the wars in Ukraine and Gaza and its interventions across the world from the South China Sea to Venezuela do not provide a convincing argument that its rise is peaceful. On the other hand, China's challenge to the US and Europe has not led to a military confrontation yet.

Should the Thucydides thesis and the subsequent "Trap" that is being discussed be relooked? China may not challenge the US and Europe militarily but may do so in other fields — economy and S&T. China's economic rise and S&T rise during the last three decades have been documented. According to an analysis by the Brookings Institution in the US, "China's rapid technological advances are playing a leading role in contemporary geopolitical competition. The United States, and many of its partners and allies, have a range of concerns about how Beijing may deploy or exploit technology in ways that challenge many of their core interests and values. While the US has maintained its position as the technologically dominant power for decades, China has made enormous investments and implemented policies that have contributed significantly to its economic growth, military capability, and global influence. In some areas, China has eclipsed or is on the verge of eclipsing, the United States — particularly in the rapid deployment of certain technologies [10]." Scholars and commentators have started writing about this. They are also discussing how in certain sectors, for example, semiconductors, artificial intelligence (AI), and space, the US and China would compete to ensure that they become the technological superpower, at the cost of building an adversarial relationship with the other [11].

The emergence of the "Tech War" narrative should fit the above-mentioned description. It describes a phase of adverse relations between the US and China, attempting to make use of technology. China is emphasising S&T to challenge the domination of the US in the post-Cold War unipolar world. On the other hand, the US also wants to use S&T to prevent China from challenging its hegemony. The US wants to contain China by containing the latter's access to S&T. According to an analysis, "Under Presidents Trump and Biden, the maintenance of US technological primacy has emerged as the leading dimension of US policy towards China. Beijing's tech development is viewed, in Washington's eyes, as a security threat of the highest order [12]." The US believes that some aspects of S&T constitute strategic technologies, "where Chinese advancement or influence could most threaten America's national security and economic interests [13]." These perceptions emerge from an apprehension that China can make use of its technological advancements to challenge the US supremacy [14]. The US wants to pursue "technological decoupling" with China. Washington also wants its allies, especially in Europe, to pursue a similar strategy vis-à-vis China.

The strategies pursued by the US and its allies in Europe vis-à-vis Huawei will underline the nature of the Tech War that is emerging.

According to Huawei, it is "a leading global provider of information and communications technology (ICT) infrastructure and smart devices," having 207,000 employees in over 170 countries, and "serving more than three billion people around the world [15]." According to the Huawei portal, the company is "committed to bringing digital to every person, home and organisation for a fully connected, intelligent world." However, for the US and its allies, Huawei is a part of China's strategy to get embedded with the strategic technologies of the West. There is a strong apprehension that the government in China could ask Huawei to share information and create backdoors.

This fear is prevalent and can be seen in multiple reports prepared in the US and Europe. The earliest report was as early as 2012. Titled, "Investigative Report on the US National Security Issues Posed by Chinese Telecommunications Companies Huawei and ZTE," the report concluded that "the risks associated with Huawei's and ZTE's provision of equipment to US critical infrastructure could undermine core US national-security interests." The report also recommended that the US intelligence community should be careful about the threat posed by telecommunications companies like Huawei and ZTE. It also recommended that the US Committee on Foreign Investment "block acquisitions, takeovers, or mergers involving Huawei and ZTE given the threat to US national security interests." The report also recommended that the "US government systems, particularly sensitive systems, should not include Huawei or ZTE equipment, including component parts. Similarly, government contractors — particularly those working on contracts for sensitive US programmes — should exclude ZTE or Huawei equipment in their systems [16]."

The above-mentioned report looked not only at the threat posed by telecommunications companies like Huawei but also at the influence that China had over them. According to the report, "Chinese intelligence collection efforts against the US government are growing in "scale, intensity and sophistication." Chinese actors are also the world's most active and persistent perpetrators of economic espionage. Thirteen US private sector firms and cybersecurity specialists have reported an ongoing onslaught of sophisticated computer network intrusions that originate in China and are almost certainly the work of, or have the backing of, the Chinese government. Further, Chinese intelligence services, as well as private companies and other entities, often recruit those with direct access to corporate networks to steal trade secrets and other sensitive proprietary data [17]."

A similar threat was observed later during Donald Trump's presidency. According to a report published by the Congressional Research Service, "Senior officials in the Trump Administration asserted that Huawei's products present an inherent security threat because the Chinese government can force Huawei to share confidential information or create "backdoors" by which the Chinese government could access Huawei systems [18]." Another independent US analysis by a leading think tank, the Council on Foreign Relations (CFR), also came to the same conclusion. A CFR analyst argued, "There are also concerns that Huawei's 5G infrastructure could contain backdoors that allow the Chinese government to collect and centralise massive quantities of data and give Beijing the necessary access to attack communications networks and public utilities. In 2022, a Federal Bureau of Investigation (FBI) investigation found that Huawei equipment can be used to disrupt US military communications, including those about the US nuclear arsenal [19]."

Clearly, the differences between the US and China over Huawei should underline the idea of a brewing Tech War. The US, Europe, and their allies are increasingly worried about China's S&T rise. The era of China known for its cheap products is over; today, China has become a technological powerhouse in manufacturing related to sensitive electronics and heavy machinery. China has also invested in electronic vehicles and high-speed trains. The rise of China's electronic vehicles should underline the emergence of a strong automobile industry. BYD and Nio are today challenging Tesla, and are believed to be ahead of their Japanese, Korean, and European counterparts [20]. During the recent Auto Show in Beijing, China's automobile industry introduced new models [21]. The rise of EVs is aided by another important breakthrough for China in battery technology, in terms of storage and speed. In the aviation sector, China has recently unveiled C919, a passenger aircraft. The Commercial Aircraft Corporation of China, referred to as COMAC, which manufactures C919, has been reported to have received more than 1000 orders. There are already reports that C919 will compete with the Airbus and

Boeing shortly [22]. In the maritime sector, in May 2024, China's third aircraft carrier — Fujian — completed its first sea trial [23]. In polar infrastructure, China has already built two ice-cutters that will enable ship movement in the Arctic [24]. In 2023, it started building its third polar vessel. In the space sector, in May–June 2024, China landed its Chang'e 6 on the far side of the moon to bring back rock samples [25]. China has been operating its space station, Tiangong, since 2022.

Clearly, China has taken giant steps in the S&T sector — from EVs to aircraft and from polar carriers to moon missions. This has caused deep apprehension in the US and Europe. Though the Soviet Union posed a challenge during the Cold War, it was more on the military field; the US and Europe had an edge in S&T over Russia. However, the threat from China is not in the military domain; rather, Beijing challenges the S&T supremacy that the West has enjoyed since World War II or even before that. The Thucydides Trap in the 2020s will not be seen in the military domain, but in S&T.

The American response to Chinese Huawei should underline the new Tech War that is emerging. Huawei may only be an example; as the Tech War deepens, it is likely to expand to broader areas, especially relating to emerging and critical technologies that would further help the big powers maintain their advantage and supremacy in other fields.

19.4 The Ukraine War and the Challenge to S&T Collaboration

Russia's invasion of Ukraine in 2022 and the ongoing war in that region have further widened the Tech War. While the Tech War was slowly evolving against China, there has been much development over the last decade, and since 2022, Russia has been under increasing sanctions in different domains, including S&T. Institutional collaborations between governmental institutions are now banned between Russia and Europe and Russia and the US. Universities in Europe and the US have also imposed restrictions on working with researchers from Russia.

The European Union wants to use technology sanctions to force Russia to abandon its Ukraine ambitions. The sanctions, according to an official, "will cover an area of export where the EU has a very clear competitive edge, and it will not be easy for Russia to substitute these technologies, at least in the short and medium term. Obviously, this will have a significant knock-on effect on Russian capacity of exporting refined oil to the European Union [26]." The same report quoted the EU Commission President Ursula von der Leyen. She is reported to have said, "We want to cut off Russia's industry from the technologies desperately needed today to build a future," and that it is aimed "to seriously degrade Russian economy in all areas in the future." The US said that, along with its partners, it "will effectively be cutting off more than half of all high-tech imports going into Russia. This includes curbing Russia's access to advanced semiconductors and other foundational technologies that Russia needs to diversify and modernise its economy." The EU Commission also suspended projects funding research and innovation in Russia. Countries have asked their universities to cut off research and innovation ties with Russia.

As Russia faces sanctions, it is increasingly relying on China and Iran for support. The war in Ukraine and the subsequent sanctions the West imposed on Russia have made Moscow more dependent on China. On the other hand, China is also closely watching how technology is playing an increasing role in the Ukraine War.

China will want to invest more in science and technology, while the West will want to prevent Beijing's ambition. Drones, satellites, and AI have become three crucial components in the Ukraine War; China will want to build its capacity further in these three fields, while the West will want to upset any developments through new restrictions.

How will the China–Russia collaboration shape the larger Tech War? How China will invest further in its S&T capabilities will have its own ramifications on the international order in the 2020s and the coming decades. Even if the Ukraine War comes to an end, the Tech War will continue. The Thucydides Trap will not take place on military ground; it will have a technology angle.

19.5 Conclusions

To conclude, we note that the global order today is in flux. During the Cold War, despite the political and military tensions between the then superpowers — the US and the Soviet Union — there were certain positive developments through collaboration in science and technology. The earlier part of this chapter, and earlier chapters of this volume, referred to multiple collaborations between the US and the Soviet Union during the Cold War.

During the 1990s, after the disintegration of the Soviet Union, collaboration between the countries witnessed a peak in various issues including science and technology. The idea of science diplomacy received a boost, and there were collaborations on multiple issues. Collaboration took place across the board — within the South, between North and South, and between East and West. Besides international forums supported by UNESCO and other similar institutions, forums like BRICS (Brazil, Russia, India, China, and South Africa) and the SCO (Shanghai Cooperation Organisation) broadened the idea and took ownership of S&T collaboration. Individual countries, for example, India, have invested in increasing collaboration in S&T. There are multiple projects involving the Department of Science and Technology and other institutions that have engaged in science diplomacy. An example of science diplomacy led by India in the Global South is COVID and supply of vaccines etc., which has been discussed elsewhere in this volume.

Will such science diplomacy continue through the 2020s into the 2030s? The answer is yes and no. This chapter looked into the emerging "Tech War," especially between the big actors — the US, China, and the EU. In areas considered "critical" and "strategic," there will be deep divisions. For example, there will be serious competition in domains relating to semiconductors, AI, cyberspace, space, energy, and critical minerals. As explained earlier, the competition might even be damaging and could pose a challenge to collaborations in science and technology. On the other hand, countries like India will lead scientific collaborations, especially in the Global South, and pressurise the North to share their technology to achieve the concept of "One World, One Science." India's chairmanship of G20 provided exactly that kind of leadership. India has also demonstrated the same enthusiasm in pursuing the "One World One Sun" initiative to build a global solar grid.

References

[1] S. Robinson *et al.*, "The globalization of science diplomacy in the early 1970s: a historical exploration," *Science and Public Policy* 50(4) (August 2023) 749–758, https://doi.org/10.1093/scipol/scad026.

[2] Address by Dwight D. Eisenhower, President of the United States of America, to the 470th Plenary Meeting of the United Nations General

Assembly (8 December 1953), https://www.iaea.org/about/history/atoms-for-peace-speech.

[3] Address by Dwight D Eisenhower, *ibid.*

[4] "History," *International Atomic Energy Agency*, https://www.iaea.org/about/overview/history.

[5] "Statute," *International Atomic Energy Agency*, https://www.iaea.org/sites/default/files/statute.pdf.

[6] O. Krasnyak, "How US-Soviet scientific and technical exchanges helped end the cold war," *American Diplomacy* (November 2019), https://americandiplomacy.web.unc.edu/2019/11/how-u-s-soviet-scientific-and-technical-exchanges-helped-end-the-cold-war. According to the author, "The exchanges had a positive impact on bilateral relations and wider implications in world politics. State leaders — Nixon and Brezhnev in the early years of détente and Reagan and Gorbachev by the end of the Cold War — whether driven by domestic or international politics or both, ultimately aimed to improve relations and sustain a peaceful coexistence."

[7] The World Academy of Sciences, https://twas.org/twas-voice-science-south.

[8] Zheng Bijian, "China's "Peaceful Rise" to Great-Power Status," *Foreign Affairs* (September/October 2005).

[9] G. Allison, "The Thucydides Trap," *Foreign Policy* (9 June 2017), https://foreignpolicy.com/2017/06/09/the-thucydides-trap/.

[10] T. Chhabra, R. Doshi, R. Hass, and E. Kimball, "Global China: Technology," *Brookings* (April 2020), https://www.brookings.edu/articles/global-china-technology/.

[11] See C. Miller, "*Chip War: The Fight for the World's Most Critical Technology*" (Scribner, United States, 2022); H. Jones, "*When AI Rules the World: China, the US, and the Race to Control a Smart Planet*" (Bombardier Books, Post Hill Press, New York, 2022).

[12] M. Ryan and S. Burman, "The United States–China 'tech war': Decoupling and the Case of Huawei," *Global Policy* (29 February 2024), https://doi.org/10.1111/1758-5899.13352.

[13] J. Bateman, "*US-China Technological Decoupling*" (Washington, DC: Carnegie Endowment, 2022). https://carnegieendowment.org/files/Bateman_US-China_Decoupling_final.pdf.

[14] Bateman explains this in his essay as follows: "Washington fears that Beijing can leverage technological linkages to steal secrets, spread disinformation, surveillance of dissidents, hold US infrastructure hostage, and leap ahead in economic competition, among other threats. As a result,

US officials of both parties have sought to substantially — though not completely — reduce the flow of technology products, services, and inputs to and from China." See J. Bateman, "*US-China Technological Decoupling*" (Washington, DC: Carnegie Endowment, 2022).

[15] "Who is Huawei," https://www.huawei.com/en/corporate-information.

[16] US House Permanent Select Committee on Intelligence, "Investigative Report on the US National Security Issues Posed by Chinese Telecommunications Companies Huawei and ZTE" (8 October 2012). https://intelligence.house.gov/sites/intelligence.house.gov/files/documents/huawei-zte%20investigative%20report%20(final).pdf.

[17] US House Permanent Select Committee on Intelligence (8 October 2012).

[18] "Huawei and US law," *Congressional Research Service* (23 February 2021), https://crsreports.congress.gov/product/pdf/R/R46693.

[19] See N. Berman, L. Maizland, and A. Chatzky, "Is China's Huawei a Threat to US National Security?" (8 February 2023), https://www.cfr.org/backgrounder/chinas-huawei-threat-us-national-security.

[20] "China's Electric Cars Keep Improving, a Worry for Rivals Elsewhere," *New York Times* (1 May 2024), https://www.nytimes.com/2024/05/01/business/china-electric-vehicles.html.

[21] "Beijing Auto Show 2024 Wraps up with a Bang, as 892,000 Visitors Chase New Technologies," *Global Times* (5 May 2024), https://www.globaltimes.cn/page/202405/1311669.shtml.

[22] "Can China's New Plane Compete with Airbus and Boeing," https://skift.com/2024/02/25/can-chinas-new-plane-compete-with-airbus-and-boeing/.

[23] "China's 3rd Aircraft Carrier Completes Maiden Sea Trials, to Greatly Empower PLA Navy," *Global Times* (8 May 2024), https://www.globaltimes.cn/page/202405/1311875.shtml.

[24] "China's Polar Icebreaker Conducts Sea Trials, to Boost Country's Research Capability in Polar Studies," *Global Times* (7 February 2023).

[25] "China Lands on Moon's far Side in Historic Sample-Retrieval Mission," *Reuters* (2 June 2024); "China's Chang'e-6 Robot Lands on Moon's far Side," *BBC*, https://www.bbc.com/news/articles/cxeejp0y2pjo.

[26] "US, EU and Allies Rachet up Moves to Isolate Russia Technologically," *Science Business* (25 February 2022). https://sciencebusiness.net/news/us-eu-and-allies-rachet-moves-isolate-russia-technologically.

Chapter 20

Epilogue

"Alone we can do so little; together we can do so much."

— Helen Keller

"If you want to go fast, go alone; if you want to go far, go together," states an African proverb. This simple but profound statement lucidly sums up everything we have discussed while covering the phenomenal scientific and technological advances in various fields. We believe that international collaboration in science is the most natural framework, as scientific results benefit and enlighten humanity, and scientific insights benefit everyone and apply to every material activity in the universe.

Collaboration was what the Neanderthals indulged in when they spread out of Africa in wave after wave, over generations. A small band of them worked together, relied on each other, and used their crude tools of stone, wood, bone, or flint to bring down the much larger and fearsome mammoths which weighed several tonnes and provided enough food for them to last weeks. This coordinated effort at "surround and kill" allowed them to take down the much larger prey, share the workload, and increase their chances of success and survival. Neanderthals lived in Europe and Asia for over 350,000 years, but disappeared around 40,000 years ago, perhaps due to climate change or possibly as they were overwhelmed by larger groups of Homo sapiens who had also started fanning out of Africa about 60,000 years ago, who were perhaps even better at adapting and collaborating, and who went on to populate the entire globe. There have also been suggestions that Neanderthals were partly subsumed by Homo sapiens.

As the Homo sapiens spread far and wide and evolved somewhat differently to adapt to the specific local conditions of climate and surroundings over millennia, distinct groups formed which were perhaps forerunners of nations. Let us also not forget that when Homo sapiens appeared on the scene, they encountered flora and fauna that had already evolved over millions of years to suit and take advantage of their environments. The different groups/nations bartered, exchanged, and traded goods found naturally or produced/developed in their respective regions. These were the earliest examples of "collaborations" across the boundaries of "nations" and benefitted all concerned. Barter economy is still not uncommon in far-flung areas and isolated communities. We have seen that archaeological evidence suggests the long-distance movement of obsidian, a volcanic glass used for making tools, as early as 100,000 years ago, confirming the existence of organised exchange networks even among the earliest nomadic groups. Moreover, the shared campsites and the presence of tools made from materials not locally available point to interactions and exchanges between different groups for essential resources.

We know that Sumerians traded barley, dates, textiles, and metalwork with neighbouring regions like Egypt and the Indus Valley. Around the same time, Egyptians traded gold, linen, papyrus, and other goods via the Nile with civilisations from East Africa and the Aegean Sea region. Similarly, the Indus Valley Civilisation people traded with Mesopotamia and other regions, exporting pottery, beads, and textiles using standardised seals, weights, and measures, facilitating long-distance trade and ensuring fair exchange. Phoenicians, who were excellent seafarers, traded purple dye, glass, wine, and olive oil throughout the Mediterranean, and established colonies like Carthage. Later, trade by Romans and Greeks and more importantly by caravans travelling along the Silk Route connected East and West. Until the sea route to India from Europe was discovered by Vasco De Gama, trade between India and Europe passed through the central Asian caravan route. This trade route stretched from China to the Mediterranean and carried goods like silk, spices, and paper. Within the historical boundaries of India, there were two major highways: Uttarapatha (the northern route) ran from present-day Kandahar through Takshashila, Lahore, Amritsar, Delhi, Kannauj, Prayag, Varanasi, and Pataliputra and ended at the port of Tamralipta on the Gangetic delta in the Bay of Bengal. Dakshinapatha (the southern route) started from

Allahabad–Varanasi and branched into two — one going towards Gujarat and the other through Kishkindha to the southern tip of India. These were already well-established routes by the sixth century BCE.

The trade along these and the Silk Road (a name given only in 1877 by the German historian Ferdinand von Richthofen, while the route had been in use for several millennia) led to the transmission of ideas, literature, religions, and technologies across vast distances, some of which we have discussed earlier. This internationalisation was alluded to by Pericles (ca. 495–429 BCE) when, during his famous Funeral Oration, he commented, "Because of the greatness of our city, the fruits of the whole earth flow in upon us; so that we enjoy the goods of other countries as freely as our own." We only need to walk into any modern garden or city today, anywhere in the world, to see how the beautiful flowers and trees which were once found only in their places of origin are now giving delight and pleasure to everyone, all over the world.

Even conquering armies and prisoners of war (mostly sold as slaves in ancient times) spread the light of knowledge, though gruesome and ghastly it was. It is a tragedy of enormous proportions that the conquerors often obliterated local knowledge.

Vigorous, organised, and large-scale collaborations across nations for scientific research are a recent phenomenon, necessitated and spurred by the complexity of problems and the vast financial, technical, and intellectual resources needed to tackle them as the solutions spiral out of the reach of any one nation. This has also had a remarkably interesting effect on the popular image of scientists.

Consider Satyendra Nath Bose. It is believed that he essentially derived his statistics while teaching a master's class at the University of Dacca a hundred years ago in 1924. He submitted his manuscript to *Philosophical Magazine*, which rejected it. Then, he sent it to Einstein, who translated it into German and submitted it for publication to *Zeitschrift fur Physik*, where it was published. The Bose–Einstein condensate, its natural corollary, could only be discovered in 1995 when, through dogged effort, a handful of scientists used laser (and magnetic evaporative) cooling to attain a temperature of less than 170 nK (nano Kelvin) in a dilute gas of rubidium atoms. A detailed study of Bose–Einstein condensates had to wait for the International Space Station (ISS) microgravity laboratory, which ensured that the condensates-undisturbed by gravity could be held together for much longer. We have seen earlier that the ISS is a glorious example of an international collaboration of several thousand engineers and scientists from several nations.

Albert Einstein, working essentially alone, predicted the existence of gravitational waves in 1916 as a ripple in the fabric of space-time triggered by the acceleration of massive objects. It took several hundred scientists from across the world another hundred years to build incredibly sensitive detectors and extremely precise monochromatic lasers to (directly) detect these waves arising from the merger of two black holes, which travelled over a billion years to reach Earth. His principle behind lasers, even though enunciated in 1917, was to take 40 years for realisation. It was to completely revolutionise almost all aspects of our everyday life and become a powerful tool for research.

Higgs boson particles were proposed more than 50 years ago by a very shy and very reclusive Peter Higgs, who shunned contact with people and liked to go on long solitary walks in the hills of Scotland. It took a massive effort lasting several decades by several thousand engineers and physicists worldwide to design and build the Large Hadron Collider and the necessary detectors of the CMS and ATLAS collaborations. The particle's discovery was announced on 4 July 2012. We saw that an earlier attempt to construct a Superconducting Super Collider by the USA to look for the Higgs boson had to be abandoned, partly because it was not a truly international effort but also because it was prohibitively expensive, which in turn threatened to soak up the funding meant for other fields.

We have seen how each of these efforts has ballooned to incorporate scientists and engineers from across the world, and sometimes even warring nations, to pool their intellectual, financial, and technical resources towards research and development, which is truly a heritage of humanity. We do not wish to repeat the many topics that we have discussed in the preceding chapters. We have seen a vibrant and sustained current of international collaboration flowing through them.

It seems to us that the sheer large number of scientists across the world and the massive and speedy information exchange assisted by the "World Wide Web" (which itself was necessitated for ease of collaboration and sharing of data in basic research) may perhaps have minimised the number and scope of "simpler" problems (low hanging fruits?) that could be tackled by a scientist working alone or in small groups.

However, human ingenuity and genius cannot be denied, which will, can, and must surely prove us wrong. Discoveries in astronomy, dark matter, dark energy, fusion energy research, climate change mitigation, rapidly changing weather forecasts, biodiversity (and its loss), depths of oceans, tackling pollution of water bodies, heat/salinity/drought/flood/pest-resistant crops in a rapidly warming world, large-scale carbon sequestration, genetics, bioinformatics, quantum computing, personalised medicine, brain research, consciousness and memory, vaccines, etc., will need massive resources, cutting-edge technologies, and a vast pool of brilliant manpower, ably assisted by powerful tools like the artificial intelligence and machine learning. Some topics of research, especially those involving genetic modification and cloning, will raise complex issues including those of ethics, which will need to be settled by international consultations.

With billions of galaxies, their suns, and their planets [1] open to study with the help of most modern telescopes, some of which we have discussed, contact with, or at least discovery of, another planet which could (have) support(ed) life cannot be categorically ruled out. This opens up how-so-ever-distant or even impossible dreams of collaboration across the universe. It may even help us unravel the mysteries of the universe, its laws, and and its formation at the microscopic level, at the macroscopic level, and at the megascopic level!

We know for sure that bringing together researchers from different countries and backgrounds with diverse perspectives and expertise leads to more innovative and creative solutions as this diversity challenges existing assumptions and biases and leads to a broader understanding of problems and potential solutions.

By combining different approaches and methodologies, international collaborations can lead to more effective problem-solving and the development of breakthrough technologies. This can be particularly beneficial for tackling complex challenges that require global solutions, such as climate change and pandemics.

We know from our experiences at CERN, ICTP, ISS, FAIR, ITER, etc., that collaboration fosters cross-cultural understanding and appreciation and leads to better communication and teamwork, as researchers learn from diverse cultural approaches to research and innovation.

The numerous successful collaborations discussed in the preceding chapters point to the building of strong personal connections, trust, and bonds, which are crucial for successful collaborations. These are in turn facilitated by cultural understanding and mutual appreciation.

We must be realistic enough to realise that nay-sayers will raise the bogey of unequal sharing, protecting intellectual property, and some irreconcilable differences in cultures, languages, and conflicts. Yet, we feel that these can be tackled by clear equitable agreements and frameworks arrived at by discussions.

We are also sure that the mutual understanding, respect, and effective communication fostered by international collaborations will contribute to a more equitable and prosperous future for all.

We once again recall the teachings of Lord Buddha, alluded to in the beginning of this book, where He talks of the futility of wars/conflicts as they lead to all-round pain and suffering, as both the victor and the vanquished suffer — one from hatred and the other from pain.

Also recall the famous incident from the life of Confucius (ca. 551–ca. 479 BCE), where he asked his disciples about their aims in life. The reply [2] which had pleased him most was from Yan Hui, or Yan Yuan (c. 521–481 BCE), who said, "... I would diffuse among the people instructions on the five great points (*sic.* Benevolence and humanity; Honesty and uprightness; Knowledge and wisdom; Faithfulness and integrity; Propriety, Good manners, and Worship), and lead them on by the rules of propriety and music, so that they should not care to fortify their cities by walls and moats but would fuse their swords and spears into implements of agriculture."

Avoidance of conflict has remained a recurring theme in all traditions. The phrase "... they shall beat their swords into ploughshares, and their spears into pruning hooks; nation shall not lift sword against nation, neither shall they learn war anymore" (Isaiah 2:3–4) has been repeatedly invoked on various occasions, e.g., the setting up of the IAEA (International Atomic Energy Agency) and the Megatons to Megawatts Program for the dismantlement of nuclear warheads.

Prof. Abdus Salam repeatedly lamented the amount of money spent on the development and deployment of aircraft carriers, submarines, destroyers, tanks, fighter planes, bombers, sophisticated and destructive weapons, and their delivery systems — many of which are never fired. Just a small fraction of their cost could sustain efforts to provide education and sustenance to all, and all the development activities of the world.

This theme is the guiding principle for the sculpture "Let Us Beat Swords into Ploughshares" by Yevgeny Vuchetich (1908–1974), gifted to the United Nations in 1959 (Fig. 20.1).

At the turn of the present millennium, Greenwood and North [3] remarked, "The scientific community has been captivated, perplexed, and excited by the array of questions and issues on the horizon of

Fig. 20.1. "Let Us Beat Swords into Ploughshares" by Yevgeny Vuchetich (Credit: Wikipedia).

scientific discovery. Each scientific discipline is considering which topics warrant study and public attention. The disciplines, as well as science, have enjoyed tremendous recent victories in discovery, innovation, and graduate education that have paid great dividends to the economy and made meaningful contributions to the state of knowledge. There are many more successes to celebrate as we look forward to a new century and new millennium of scientific innovation." We only add that international collaboration has played and will continue to play a pivotal role into these successes.

His Holiness the Dalai Lama once narrated a very touching incident, which tells of the strong conviction of His followers in the teachings of Lord Buddha. A monk released from a Chinese prison, where he was brutally tortured for 18 years, came to see Him. At some stage, the monk said that he occasionally faced some danger during his days in prison. In reply to a question from the Dalai Lama about the danger, the monk added that, many times, he was in danger of losing compassion for the Chinese! The futility of conflict and the healing and calming effects of empathy cannot be explained more impressively. It is pivotal to human civilisation.

Let us make the following very clear, even though we have repeatedly stated it throughout the preceding pages. *Climate Change* is one problem that the entire world is facing right now. Even at the risk of being described as melodramatic, we wish to emphasise that the worst possible predictions of scientists due to global warming have started coming true, and even much more rapidly than their worst expectations. This affects every country and every corner of the Earth, its air, its land mass, and its seas, and every living being, humans, animals, marine life, insects, and plants, and severely affects the entire biodiversity as well as the production of food, availability of potable water, and clean air to breathe.

Every country must contribute to the measures suggested by our scientists, which are based on observations and solid scientific arguments. Unfortunately, their early cries of alarm have turned to cries of painful despair, without enough being done. Industrialised countries have created a problem which affects everyone, though the developing countries in Asia, Africa, and Latin America face its severity most strongly [4, 5]. This can only be solved by concerted international collaboration and perseverance. The struggle for the arrest and reversal of climate change will be long and hard, and we must and can win it, only if we act together. We would like to borrow a phrase made famous by Ernest Hemingway (1899–1961 CE) — "*now, now, now, the only now, and above all now, and there is no other now*" — to describe the present situation and emphasise that we need to seize the present moment and start action immediately, else the very survival of humanity will be at stake.

We close this entire discussion by recalling that Rabindranath Tagore adopted "“यत्र विश्वं भवत्येकनीडम्”" (where the entire world becomes a nest) from the Shukla Yajurveda when he established his university Vishwabharati for exchange and exploration of knowledge by all nations, which we mentioned earlier too.

We would further like to recall the famous comment from Maha Upanishad:

"अयं निजः परो वेति गणना लघुचेतसाम्।
उदारचरितानां तु वसुधैव कुटुम्बकम्॥"

"Those with small minds consider what is theirs and what is another's.
For those with noble character, *the whole world is their family*."

With a strong conviction in the futility of conflict and a stronger conviction in the fact that the entire world is one family, we can proceed to reap the fullest benefits of international collaboration in sciences and the exchange of knowledge from across borders and usher in everlasting peace and exult in ever-deeper scientific knowledge in the universe.

References

[1] Recall the famous comment generally attributed to Dr. A. P. J. Abdul Kalam, "Look at the sky. We are not alone. The whole universe is friendly to us and conspires only to give the best to those who dream and work". He often told this to students while discussing science with them.
[2] "Chapter V, Section III: His Immediate Disciples", *Confucian Analects, The Great Learning, and The Doctrine of the Mean*. Translation and commentary by James Legge, (Second Edition, Clarendon Press, Oxford, 1893).
[3] M. R. C. Greenwood and K. K. North, "Science through the Looking Glass: Winning the Battles but Losing the War?", *Science* 286 (1999) 2072–2078.
[4] D. K. Srivastava and V. S. Ramamurthy, "*Climate Change and Energy Options for a Sustainable Future*" (Singapore: World Scientific, 2021).
[5] V. S. Ramamurthy, D. K. Srivastava, and S. Nayak, "*Art and Science of Managing Public Risks*" (Singapore: World Scientific Publishing Company, 2022).

Index

E

F

G

Y

Z

www.ingramcontent.com/pod-product-compliance
Lightning Source LLC
LaVergne TN
LVHW021055220926
848484LV00021B/846